T0401890

ALLE ZEIT WACH
1842

W. Pellet M. Cannoni A. Pech

Otoneuro-surgery

In Collaboration with
Ph. Farnarier S. Malca S. Valenzuela J.-M. Thomassin
J.-M. Triglia M. Zanaret C. Lacroix Ph. Querruel

Foreword by
W. House and W. Hitselberger

With 139 Figures

Springer-Verlag Berlin Heidelberg New York
London Paris Tokyo Hong Kong Barcelona

Professor William Pellet
Service de Neuro-Chirurgie (Pr. A. Combalbert), Hôpital de Sainte Marguerite
270 Bd de Sainte Marguerite, F-13009 Marseille

Professor Maurice Cannoni
Service d'Oto-Rhino-Laryngologie, Hôpital de la Timone
Rue Saint Pierre, F-13385 Marseille Cedex 5

Professor André Pech
Clinique d'Oto-Rhino-Laryngologie, Hôpital de la Timone
Rue Saint Pierre, F-13385 Marseille Cedex 5

Translation by
David Le Vay, MD
6, Garstons, High Street, Burwash, East Sussex TN19 7EZ, England

Drawings by
J. P. Jacomy
CRI, 39 Allée Albéniz, F-13008 Marseille

The authors greatly appreciate Doctor Jean Bernard Causse's editorial assistance.

ISBN 978-3-642-48694-4 ISBN 978-3-642-48692-0 (eBook)
DOI 10.1007/978-3-642-48692-0

Library of Congress Cataloging-in-Publication Data
Pellet, W. (William), 1938- [Oto-neuro-chirurgie. English] Otoneurosurgery / W. Pellet, M. Cannoni, A. Pech ; in collaboration with Ph. Farnarier . . . [et al.] ; foreword by W. House and W. Hitselberger. p. cm. Translation of: Oto-neuro-chirurgie. Includes bibliographical references.
ISBN 978-3-642-48694-4
1. Petrous bone-Surgery. 2. Nerves, Cranial-Tumors-Surgery. I. Cannoni, M. II. Pech, A. (André) III. Title. [DNLM: 1. Cranial Nerve Neoplasms-surgery. 2. Petrous Bone-surgery. WL 330 P380o] RF 126.P4613 1990 617.5'14-dc20 DNLM/DLC for Library of Congress 90-9722 CIP

2127/3145-543210 - Printed on acid-free paper

Foreword

Doctors Pellet, Cannoni and Pech have joined forces to write a masterly work on the subject of otoneurosurgery and tumors of the base of the skull. The anatomic illustrations, derived from their meticulous dissections, are of excellent quality and a credit to the famed tradition of French neuroanatomists of the past. It is evident that the operative techniques described by these surgeons indicate that they have spent a great deal of time in the dissecting-room and that they have completely mastered the temporal bone. A perfect knowledge of the fundamental anatomy of the petrous bone is essential before embarking on the difficult surgical procedures in this region of the brain and base of the skull.

Reading this work makes it abundantly clear that the results of this type of surgery are vastly improved when neurosurgeons and otologists combine their skills. Each of these specialists brings his unique experience and entire knowledge to the operation for the greater benefit of the patient.

We wish to congratulate Doctors Pellet, Cannoni and Pech on the publication of this excellent work.

March 1990

William F. House, MD
William E. Hitselberger, MD

Contents

List of Coauthors

Farnarier, Philippe
Neuroradiologist, Head of Radiology Department, C. R. A. C. M., 232 Bd de Sainte-Marguerite, F-13009 Marseille

Lacroix, Claudine
Anesthetist-Specialist in Intensive Care, Staff-member, Department of Anesthesia and Intensive Care (Prof. G. Francois), Hôpital de la Timone, Rue Saint-Pierre, F-13005 Marseille

Malca, Samuel
Neurosurgeon, Staff-member, Department of Neurosurgery (Prof. A. Combalbert), Hôpital de Sainte-Marguerite, 270 Bd de Sainte-Marguerite, F-13009 Marseille

Querruel, Philippe
Anesthetist-Specialist in Intensive Care, Staff-member, Department of Anesthesia and Intensive Care (Prof. G. Francois), Hôpital de la Timone, Rue Saint-Pierre, F-13005 Marseille

Thomassin, Jean-Marc
Professor of Otorhinolaryngology, Department of Otorhinolaryngology, Hôpital de Sainte-Marguerite, 270 Bd de Sainte-Marguerite, F-13009 Marseille

Triglia, Jean-Michel
Otorhinolaryngologist, Staff-member, Department of Otorhinolaryngology (Prof. A. Pech), Hôpital de la Timone, Rue Saint-Pierre, F-13005 Marseille

Valenzuela, Santiago
Assistant in Neurosurgery, Department of Neurosurgery (Prof. A. Combalbert), Hôpital de Sainte-Marguerite, 270 Bd de Sainte-Marguerite, F-13009 Marseille

Zanaret, Michel
Otorhinolaryngologist, Staff-member, Department of Otorhinolaryngology (Prof. Ag. Cannoni), Hôpital de la Timone, Rue Saint-Pierre, F-13005 Marseille

Introduction

W. Pellet

Because its anatomy is relatively simple, the cranial vault provides the ideal zone of access to the cranial cavity. The hemispheric convexity, whether cerebral or cerebellar, is readily so exposed. The structures of the base, on the other hand, are far more difficult to approach through this opening. In such cases the neurosurgeon endeavours to cut his flap as low as possible, at the level of the base, on the supraorbital margins, on the zygoma or on the mastoid. Though he may sometimes risk encroaching on these borders, the complexity of the anatomic structures of the skull base induces him to limit his opening to this level. In order to "come down" on to the base, he then has to separate the brain from the osteodural plane. The ease, efficacy and especially the aggressiveness of this maneuver vary greatly with the topography of the approach. The relative narrowness of the posterior fossa compared with the size of the cerebral compartment, the vertical nature of the posterior surface of the petrous, especially as it is buried in the dihedral angle it forms with the tentorium cerebelli, the presence of the cranial nerves stretching towards their respective foramina, the proximity of the brain-stem, the course of the dural sinuses: all these factors combine to make exposure of the peripetrous regions particularly restricted and difficult, especially the way down along the posterior surface of the petrous. In the absence of an expanding lesion, the neurosurgeon, if he has some degree of experience and if the necessary instruments are available, may attain his objectives without excessive difficulty or risk. On the other hand, the presence of a tumor which reduces the available space, stretches the nerves and compresses the brain-stem seriously complicates the problems and increases the risks. If this tumor remains strictly intradural, such as the epidermoid cyst of the cerebellopontine angle, many meningiomas of the posterior surface of the petrous or rarer tumors (in our experience, metastases or sarcomas), complete excision is still possible but only at the cost of operative difficulties and technical achievements which only experienced surgeons can tackle without fear of causing intimidating damage. But, in fact, this is the least likely possibility; and because of the presence at this level of the internal acoustic meatus and the posterior foramen lacerum, and because the commonest tumors, neurinomas of the vestibulocochlear nerve and tumors of the jugular bulb, arise at this level and are quite able to cross these orifices, the neurosurgeon usually finds himself faced with the double problem of expansion in the cerebellopontine angle with intrapetrous, or even infratemporal, extension. To expose properly the entire region without abuse of the retractor, to completely define the intrapetrous extension and proceed to a really total excision of the tumor, to identify the nerves in a healthy area and thus to respect more easily their continuity, it would seem logical to remove the osteodural wall, which, under the circumstances, means resecting the whole or part of the petrous pyramid. This is certainly more difficult than cutting a flap and calls for knowledge and training acquired only by otologists. This is the justification for otoneurosurgical collaboration where the surgery of petrous and peripetrous tumors is concerned.

The petrous pyramid, a wedge embedded in the base of the skull between the supra- and subtentorial levels and the uppermost latero-cervical regions, constitutes a massive obstacle which blocks approaches to the middle cerebral fossa below, the classical suboccipital approach in front and the highest cervicotomies above. This mass is hollowed out by three main sinuous excavations: the cochleo-vestibular cavities, the somewhat complex course of the facial canal crossed by the facial nerve, and the carotid canal. Its surface is marked by impressions of varying depths created by the passage of venous sinuses and nerves. The otoneurosurgical routes of approach are intended to ablate tumors developing at this level by resecting whole or part of this bony pyramid by means of microsurgical abrasion. While so doing, and when the surgeon intends to open the facial or carotid canal, the facial nerve or internal carotid must be respected. The problem is whether to sacrifice or preserve the cochleovestibular apparatus, knowing that any breach of the labyrinthine cavities will be penalized by immediate and permanent deafness. Fortunately, this problem is very often solved before operation because of spontaneous deterioration of hearing due to tumoral growth. In such a case, there is no harm in resecting the posterior labyrinthine massif (translabyrinthine route), the entire labyrinthine massif (transcochlear route) or even more (extended transcochlear route). On the other hand, the problem arises most keenly when preoperative hearing remains functional. If the size of the tumor is small, trephining confined to the roof of the internal auditory meatus (the suprapetrous or middle cerebral fossa route) may permit complete resection of the tumor with preservation of hearing. If the tumor is larger, and this case is fortunately quite rare, the use of these approaches makes sacrifice of hearing inevitable. Several factors will be involved in arriving at a decision: the tumor itself (nature, site and size), the patient (age, occupation, risk factors), the contralateral ear (normal or diseased) and the surgeon (especially his customary procedures). We feel it important to stress that these otoneurosurgical approaches offer the neurosurgeon an angle of attack and an exposure which none of the classical neurosurgical approaches can provide. It also has to be said that these approaches can no longer be criticized for being so narrow, provided of course that the otologist is perfectly familiar with the technique and knows how to push the resection within the realms of possibility.

Collaboration between otologists and neurosurgeons, if it is to be actually fruitful, must be engaged in candidly and completely and at all times, from when the patient is taken care of, during the diagnostic process and the discussion of operative indications, during the operation itself and throughout postoperative surveillance. There should be no question of preeminence of one team rather than the other; otherwise relationship will deteriorate into a sordid coexistence of precarious nature. It will be neither constructive nor stimulating for the two surgeons if, as Derome* complained, one of them is "co-opted in this surgery in a purely ancillary role". On the contrary, sharing knowledge, concepts and ideas is essential to promote the progress of the enterprise. And this naturally requires, from the onset, a state of mind open to such collaboration and major exertions to develop it. Thus, the neurosurgeon must devote himself to the acquisition of purely otologic information, and conversely. To secure progress along these lines, we feel that it is very important to outline very adequately, in the first chapter, those anatomic concepts required for a proper understanding of the purely technical problems, still mainly otologic, to be discussed in the second chapter. The assimilation of this information, remote only at first glance from the normal preoccupations of neurosurgeons, seems of primary importance for those of our colleagues who may be attracted by this undertaking. On the other hand, our third chapter is centered on the problem of operative indications and marks a return to problems of a more specifically neurosurgical nature, since it discusses the tumors usually dealt with habitually by neurosurgeons (neurinomas, meningiomas, cholesteatomas, tumors of the jugular bulb and some rarer conditions); though it should be understood that this report, presented to the Society of French-speaking Neurosurgeons, does not pretend to deal with such purely otologic tumors as carcinomas of the middle ear or certain tumours of the superior jugular bulb for example (type A, tympanic, or B, tympanomastoid of Fisch). We have always borne in mind the words of Andre Gide: "To understand something is to feel capable of doing it", placed by Pertuiset at the head of the conclusions of his report on acoustic neurinomas presented before this same Society in 1970. It is our desire that this book should bring to those of our colleagues who have not yet considered it, an understanding of the full importance of this otoneurosurgical collaboration.

---

* P. Derome (1972) Les tumeurs sphéno-ethmoïdales. Possibilités d'exérèse et de réparation chirurgicales. Neurochirurgie 18 (suppl 1): 4

Spheno-ethmoidal tumors. Possibilities of excision and surgical repair.

Basic Anatomy

W. Pellet

In Collaboration with
Ph. Farnarier,
J. M. Triglia, S. Valenzuela,
and M. Zanaret

The most difficult thing in the world is to say with due thought what everyone says without thinking.
(Alain: *Histoire de mes pensées*)

The Petrous Pyramid

The petrous pyramid or petrous bone is only one of the three portions, the petrous portion, of the temporal bone, which also comprises the squamous and the tympanic portions. These three bony components are separate in the fetus and the newborn, but fuse completely so as to form what appears to be a single bone in the adult though certain sutural vestiges remain perfectly identifiable. We cannot deal at length here with the description and arrangement of these three portions, details of which will be found in the classical studies [3, 56, 88, 89]. The squama, incorporated in the cranial vault, is a very simple structure well-known to neurosurgeons.

The petrous portion, on the other hand, with the closely attached tympanic portion, is, as stressed by Y. and B. Guerrier [41], the most complex bony component of the body. It is incorporated in the base of the skull. Paired and symmetrical, on either side of the body of the sphenoid, the two pyramids seen intracranially form the step separating the middle and posterior fossae. They run obliquely forward and inward to form an obtuse angle of about 100°. Each therefore makes an angle of about 40° with the frontal plane, an important consideration when positioning the head for an operation (Fig. 1).

Classically, the petrous is described as a quadrangular pyramid with two intracranial surfaces and two extracranial surfaces. However, the standard lateral view of the skull gives the impression that it is roughly triangular. This is why, as already suggested by Y. and B. Guerrier [41] and as done by Anson and Donaldson [3] in their monumental work, we shall describe three surfaces, superior, posterior and inferior, the base being the mastoid. Schematically, this pyramid may be regarded as composed of a superficial shell made of compact bone, hollowed out by specific cavities; the cochleovestibular apparatus, the facial canal and the carotid canal, the remaining spaces being occupied by a packing system: the petrous pneumatic cells.

The Petrous Cortex

This is, as it were, the frame-work whose contours and details are wellknown and described in the anatomical texts. We review here the essential features which guide the surgeon throughout the otoneurosurgical procedure.

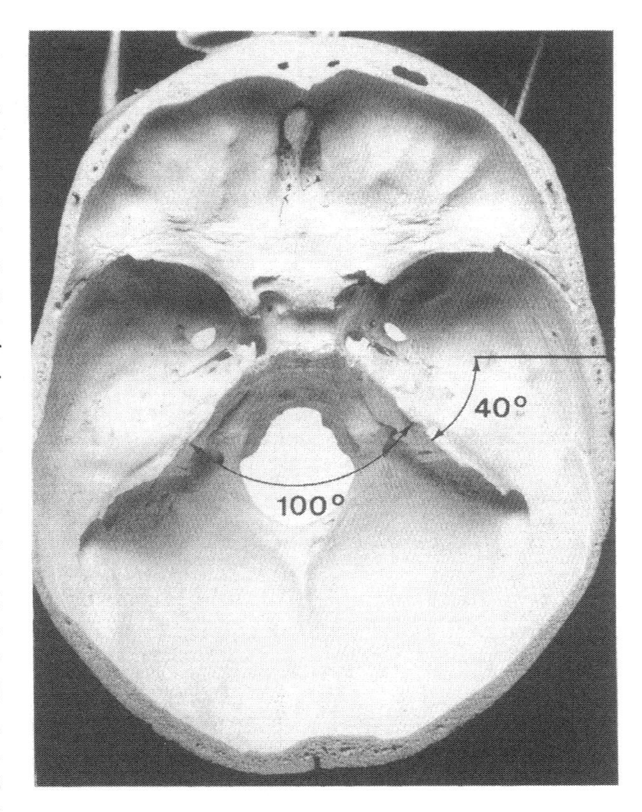

Fig. 1. Base of skull, superior aspect. Note the obliquity of each of the pyramids in relation to the frontal plane and the angle formed between them

Surfaces

Superior Surface

This is the anterosuperior surface of the classical texts (Fig. 2), particularly well described from the otoneurosurgical aspect in the theses of Gitenet [38] and Zanaret [107].

Triangular in shape, with the base laterally and the apex medially, it is bounded posteriorly by the superior petrosal margin, which, in the dried specimen, constitutes the posterior border of the middle cerebral fossa.

Laterally, it is continuous without abrupt transition with the intracranial aspect of the vertical portion of the squama. Fused in front over its lateral two-thirds with the horizontal portion of the squama and over its medial third with the posterior margin of the greater wing of the sphenoid, it forms with these structures part of the composition of the floor of the middle cranial fossa.

While it is almost horizontal laterally, it becomes distorted by sloping forward and downward as it approaches the apex. At the same time the horizontal portion of the squama and especially the greater wing of the sphenoid slope backward and downward to the extent that the middle fossa has an excavated shape whose base is more or less centered on the spinous foramen.

Smooth enough at first glance, its surface shows several irregularities which it is important to detect since they constitute very useful landmarks during operations via the middle cranial fossa (hereafter called the suprapetrous approach).

The Arcuate Eminence (Figs. 2a and 2b). This is the best known if not the most definite, since it is very variable from one subject to another. As emphasized by Y. and B. Guerrier [41], allowing for the tangential approach to this superior surface, it appears to the surgeon engaged in stripping the dura mater as a major eminence even when it is minimal. This eminence is oval-shaped, with its long axis more or less sagittal. It is situated at the junction of the outer third and inner two-thirds, close to the superior petrosal margin. The precise cause of this eminence is of no great import. Some regard it as caused by the convexity of the anterior semicircular canal, others as due to the presence of an air-cell or a simple mamillary eminence determined by the posterior part of the third temporal sulcus. What is important is that it represents the best surface landmark for locating the loop of the anterior semicircular canal. It is well shown by the scanner (Fig. 2e), and Cohadon and Castel [22] have specified that this loop is to be found at between 1 and 3.5 mm depth under the anteromedial slope and usually at the middle third.

Before reaching the arcuate eminence, the surgeon will cross a plane surface in continuity with the intracranial surface of the squama: the tegmen tympani, the roof of the tympanic cavity. This is an often very thin bony layer, very easily staved in, sometimes even exhibiting small zones of dehiscence connecting the cranial and tympanic cavities. This surface is crossed obliquely by the superior petrosquamous fissure, the intracranial boundary between the petrous pyramid and the squama. This fissure, particularly marked in its posterior part, is situated just above the tympanic cavity.

Once past the arcuate eminence, the surgeon descends again on another plane surface of triangular shape, based laterally on the eminence and situated along the superior petrosal margin; this is the meatal area, and it is at this site that it is necessary to excavate to uncover the internal acoustic meatus. It narrows progressively to its apex, which is marked by the second irregularity.

The Tubercle of Princeteau (Figs. 2b and 2c). This is a small prominence, more or less acuminate, situated very close to the posterior margin and dominating, as a strong-point dominates a valley, a dome-shaped depression situated just medial to it and extending over the apex of the petrous: this is the fossula marked out by the inferior aspect of the trigeminal ganglion (the trigeminal depression).

The Hiatus of the Canal for the Greater Petrosal Nerves (Fig. 2d). This is the third irregularity, but a hollow one. It is a small orifice about 1 mm in diameter, extended towards the tip of the petrous by a small groove for the passage of the petrosal nerve. This orifice is situated in front of the arcuate eminence and more or less medial to it. Its position depends on the length of the canal that opens there, and this varies between 0 and 7 mm according to the measurements of Hall and Rhoton [44, 82], of Cohadon and Castel [22] or those of Parisier [69]. This variability means that it is not to be regarded as a landmark but rather as a warning of the proximity of the geniculate ganglion and an indication for surgical prudence, all the more so as in 15% of cases, according to Hall and Rhoton, the geniculate ganglion is itself more or less unprotected under the dura mater. Classically, the greater superficial and deep petrosal nerves emerge from the hiatus, but according to Portmann [78] there is only a single nerve, the petrous nerve, stemmed from the genicu-

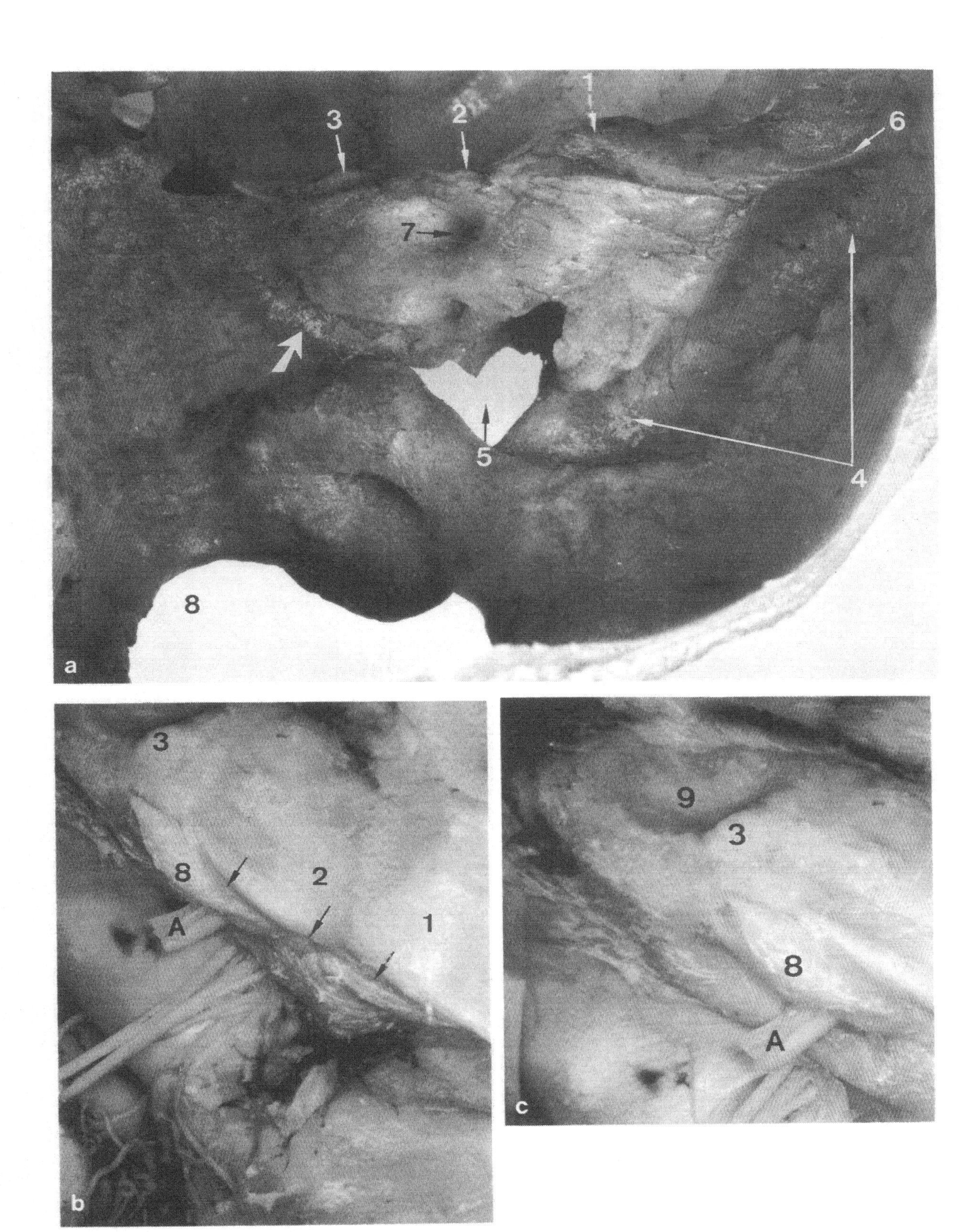

Fig. 2 a–e. Superior aspect of the petrous. **a** Posterior view of right petrous: prominence of arcuate eminence *(1)*, plane of meatal area *(2)*, tubercle of Princeteau *(3)*. Note that the groove for the sigmoid sinus *(4)* which arrives at the posterior lacerate canal *(5)* passes under the superior margin of the petrous *(6)* – the aperture of the internal acoustic meatus *(7)* – foramen magnum *(8)* – groove for inferior petrosal sinus: *white arrow.* **b** Superior aspect of right petrous: arcuate eminence *(1)*, meatal area *(2)*, tubercle of Princeteau *(3)*, supra-acoustic eminence *(8)*. The *arrows* indicate the groove of the superior petrosal sinus. Acousticofacial bundle entering IAM *(A)*. **c** Upper aspect of right petrous: tubercle of Princeteau *(3)*, trigeminal depression *(9)*, supra-acoustic eminence *(8)*. Acousticofacial bundle *(A)*

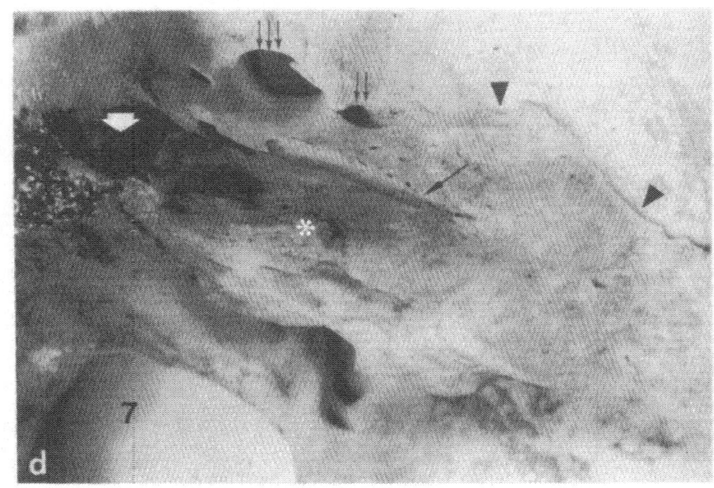

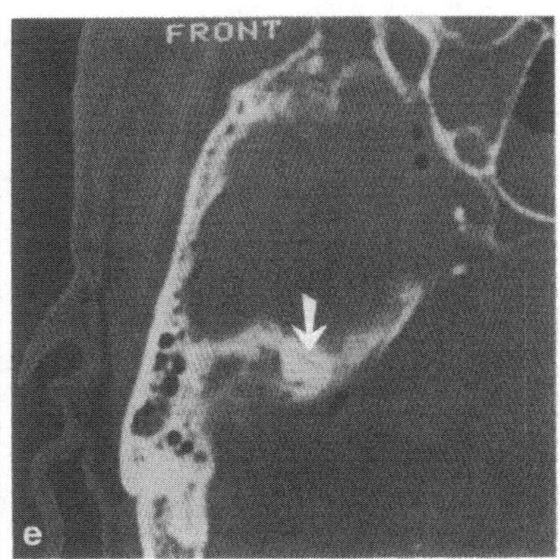

Fig. 2. d Upper aspect of right petrous: tubercle of Princeteau *(white star)*. Anterior lacerate foramen *(white arrow)*. Hiatus of canal for petrous nerve *(black arrow)* prolonged inward by the groove for the petrous nerve. In front, spinous foramen *(double black arrows)*, foramen ovale *(triple black arrows)*. Foramen magnum *(7)*. **e** Transverse CT section of left petrous passing through the arcuate eminence. Loop of anterior semicircular canal *(arrow)*

late ganglion. The hiatus is often associated with a second orifice situated slightly in front and medially, the accessory hiatus, through which classically emerge the lesser petrosal nerves and, according to Portmann, the tympanic nerve itself.

Posterior Surface (Figs. 2a and 3)
This is practically vertical, especially laterally, and under the superior petrosal margin it constitutes the anterior wall of the posterior cranial fossa. It is triangular, too, its lateral base bearing the impression of varying depth but always marked, of the lateral sinus as it describes its sigmoid curve descending to the posterior foramen lacerum. Medial to this last and towards the apex, the groove of the inferior petrosal sinus marks the anterior petro-occipital fissure more or less deeply. This surface, even while covered with dura mater, bears a major landmark:

The Aperture of the Internal Acoustic Meatus (Porus Acousticus) (Fig. 3a). This is situated at the junction of the medial third and lateral two-thirds and is usually an oval orifice, opening slant-wise on the posterior surface. Its average measurements, according to Paleirac [in 41], are 4.3 mm in vertical diameter and 8.6 mm in transverse diameter. It is situated closer to the superior petrosal margin than to the inferior border, which here constitutes the superior margin of the intracranial orifice of the posterior foramen lacerum. Close to the aperture, in a zone no larger than a French 50 centimes coin or than an American penny coin [Anson and Donaldson, 3], incision of the dura mater, as for trephining of the meatus, exposes three other smaller orifices which also constitute important landmarks to avoid opening the labyrinthine cavities during drilling.

The Subarcuate Fossa (Fig. 3a). This is a small depression, very variable in shape and size, situated above and lateral to the aperture of the meatus, so close up against the superior petrosal margin as to notch the latter markedly and render the adjacent segment of the margin more prominent. As this overhangs the aperture, it gives the appearance of a small bony rim, the supra-acoustic eminence. In the base of this small fossa, in the adult a very small foramen opens, of pinhead-size according to Anson and Donaldson [3], a vestige of what was a sizeable opening in the fetus. Here opens a small canal (Fig. 3b), the petromastoid canal, which goes through the loop of the anterior semicircular canal at its center and ends at the mastoid antrum. This canal is occupied by the subarcuate artery destined to supply the mastoid antrum and also the posterior labyrinth [64]. This is one reason justifying its detection during operation. According to Gueurkink [43], its average distance from the posterior margin of the aperture of the internal acoustic meatus is 5.5 mm, but there is great variability from 1 to 12.5 mm.

The Endolymphatic Fossula (Fig. 3a). This is also known as the ungual fossula since it typically has the shape of a nail-scratch. It is a slit generally run-

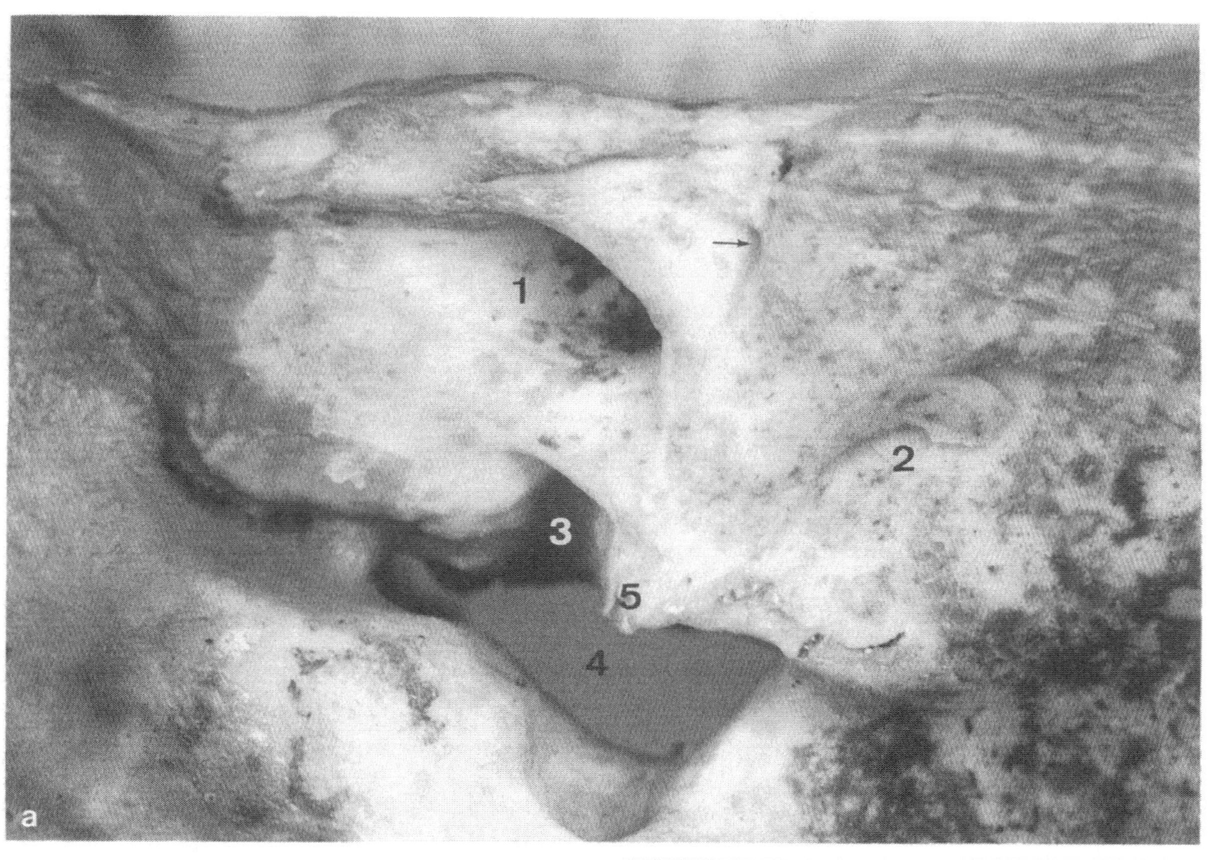

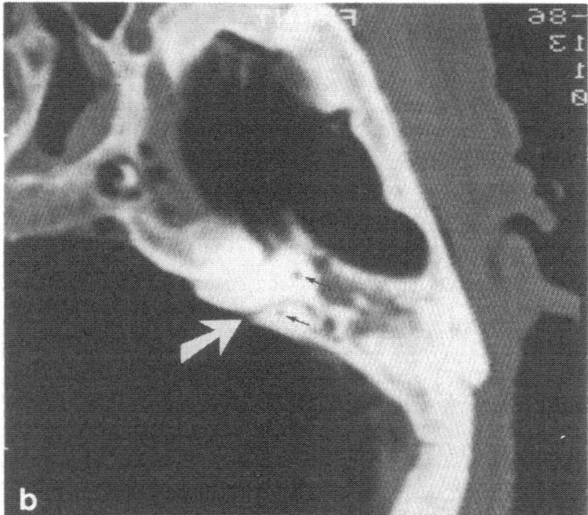

Fig. 3 a, b. Posterior aspect of petrous pyramid. **a** Aperture of IAM *(1)*, subarcuate fossa *(black arrow)*. Endolymphatic (ungual) fossula *(2)*. Pyramidal fossula *(3)*. Posterior lacerate canal *(4)*. Jugular spine of temporal *(5)*. **b** Transverse CT section of right petrous passing through the subarcuate fossa *(white arrow)* which is prolonged by the petromastoid canal which travels in the loop of the anterior semicircular canal *(black arrows)*

ning obliquely downwards and inwards, more or less parallel to the margin of the groove of the sigmoid sinus. It is often more evident in its upper portion. Anson and Donaldson [3] give its dimensions as between 2 and 6 mm in 9 cases out of 10, but the extremes of range go from a pinhead to 10 mm. It is situated halfway along the superior petrosal margin and the upper border of the posterior foramen lacerum, and between the aperture of the internal acoustic meatus medially and the groove of the sigmoid sinus laterally and below, always closer to the latter according to Anson and Donaldson [3]. According to Gueurkink [43], the average distance separating it from the posterior margin of the aperture of the IAC is 10.7 mm (11 mm for Anson and Donaldson [3] with a range of 7 to 19 mm). Gueurkink also notes that its average distance from the subarcuate fossa is 9.5 mm with a range of 6 to 15 mm. During trephining of the meatus, the flap of dura mater must not exceed the size indicated by these two measurements. In fact, this ungual fossula is the flared orifice of the acqueduct of the vestibule which expands terminally to accomodate the endolymphatic sac, not to be damaged or obstructed during drilling if hearing is to be preserved.

The third landmark is the external orifice of the canaliculus or aqueduct of the cochlea. This is actually on the inferior aspect of the petrous, close against the margin separating it from the posterior surface (Fig. 3a) at the summit of the pyramidal fossula.

Inferior Surface (Figs. 4a and b)
This is the extracranial aspect, classically subdivided into two surfaces, postero- and antero-inferior; but many authors, such as Sterkers [96], find these poorly demarcated which inclines us to regard them as one surface. Similarly, the purists distinguish between the tympanal and the pyramid, but these bones are so intimately fused in the adult that we shall regard them as one and the same. We describe this surface as having 3 portions:

The Posterolateral Portion. This is the internal aspect of the mastoid. Roughly, it occupies the posterior half of the inferior surface. It is large and goes down almost 2 cm under the skull-base. This medial mastoid surface is deeply scored by the digastric groove, the course of which is almost sagittal but slightly oblique forward and inward (Fig. 4a). The somewhat jutting inner margin of this groove constitutes the juxtamastoid eminence, sometimes notched by the groove of the occipital artery. The digastric groove is an important landmark during the translabyrinthine approach; during curettage of the mastoid it is opened progressively from the base (Fig. 48) and uncovering of the muscle marks the inferior limit of the approach. On the other hand, we know that at its anterior extremity there is the stylomastoid foramen, crossed by the facial nerve emerging from the facial canal. It is therefore possible to locate the facial nerve by following the groove, which creates a problem, mainly when there is a bulge of the jugular bulb or of the lateral sinus.

Middle Portion. This is a complex region, as it juxtaposes a series of characteristic anatomic irregularities in an area of less than 2 cm^2 (Figs. 4a and b).

A bony escarpment prolongs the prominence of the mastoid forward and inward in the axis of the pyramid. It consists successively of:

- The most inferior part of the tympanic sulcus, fused to the anterior vertical border of the mastoid by the tympanomastoid or posterior tympanosquamous fissure
- The styloid process
- The vagina of the styloid process which clothes the anterior aspect of the styloid and extends well beyond it medially in the direction of the internal carotid.

In front, the anterior aspect of the tympanal, flat, smooth, practically vertical and in the same oblique forward and inward plane as the pyramid, descends from the interior surface of the horizontal portion of the squama, to which it is attached by means of the petrotympanic fissure. This plane surface enters into the structure of the posterior wall of the mandibular fossa, the greater part of which is hollowed out in the inferior surface of the horizontal portion of the squama. This fossa is dominated in front by the double temporal articular tubercle, laterally by the anterior zygomatic tubercle and behind by the posterior zygomatic tubercle. It is crossed in its posteromedial portion by the petrotympanic fissure, a simple anterior tympanosquamous fissure in its lateral part but double medially because of the interposition of the inferior extension of the tegmen tympani into the interstice of this anterior tympanosquamous fissure (Fig. 6c). The chorda tympani emerges from the midst of this fissure. The mandibular fossa bars access to the posterior lacerate canal and the carotid canal, and its excision is essential in approaching these structures by the infratemporal route or by the extended transcochlear route employed for excision of a tumor of the glomus jugulare or of a neurinoma of the mixed nerves.

Posteriorly, there are several orifices:

- *The stylomastoid foramen* (Fig. 4). This is situated just behind the styloid process and at the anterior end of the digastric groove. It is the 2 mm diameter external orifice of the facial canal, through which the facial nerve emerges and the stylomastoid artery enters. Just in front and medial to it, at the inner aspect of the base of the styloid, there is another, much smaller, orifice, the ostium exitus, the external orifice of the innominate canal conveying the anastomosis of the vagus nerve to the facial (the auricular branch of X).

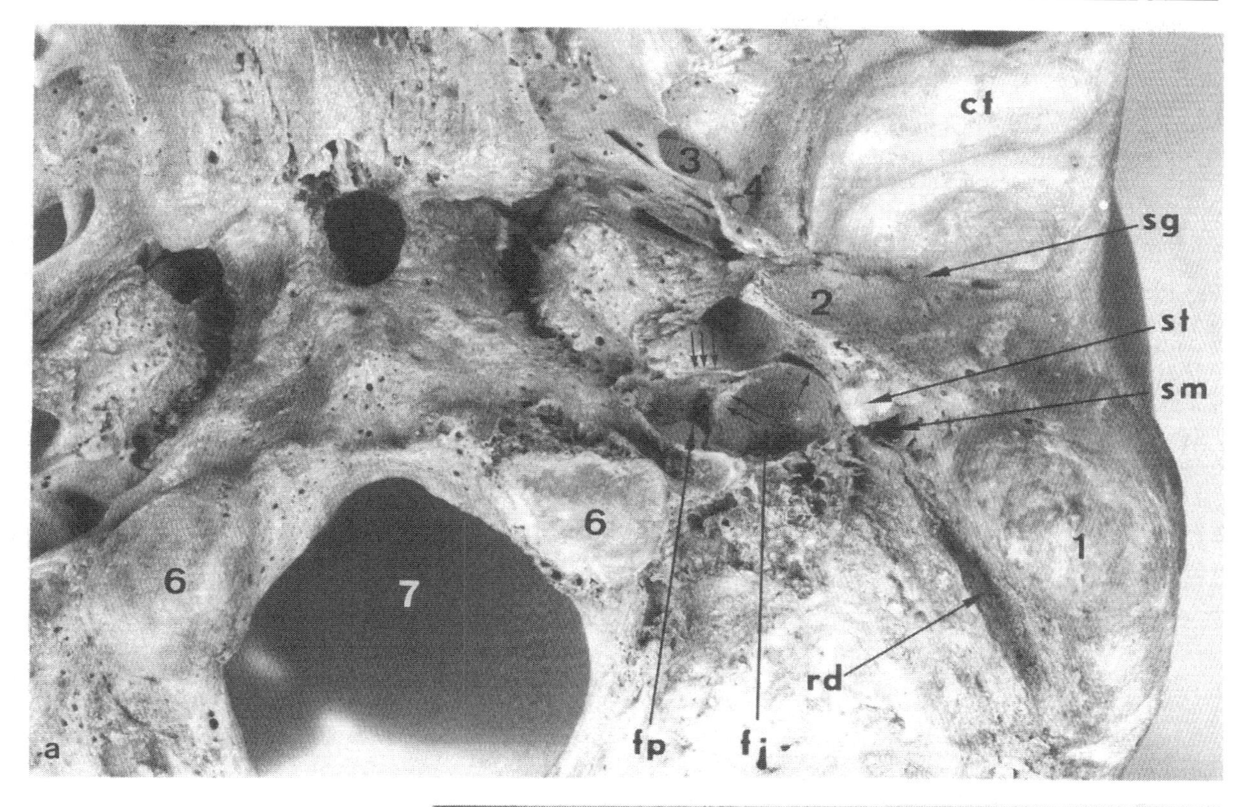

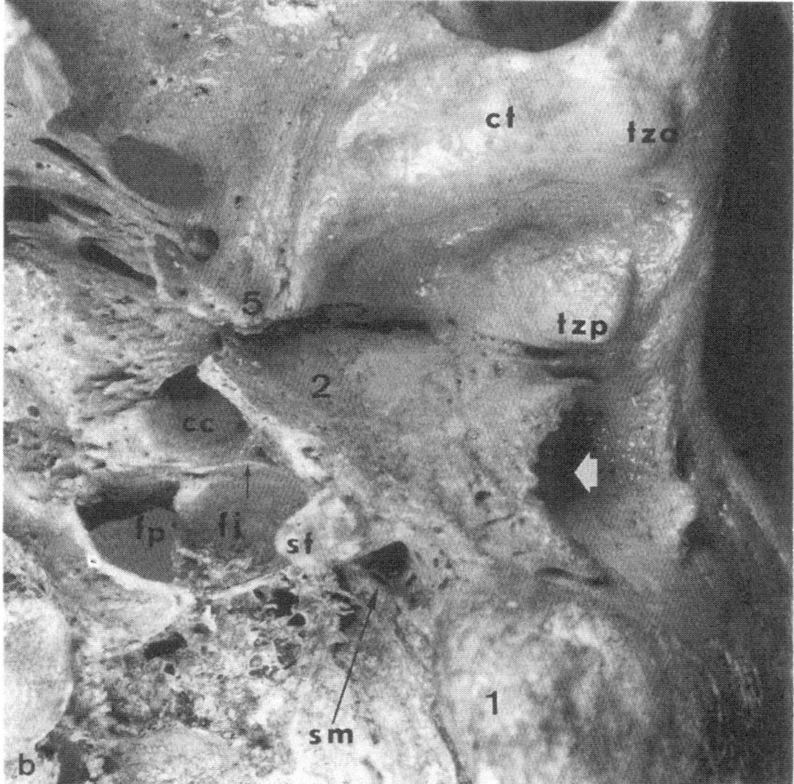

Fig. 4 a, b. Inferior aspect of petrous pyramid **a** and **b**. Mastoid apophysis *(1)* - digastric groove *(rd)*. Styloid process *(st)*. Stylomastoid foramen *(sm)*. Vagina of styloid process *(2)*. Petrotympanic fissure *(sg)*. Temporal condyle *(ct)*. Anterior *(tza)* and posterior *(tzp)* zygomatic tuberosities. Jugular fossa *(fj)*. Pyramidal fossula *(fp)*. Curved retrotympanic plate *(black arrow)*. Petroso-jugular crest *(double arrows)*. Pyramido-carotid crest *(triple arrows)*. Carotid canal *(cc)*. External acoustic meatus (white arrow). Foramen ovale *(3)*. Spinous foramen *(4)* sphenoidal spine *(5)* occipital condyles *(6)* occipital foramen *(7)*

- *The jugular fossa of the temporal* (Fig. 4a and b). This is the anterosuperior roof of the posterior lacerate canal. It is an oval excavation with its long axis transverse, situated just behind the vagina of the styloid and immediately medial to the base of the styloid process. The anterior margin of this hollow is not fused with the tympanic portion of the petrous and thus appears as a slender crest known as the curved retrotympanic plate. This curved plate divides medially into two secondary crests which embrace the petrosal fossula between them. The more lateral one, the petroso-jugular crest, continues to outline the circumference of the jugular fossa. It ends at the margin separating the inferior and posterior surfaces by raising a small bony spine, the jugular spine of the temporal, which appears the more prominent since just lateral to it. The posterior margin of this inferior surface is notched by the jugular excavation (Fig. 3a). The base of this hollow, i.e., the dome which overhangs the posterior foramen lacerum, forms the floor of the tympanic cavity. The higher this roof, the thinner it becomes and the more it elevates the floor of the cavity. It sometimes happens that this dome is so thin that there is an infraction permitting direct contact of the apex of the jugular bulb and the mucosa of the cavity, or even producing a "hernia" of the bulb into the tympanic cavity. This contiguity explains the extension of tumors of the glomus jugulare into the tympanic cavity. The depth of the jugular fossa is very variable, from 0 to 14 mm according to Anson and Donaldson [3], 70% of cases being between 2 and 6 mm. Di Chiro [25] pointed out the frequent asymmetry between the two sides, the right usually being larger than the left. Further, the higher the bulb, the closer it approaches the internal acoustic meatus, which lies above it at an average interval of 6.5 mm [Gueurkink, 43]. Cohadon et al. [23] state that it even exceeds the floor of the meatus in height in 15% of cases. Although finding a slightly lower incidence, our operative findings completely confirm the assertions of the Bordeaux school [23]. When it exists, this prolapse of the jugular bulb complicates the translabyrinthine approach and calls for careful abrasion of all the bone covering the dome so as to free it and displace it downwards, giving an adequate view of the cerebellopontine angle and especially of the mixed nerves.
- *The carotid foramen* (Fig. 4b). This is an oval orifice, 7-9 mm in diameter, tangential to both the base of the tubal process of the tympanic portion in front and the anterior circumference of the jugular fossa in the back. The second, more medial, crest of division of the curved retrotympanic plate is tangential to its posterior circumference. The tympanic canal containing the tympanic nerve opens on this pyramido-carotid crest.
- *The petrosal fossula* (Figs. 3a and 4). This is a depression of pyramidal shape, situated right against the posterior edge of the inferior surface of the petrous and deeply notching this edge, just medial to the intrajugular process of the temporal, at the level of the aperture of the internal acoustic meatus. This is the third landmark of the "penny field" of Anson and Donaldson [3]. At the apex of this fossula there opens the cochlear canaliculus, which connects the scala tympani with the subarachnoid spaces (Fig. 13).

The Anteromedial Portion (Fig. 4a). This is convex transversely and appears fairly uniform, though roughened by some asperities indicating the attachments of the levator palati muscle and the suspensory ligament of the pharynx. On its anterior flank there are two small grooves for the tensor tympani muscle and the auditory tube which, as they pass under the tubal process of the tympanic portion attached to the petrous, become the musculo-tubar canal and the bony segment of the auditory tube. Its posterior flank, attached to the corresponding margin of the basilar apophysis, is excavated by a groove straddling these two margins and occupied by a venous plexus.

Base

Roughly speaking, this is the outer aspect of the mastoid; but since we have decided to consider the petrous and the tympanic portions as one and the same component in the adult we include here the aperture of the external acoustic meatus (EAM) and also the posterior root of the zygoma. This is the commonest plane of attack in otoneurosurgery. This base is continuous above with the vertical surface of the outer aspect of the temporal squama, and behind with the parieto-occipital convexity at the level of the asterion. It is roughly triangular and may be described as follows (Fig. 5):

Three Margins

The Anterior Margin (Fig. 5a). In its upper half it is formed by the anterior margin of the EAM. It commences at the point of attachment of the anterior tympanic horn on the posterior slope of the posterior zygomatic tubercle, the most lateral point of the petrotympanic fissure. It descends vertically, forming an anterior convexity, and terminates at the lowest point of the tympanomastoid fissure. Below, its inferior half is formed by the blunt anterior vertical border of the mastoid.

Posterior Edge. Also blunt, it runs obliquely forward and downward (Fig. 5a).

Superior Edge (Fig. 5a). It is marked by a crest slightly concave upwards and in general oblique in an upward and backward direction: the supramastoid crest or linea temporalis (inferior temporal line). This crest is the posterior continuation of the upper border of the zygoma and its posterior root. It is situated on the outer table, more or less where the squama is attached to the petrous, and is a fairly good landmark for the level of the upper aspect of the latter. It is more or less marked, sometimes fading out towards the parieto-mastoid suture, sometimes, when more prominent, continuous with the curved inferior temporal line of the parietal. It marks the limit of attachment of the lowermost fibers of the temporalis muscle and becomes very obvious when these have been detached. Behind this crest, the limit of the mastoid is marked by the end of the parietomastoid suture, from the parietal fissure of the temporal to the asterion.

Three Angles

Only the inferior angle is obvious; this is the tip of the mastoid, often jutting out, large but always quite rounded and also roughened by the attachments of the sternocleidomastoid muscle (Fig. 5b).

The anterior angle is marked by the posterior zygomatic tubercle, arising from the posterior root of the zygoma just above and in front of the aperture of the EAM (Fig. 5b).

The posterior angle is situated on the asterion, limited above by the temporoparietal branch and in the back by the temporo-occipital branch of this suture (Fig. 5a).

Lateral Surface

This is immediately subcutaneous, convex in the anteroposterior direction and flat in the vertical direction, except towards the posterior edge and the tip, both very blunt, where it becomes clearly convex. It is divided into two zones, anterosuperior and posteroinferior, by the lateral petrosquamous fissure, which begins at the anterior edge just above the tympanomastoid fissure and crossed the entire lateral surface to end at the parietomastoid fissure.

The anterosuperior zone is the squamous zone, or rather, from our viewpoint, the tympanosquamous zone.

The Aperture of the EAM (Fig. 5). This is an oval orifice with its long axis vertical. Its sharp anterior margin is limited by the outer border of the tympanic portion. Its posterior blunt edge is formed by the anterior border of the mastoid. Below, there is the start of the petrotympanic fissure. The upper border of the external acoustic meatus is formed by the most lateral part of the horizontal portion of the squama, topped by the posterior root of the zygoma, behind the posterior zygomatic tubercle. Just behind, its upper pole is marked by two features:

The Suprameatal Spine (Fig. 5b). This is a bony crest which separates from the posterior root of the zygoma and forms the posterior margin of the upper pole of the meatus. It shows a quite sharp peak which is an excellent landmark for the posterior rim of the EAM when the superficial planes are stripped to expose the mastoid. Just above and behind it is:

The retromeatal cribriform zone (Fig. 5b), a small triangular zone between the suprameatal spine below and the beginning of the inferior temporal line above, pierced by numerous vascular apertures that are clearly visible, especially in childhood.

The posteroinferior zone is the petrous zone. It is rather roughened by more or less marked spicules for the attachment of the sternocleidomastoid, and behind this the splenius and longissimus. Near the posterior border, almost at its middle, is an orifice: the mastoid foramen (Fig. 5b), the external orifice of a small canal, the mastoid canal, which opens into the groove for the sigmoid sinus, often in the region of its bend. This canal contains the mastoid emissary vein, often quite large, connecting the sigmoid sinus and the superficial veins. This vein always bleeds freely during exposure of the posterior border of the sigmoid sinus.

Edges

Anterior. Fused over its greater extent with the squama and the greater wing of the sphenoid, it is obvious only in its most medial portion where it forms the posterior boundary of the anterior lacerate foramen (Fig. 2d).

Posterior (Figs. 1 and 2a). This is the petrous ridge which separates the superior and posterior surfaces. It is the posterior border of the middle cranial fossa, to which is attached the greater circumference of the tentorium cerebelli. Classically, it is crossed by the groove of the superior petrosal sinus, which in

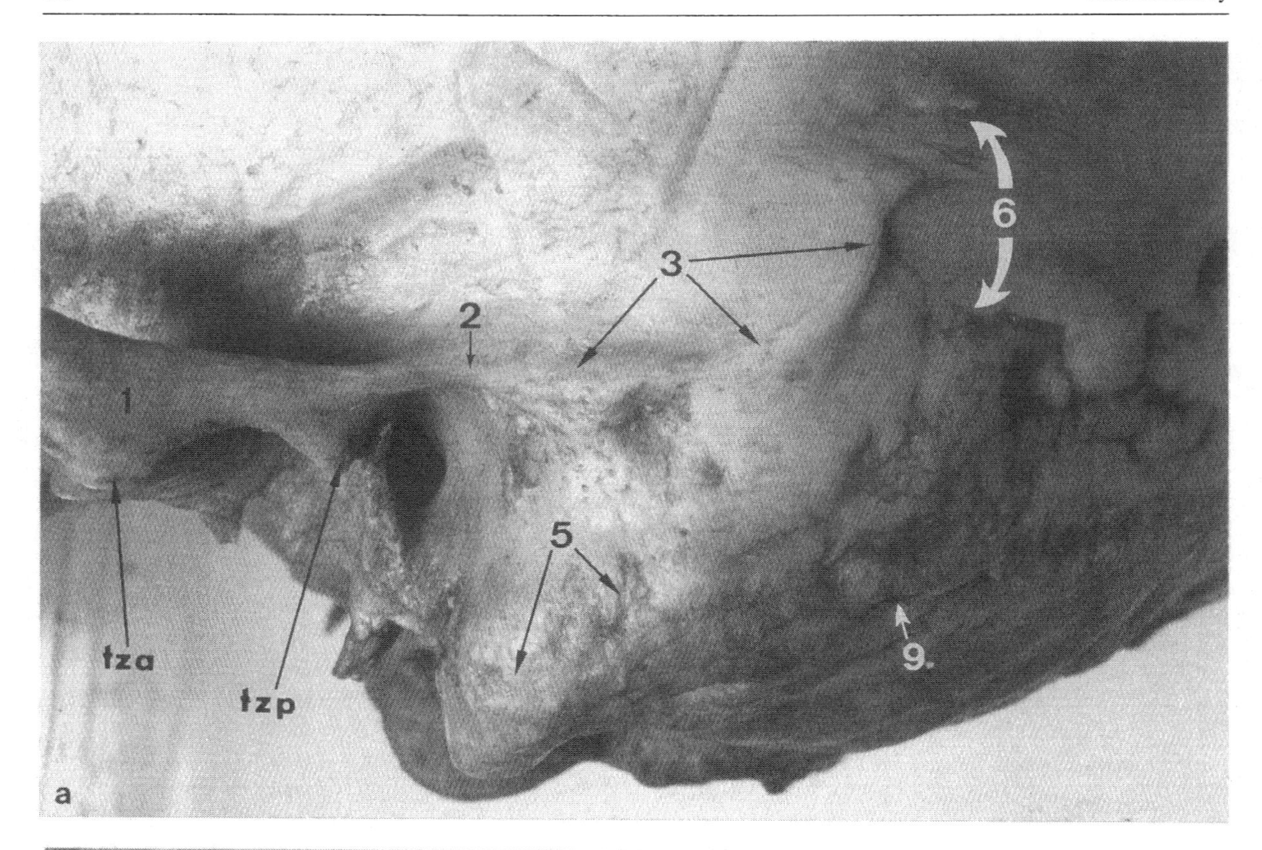

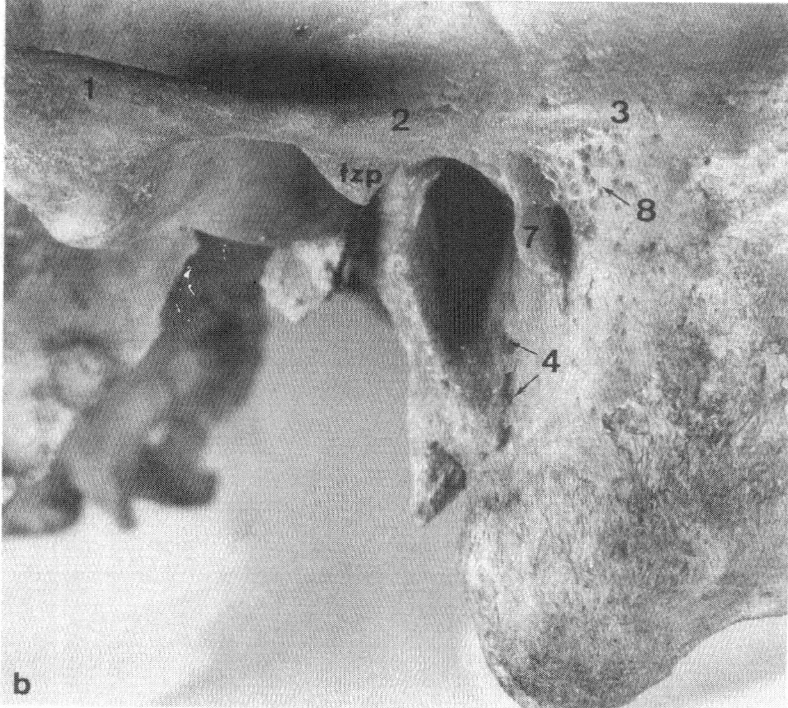

Fig. 5 a, b. Outer aspect of petrous **a** and **b**. Zygoma *(1)*, posterior root of zygoma *(2)*, temporal line *(3)*, tympanomastoid suture *(4)*, anterior zygomatic tuberosity *(tza)*, posterior zygomatic tuberosity *(tzp)*. Lateral petrosquamous suture *(5)*, temporo-parietal suture (parietal incisura) *(6)*. Suprameatal spine *(7)*, retromeatal cribriform zone *(8)*, mastoid foramen *(9)*

fact crosses this ridge in a very elongated X. Laterally, the ridge is situated beneath the edge of the superior surface, actually well above the posterior surface. It crosses this edge at the level of the notch created by the subarcuate fossa and then skirts the posterior border of the superior surface. This posterior border and its satellite sinus constitute essential landmarks during the approach to the internal acoustic meatus by way of the middle cranial fossa (suspetrous route).

Inferior (Fig. 2a). This is free only at the level of the posterior lacerate canal, where it forms the superior border, marked by the jugular spine of the temporal bone; this is determined mainly by the presence of two notches, the jugular laterally and the petrous medially (Fig. 3a).

The Petrous Cavities

These form the complexity of the petrous and account for the wariness of neurosurgeons towards petrosal surgery; but his perfect knowledge of them guides the otologist throughout the drilling procedure.

Cochleovestibular Cavities

These both constitute and contain a composite organ sensitive to sound and to the effects of gravity and movement. Classically, it is described under three heads.

External Ear

This is composed of three parts: the external ear itself, the external acoustic meatus and the tympanic membrane. We recommend those readers desiring a detailed description of the external ear or tympanic membrane to refer to the basic texts on otorhinolaryngology, particularly the *Encyclopédie Médico-chirurgicale* [89] or to the manual of Legent, Perlemuter and Vanderbrouck [56].

However, the external acoustic meatus must be considered in some detail (Fig. 6). It is an osteocartilaginous channel, slightly flattened in the anteroposterior direction, 8 to 10 mm in diameter at its external orifice and narrowing progressively towards the junction of the inner fourth and outer three-fourths where it is only 5–6 mm across. This is the isthmus of the meatus and subsequently its caliber expands again. Its direction is virtually transverse, with a slight backward inclination of 10° in relation to the frontal plane. If this obliquity is not borne in mind, there is a risk during drilling of the posterior wall of this meatus to make a breach which leads to the subsequent formation of a postoperative cholesteatoma. Since the tympanic membrane, which occludes its inner end, is arranged in a plane that is oblique downwards, outwards and backwards, the posterior wall of the meatus is shorter than the anterior wall (about 25 and 31 mm respectively) (Fig. 6b). This meatus is formed by the juxtaposition of two portions, the outer fibrocartilaginous and the inner osseous.

The wall of the outer portion is cartilaginous in its anteroinferior part and fibrous over the rest of its circumference. The cartilaginous groove so formed is wider laterally (half-circle) than medially (one-third of a circle) and joins with the circumference of the bony meatus at the level of the outer border of the tympanic portion and the tympanomastoid fissure. The fibrous circumference is strongly adherent to the bone, particularly to the suprameatal spine. This has to be stripped somewhat when exposing the posterior margin of the bony meatus before starting to drill of the mastoid. The bony portion is slightly longer than the fibrocartilaginous portion. It is formed by the tympanic portion and by the squama, the horizontal portion for the roof of the meatus and the mastoid portion for the posterior wall. The latter is diagonally crossed by the petromastoid (tympanomastoid) fissure, clearly visible under the microscope after stripping of the fibrous meatus (Fig. 5b). The meatus is lined with skin. This is thick (1 mm) in the fibrocartilaginous portion. It is thin (0.2 mm) and very adherent to the bone in the bony portion, calling for extreme caution during its stripping to avoid making a breach in the posterior wall of the meatus that might favor the formation of a postoperative cholesteatoma. The internal orifice is closed by the tympanic membrane, fixed in a circumferential groove, the tympanic sulcus.

Middle Ear

This consists of the tympanic cavity, classically linked with the mastoid air cells, to be described separately, and the auditory tube.

Tympanic Cavity. Still known as the tympanum, this is described as a small cavity of roughly parallelepiped shape with 6 faces, interposed between the external and internal ear. In fact, with a length and depth not exceeding 15 mm at the most and especially a width varying from 6 mm (above and below) to 2 mm (at its middle), it more resembles a slit whose anteroposterior direction is oblique forward

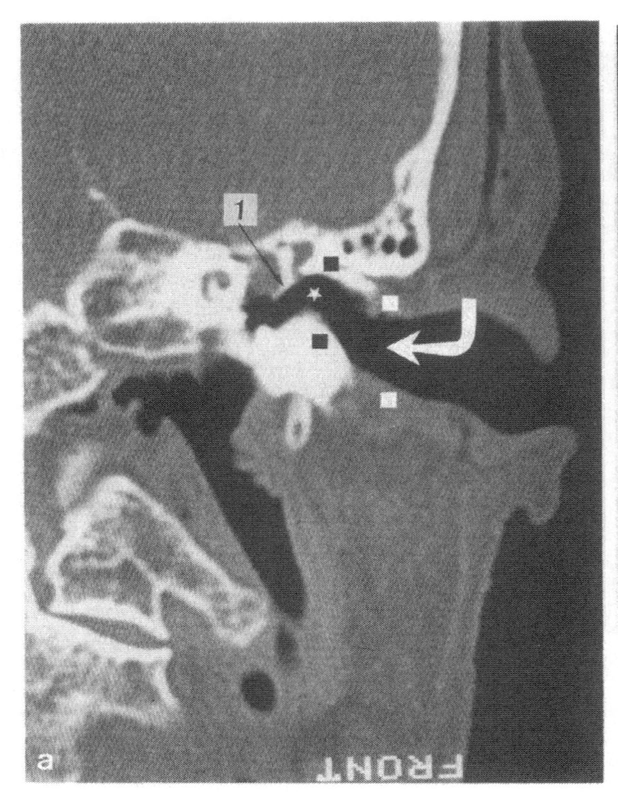

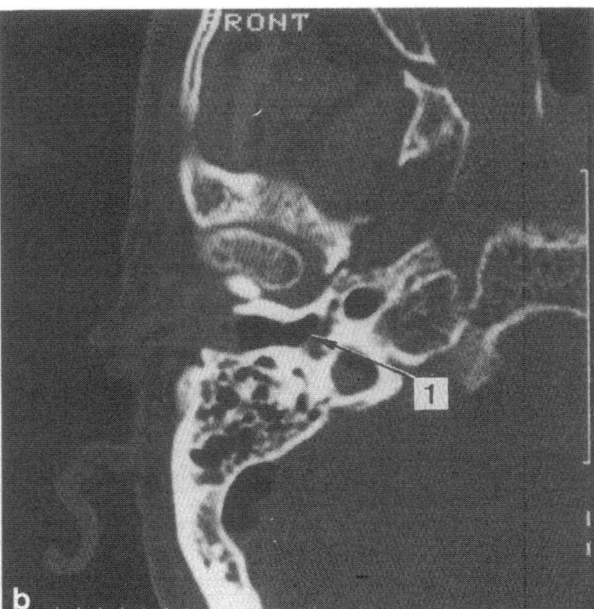

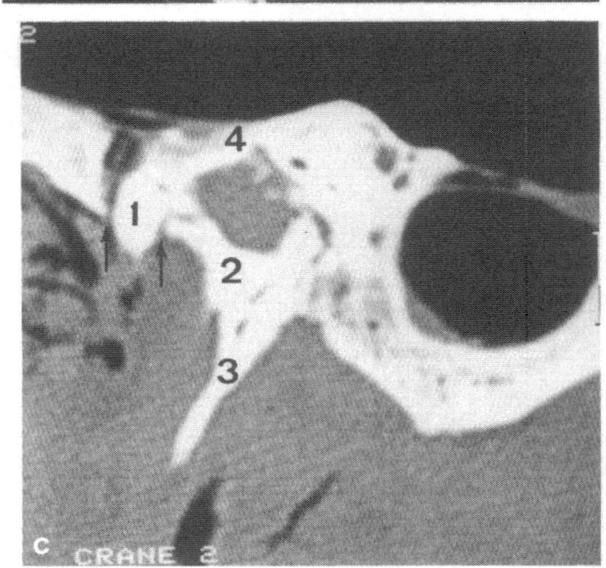

Fig. 6 a–c. External acoustic meatus. **a** Frontal CT section passing through the EAM *(white arrow)* of a right petrous. Note the two portions, external fibrocartilaginous *(white squares)* and internal bony *(black squares)*. Note also the obliquity of the handle of the malleus *(1)* indicating the direction of the tympanic membrane. The white star marks the isthmus. **b** Transverse CT section of a right petrous (seen from below). Note again the obliquity of the tympanum *(1)*. **c** Sagittal CT section. Inferior prolongation of tegmen tympani *(1)* splitting the petrotympanic fissure *(arrows)*. Tympanal *(2)*. Styloid process *(3)*. Tegmen tympani *(4)*

and inward, forming an angle of 20° with the sagittal plane [42] and thus of about 30° with the axis of the petrous (Fig. 10a, b). It is not essential for the neurosurgeon to know every detail of the six faces described in the fundamental texts [89], but he should be aware of some of them serving as landmarks or dangers to be avoided.

Outer Wall (Fig. 7). This is the tympanic wall, as the tympanum forms three-fifths of it, the remainder consisting mainly of the bony framework in which this membrane is inserted. This framework is very narrow over almost all its circumference, measuring only about 2 mm, save in its uppermost part where there is a space of about 4 mm deep according to Savié and Djerié [90] between the tympanum and the roof; this is the wall of the compartment. It is oblique inwards and its outer face corresponds to the superior wall of the external acoustic meatus (Fig. 10c).

The tympanum, fixed in its peripheral frame, forms a fibrous membrane occluding the base of the external acoustic meatus. This membrane is not flat but inwardly convex, having the shape of a cone whose apex is 2 mm more medial than its base. To its internal aspect the handle of the mal-

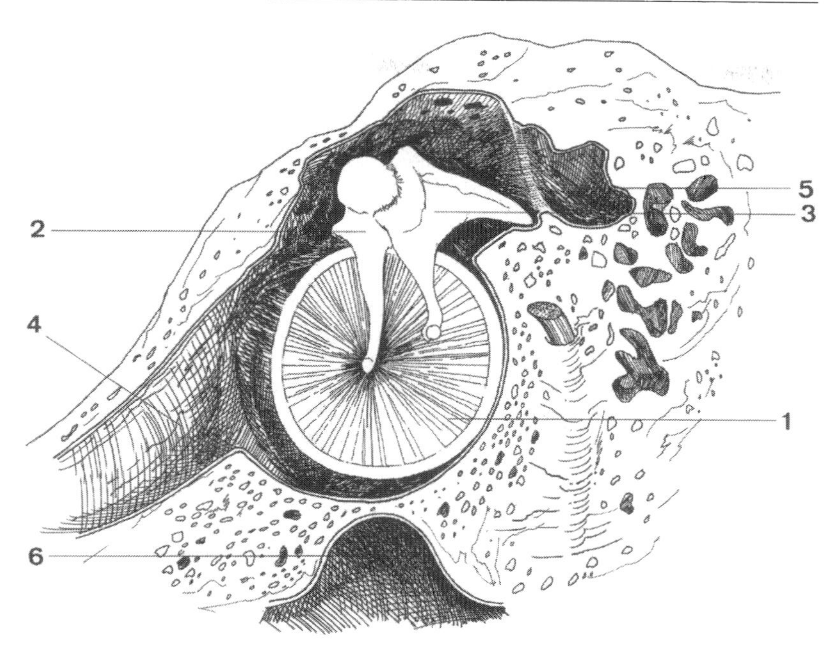

Fig. 7. Outer aspect of tympanum, drawing. Tympanic membrane *(1)*, malleus *(2)*, incus *(3)*, auditory tube *(4)*, antrum *(5)*, jugular bulb *(6)*

leus is attached, the inferior extremity of which corresponds to the apex of the cone (umbilicus). It is invested on its inner surface by the mucosa that clothes all the walls of the tympanic cavity and on its outer face by a cutaneous layer continuous with the skin of the external acoustic meatus.

Inner Wall. This is the labyrinthine wall, as it is formed by that portion of the labyrinthine massif which emerges from the mastoid cells. Several features are to be recognized (Fig. 8):

- the promontory. This is a smooth rounded prominence, flush with the wall like a sphere that is three-quarters buried. It is raised by the first turn of the cochlear spiral, which is nearly at the center of the wall. It bears the fine radiating impression of the tympanic nerve and its branches. Its anterior slope fades in a gentle incline towards the anterior wall. Its posterior slope, however, is abrupt. It is umbilicated by a fossula of variable shape (a gothic arch, triangular, trapezoidal, comma-shaped according to Anson and Donaldson [3]): the fossula of the cochlear window, at the base of which is the cochlear (round) window which faces backwards and slightly outwards. This cochlear window opens into the scala tympani of the cochlea. It is occluded by a fibrous membrane, the secondary tympanic membrane. The posterior margin of the fossula of the cochlear window is bordered by a small bony crest running obliquely backwards: the subiculum of the promontory. A depression, the tympanic sinus opens above and behind this crest. This depression is usually embedded 3 to 4 mm beneath the facial canal and the lateral semicircular canal, but it may be much more extensive, even sometimes penetrating as far as the posterior semicircular canal. The base of this tympanic sinus may be opened during drilling of the posterior labyrinthine massif in the course of the translabyrinthine approach. Its careful occlusion will prevent the risk of a CSF fistula towards the tympanic cavity. At the level of the cochlear window and above it, there is an elliptical fossula the long axis of which runs obliquely backward and downward, laterally orientated, 3 mm long and 1.5 mm high. This is the fossula of the vestibular (oval) window, which is situated at its base and opens into the scala vestibuli of the cochlea. This vestibular window is obstructed by the base of the stapes, held in place by the annular ligament. It opens into the vestibular cavity, which is widely broken open in the course of the translabyrinthine approach. Thus the base of the stapes is quite often mobilized, either during the drilling or when the middle ear is reached, and the gap so opened connects the operative cavity and the middle ear, allowing exit of CSF towards the tympanic cavity and then into the auditory tube. To prevent this possible postoperative rhinorrhea, it is proper to safeguard the base of the stapes or, if it has been mobilized, to carefully block the connexion with muscle and biologic glue.

- The canal of the tensor tympani muscle (Fig. 8). This small canal, containing the belly of the muscle, half-buried in the labyrinthine surface, raises a

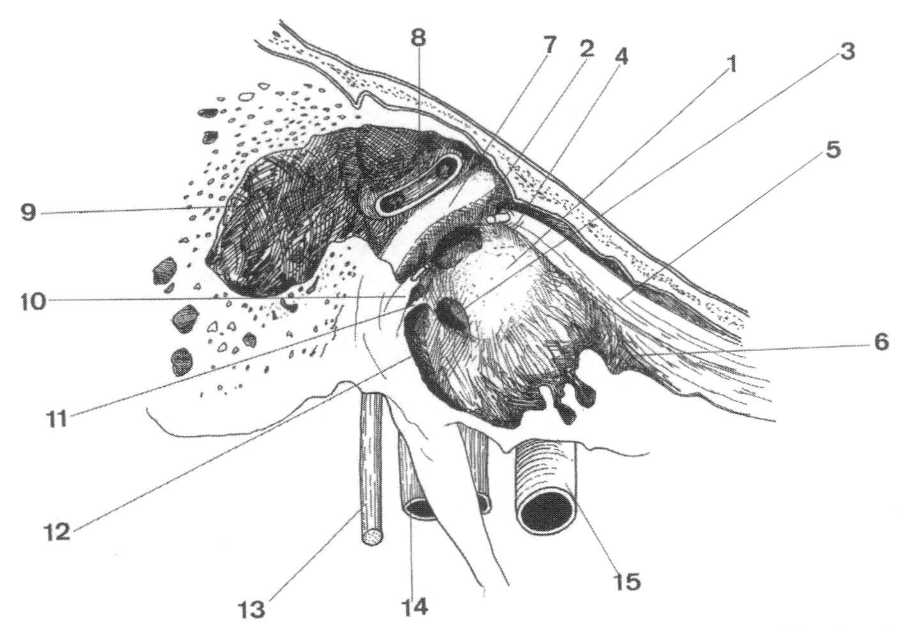

Fig. 8. Internal aspect of tympanum, drawing. Promontory *(1)*, fenestra cochlea (round) *(3)*, cochleariform process *(4)*, canal for tensor tympani *(5)*, auditory tube *(6)*, second part of facial nerve *(7)*, lateral semicircular canal (opened) *(8)*, antrum *(9)*, pyramid and tendon of stapedius muscle *(10)*, ponticulus *(11)*, sinus tympani *(12)*, extracranial facial nerve *(13)*, jugular vein *(14)*, internal carotid artery *(15)*

small horizontal ridge just above the promontory. This ridge, beginning in front of the promontory and about 5 mm long, ends more or less at the level of the center of the latter as a small prominence, the cochleariform process, at the apex of which there opens a small orifice through which the tendon of the tensor tympani emerges at a right angle.

- The second part of the facial nerve (Fig. 8) creates a prominence appearing just above the cochleariform process, directed backwards and downwards at an angle of 35 to 40° with the horizontal, skirting the upper border of the fossula of the vestibular window to disappear in the posterior wall. The shell overlying the facial nerve is more or less thin, sometimes dehiscent so as to leave the nerve exposed on the wall of the tympanic cavity.
- The lateral semicircular canal (Fig. 8) forms another prominence just above the former one. It is smooth, convex, also directed downwards and backwards, but less inclined to the horizontal, at 10°, than the prominence of the facial canal from which it progressively distances itself. It too plunges into the posterior wall of the tympanic cavity. Directly visible after the antrum has been opened, in the first stages of the translabyrinthine approach, this prominence of the lateral semicircular canal is an excellent landmark for locating the second part of the facial nerve (the tympanic portion).

Posterior Wall. This is the mastoid wall and is very narrow, especially in its inferior portion. It is divided into two parts:

- *The aditus ad antrum* (Fig. 9) occupies the upper two-fifths. It is a roughly triangular orifice with the apex below, 4 to 5 mm high, connecting the antrum and the upper part of the tympanic cavity known as the epitympanic recess (Fig. 10c). Its inner border corresponds to the prominence of the lateral semicircular canal. Its lateral border is part of the mastoid. Just under its apex there is excavated a small fossula, the fossa of the incus, to which is applied the short or horizontal limb of the incus. The posterior ligament of the incus fixes this limb in the fossa of the incus.
- The retrotympanum is formed by the inferior three-fifths. According to Proctor [79], it is always possible in the adult to make out three prominences (Fig. 9):
  - *The pyramidal eminence:* this is an almost vertical conical eminence whose apex shows a small orifice from which emerges the tendon of the stapedius muscle. This, the smallest muscle in the body, is lodged within a small canal 6 to 7 mm long hollowed in the pyramid. Its medial border constitutes the outer margin of the entry to the tympanic sinus. A bony bridge, the ponticulus,

often connects this pyramid or pyramidal eminence to the promontory; it then divides the sinus tympani into 2 cavities, the sinus tympani properly so-called below and the posterior sinus tympani of Proctor [79] above.

- *The chordal eminence*, situated laterally to the pyramid, may be a simple ridge or a true conical prominence from the apex of which the chorda tympani emerges. It is often joined to the pyramidal eminence by a transverse ridge, the chordal ridge.
- *The styloid eminence*, often minor, sometimes more marked, is a small prominence situated at the base of the pyramid and just in front of it. Within the tympanic cavity it marks the base of the styloid process. This eminence is prolonged inward by a ridge which is no more than the termination of the subiculum which joins it to the promontory.

Between the chordal eminence and the pyramid there is a small depression: the posterior sinus or facial sinus.

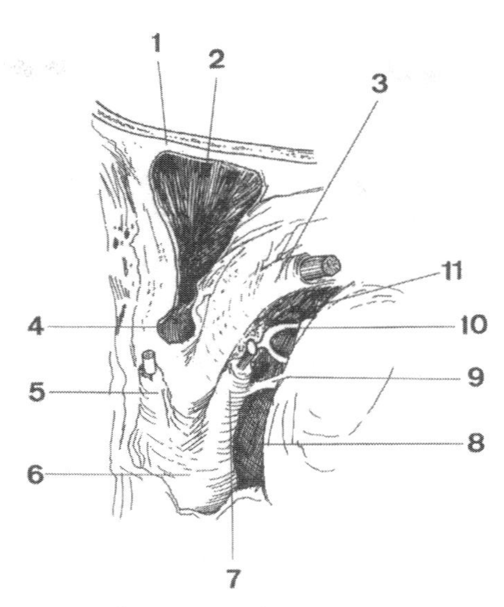

Fig. 9. Posterior aspect of tympanum, drawing. Tegmen tympani *(1)*, aditus ad antrum *(2)*, second part of facial canal *(3)*, fossa incudis *(4)*, chordal eminence *(5)*, styloid eminence *(6)*, pyramid *(7)*, sinus tympani *(8)*, ponticulus *(9)*, stapes *(10)*, posterior tympanic sinus of Proctor *(11)*

Anterior Wall (Fig. 8). This is the carotid wall; like the posterior wall it is very narrow. In its inferior third it is in direct relation with the vertical portion of the carotid canal. The often very slender bony wall is pierced by openings through which pass nerve strands anastomosing the tympanic nerve and the pericarotid plexus, as well as minor caroticotympanic branches of the internal carotid. The middle third is occupied by the tympanic orifice of the auditory tube. This is an oval orifice with a long axis of 4 to 5 mm and a small axis of at most 2 mm. Just above and laterally to it is the exit orifice of the chorda tympani. The upper third, according to Savié and Djerié [90], is usually (70%) pneumatized and in half the cases presents a depression, the anterior epitympanic sinus.

Superior Wall. This is the tegmen tympani (Fig. 8). This roof becomes progressively lower from behind forward, to the extent that the height of the tympanic cavity decreases of about 15 mm behind to 7 mm in front. According to Savié and Djerié [90], this wall is concave downwards and has an average thickness of 2.5 mm, but may range from the thickness of a sheet of paper to 5 mm. In 42% of cases it is made of compact bone (1 to 3 mm), in 31% it is thicker (2 to 5 mm) and is pneumatized but is compact on its intracranial aspect, while in 16% of cases the cells open into the endocranium. Finally, in 10% of cases it has the thickness of a sheet of paper. It is important to be aware of these findings during stripping of the dura mater in the approach via the middle cranial fossa, especially as there may exist spontaneous dehiscences, the incidence of which varies between 20 and 34% according to Kapur [52].

Inferior Wall (Fig. 7). This is the jugular wall, corresponding to the jugular bulb. Graham [39] states that it may consist of compact or pneumatized bone. Its thickness varies, depending mainly on the height of the jugular fossa. It is not unusual (6.7% of cases according to Korner [54]) for a dehiscence in the floor to bring the jugular bulb and the tympanic cavity into contiguity.

The Chain of Ossicles. This is formed by the malleus, firmly attached by its handle to the tympanic membrane, the intermediate element or the incus, and the stapes whose base is embedded in the vestibular window (Fig. 9). These three ossicles articulate between themselves and are additionally suspended from the walls of the tympanic cavity by ligaments: superior ligaments suspending the malleus and the incus from the superior wall, the anterior ligament attaching the malleus to the anterior wall, the posterior ligament fixing the incus to the posterior wall, and the lateral ligament fixing the head of the malleus to the wall of the epitympanic recess. Further, two muscles act on this chain of ossicles: the tensor tympani pulls the malleus inward

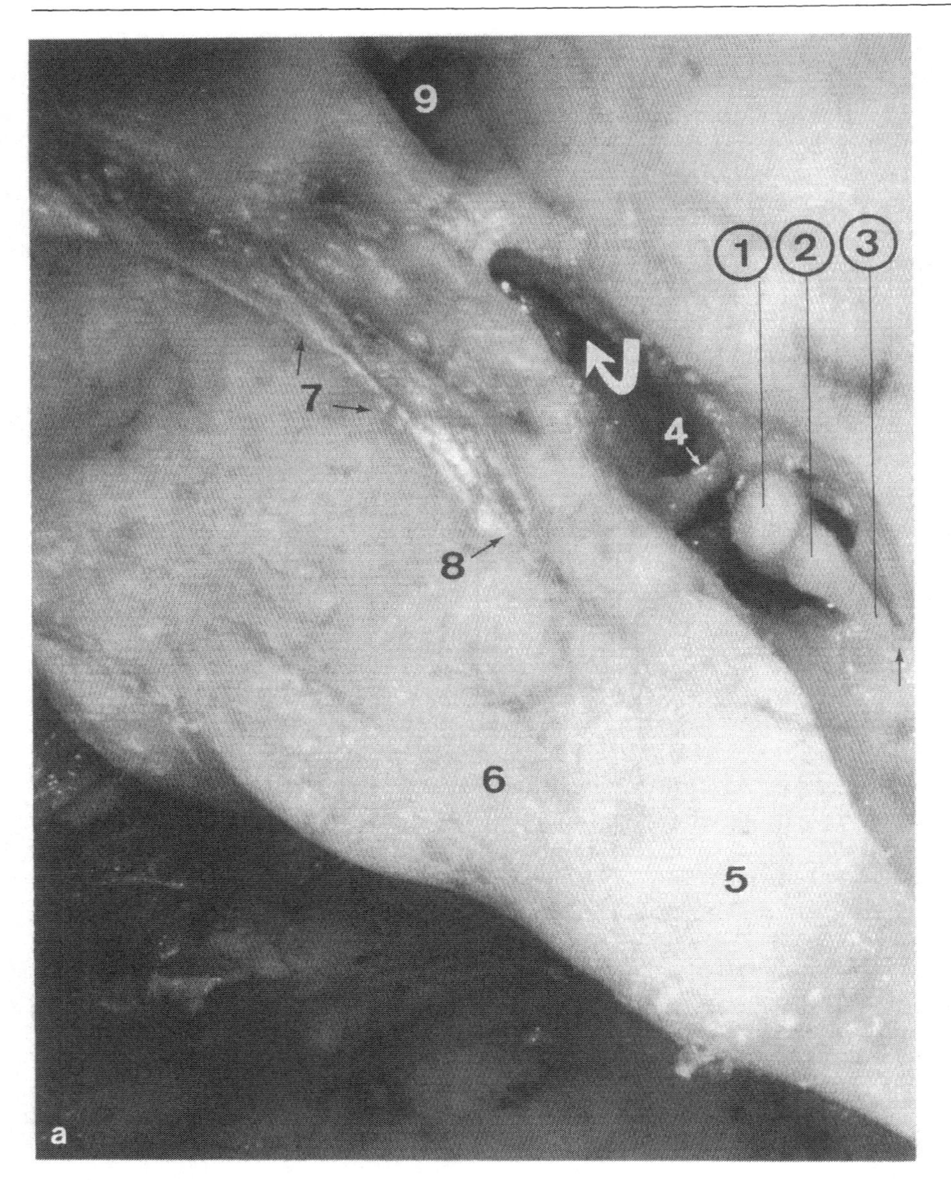

Fig. 10 a–c. The tympanum. **a** Superior aspect of right petrous pyramid. Superior view of tympanum after trephining of tegmen tympani. Visible are: the head of the malleus *(1)*, articulating with the incus *(2)*, whose short limb *(3)* is applied to the fossa incudis *(arrow)*. More deeply: the tensor tympani tendon *(4)*, arcuate eminence *(5)*, meatal field *(6)*, petrous nerve *(7)*, emerging from the hiatus of its canal *(8)*. The white arrow indicates the entrance to the auditory tube. Spinous foramen *(9)*

and tightens the tympanic membrane, while the stapedius has the converse effect. Only the incus deserves special mention, since it is always exposed during the translabyrinthine approaches and must be recognized and handled gently so as not to risk tearing the tympanic membrane via the intermediate malleus (Fig. 7). Anson and Donaldson [3] liken its shape roughly to that of a tooth with two roots, the short and long limbs. The body articulates with the head of the malleus. The long (vertical) limb, virtually parallel to the handle of the malleus, articulates with the head of the stapes. The short (horizontal) limb extends backwards and its tip is applied to the fossa incudis to which it is attaches by the posterior ligament. The tip of this short limb is an important landmark for locating the facial nerve and the lateral semicircular canal. According to Anson and Donaldson [3], this tip is at about 1.25 mm laterally to the lateral semicircular canal and the genu of the facial nerve, and is situated at about

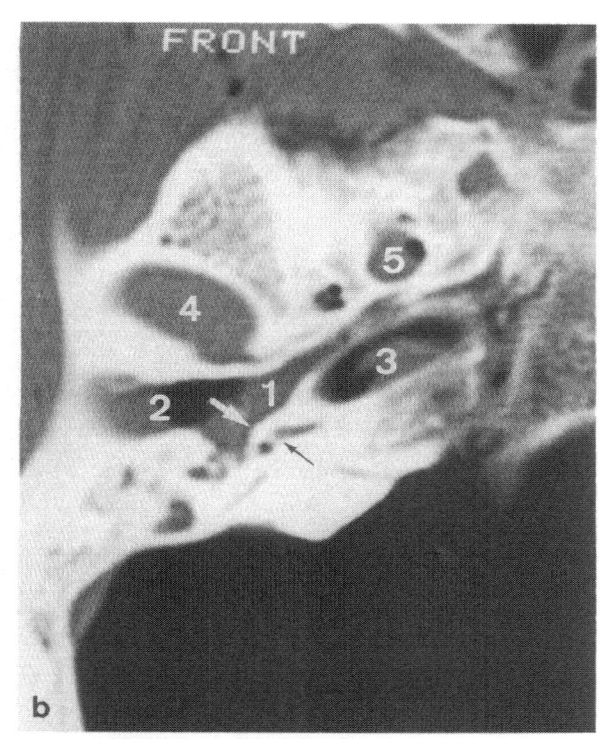

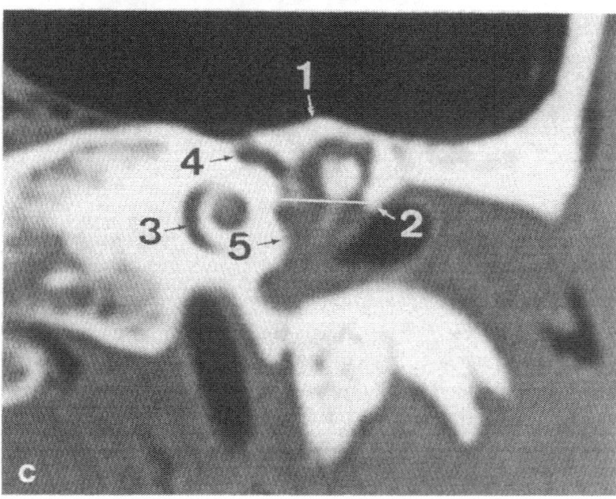

Fig. 10. b Transverse CT section of a left petrous passing through the tympanum *(1)* EAM *(2)*. Promontory *(white arrow)*. Start of first turn of cochlear canal *(black arrow)*. Carotid canal *(3)*. Fundus of glenoid fossa *(4)*, foramen ovale (5). **c** frontal CT section of a right petrous: tegmen tympani *(1)*, anterior wall *(2)*. The *horizontal line* (in white) starting from this is the lower limit of the attic. Cochlea *(3)*, compartment of geniculate ganglion *(4)*, promontory *(5)*

1.77 mm beneath the lateral semicircular canal and 2.36 mm below and a little medial to it.

The tympanic cavity is entirely clothed with mucosa, as are the ossicles, which raise this mucosa into folds comparable to the peritoneal mesenteries. The shape of the cavity and of these folds bring about a division into compartments. In the vertical direction the prominence on the internal wall of the canal for the tensor tympani and of the second part of the facial nerve and the prominence of the outer wall due to the obliquity of the wall of the recess, produce a constriction. Above, there is the epitympanic recess (Fig. 10c) which contains the head of the malleus and the body and short limb of the incus, and which communicates behind via the aditus with the mastoid antrum. Below is the mesotympanum or atrium, which contains the handle of the malleus, the long limb of the incus and the stapes, and which communicates with the upper pharynx by the auditory tube. Further, as the floor of the tympanic cavity if often lower than the lower border of the tympanic membrane, there exists a deep recess called the hypotympanum. Again, in the anteroposterior plane, the posterior part of the cavity is called the retrotympanum as opposed to the anterior part which is the protympanum, with the mesotympanum as the intermediate part.

The Auditory Tube (Figs. 7, 10 and 20a). This is an osteocartilaginous channel passing obliquely downwards, forward and inward, with an average length of 37 mm, connecting the tympanic cavity to the upper pharynx. It consists of two cones, one lateral and bony, the other medial and fibrocartilaginous, joined at their apices at a constriction, the isthmus of the tube, which measures 2 mm high and 1 mm wide at the most. The bony tube is no more than the protympanum, formed by the adhesion of the tubal process of the tympanal portion of the temporal bone to the petrous pyramid. It is subjacent to the canal of the tensor tympani and corresponds to the carotid canal medially, from which it is separated by an often very thin bony wall (Fig. 20a). The auditory tube must be carefully blocked with muscle, at the same time as the tympanic cavity, at the end of the translabyrinthine approach in order to avoid any risk of postoperative cerebrospinal rhinorrhea.

Internal Ear

The internal ear is a complex organ specialized in the functions of hearing and balance, and consists of two intricate systems: the bony labyrinth, which is a maze of cavities hollowed out in the petrous pyramid, intercommunicating and filled with perilymphatic fluid, and the membranous labyrinth, an equally complex system made up of spaces and epithelial tubes, differentiated by site into sensory systems specialized for audition (cochlear canal) and balance (vestibule), roughly adapted to the shape of the bony labyrinth but smaller than it, floating in the perilymphatic fluid and filled with a slightly different fluid, the endolymphatic fluid. Each of these two systems communicates with the subarachnoid spaces by its own canal: the cochlear duct for the bony labyrinth and the perilymphatic system, the aqueduct of the vestibule for the membranous labyrinth and the endolymphatic system. This labyrinth is connected with the internal acoustic meatus, which channels its nerves and vessels.

The membranous labyrinth, though cardinal in terms of function, is actually negligible in the field of otoneurosurgery. A detailed description is not required here and may be found in such books as those of Anson and Donaldson [3] or Legent et al. [56], or in the article by Tran Ba Huy et al. [98]. On the other hand, a detailed knowledge of the bony labyrinth is essential when the surgeon undertakes the drilling burring of the pyramid, either to avoid opening it, or on the contrary, to resect all or part of the labyrinthine massif while referring to the pattern of the progressively eroded cavities. This bony labyrinth is limited (Figs. 12 and 14) by a cortex of compact bone 2-3 mm thick, usually quite distinct from the spongy bone which forms the rest of the pyramid; but in the adult there exist zones of fusion of variable extent with the layer of compact bone which forms the cortex of the pyramid. This bony labyrinth is about 2 cm long. It is located roughly in the axis of the pyramid, but also slightly transversely, as noted by Girard [37], since its anteromedial portion, the cochlea, adheres to the cortex of the anterior slope of the petrous, while its posterolateral portion, the posterior semicircular canal, is close to the posterior wall. The bony labyrinth occupies the middle portion of the pyramid. It is divided into two parts.

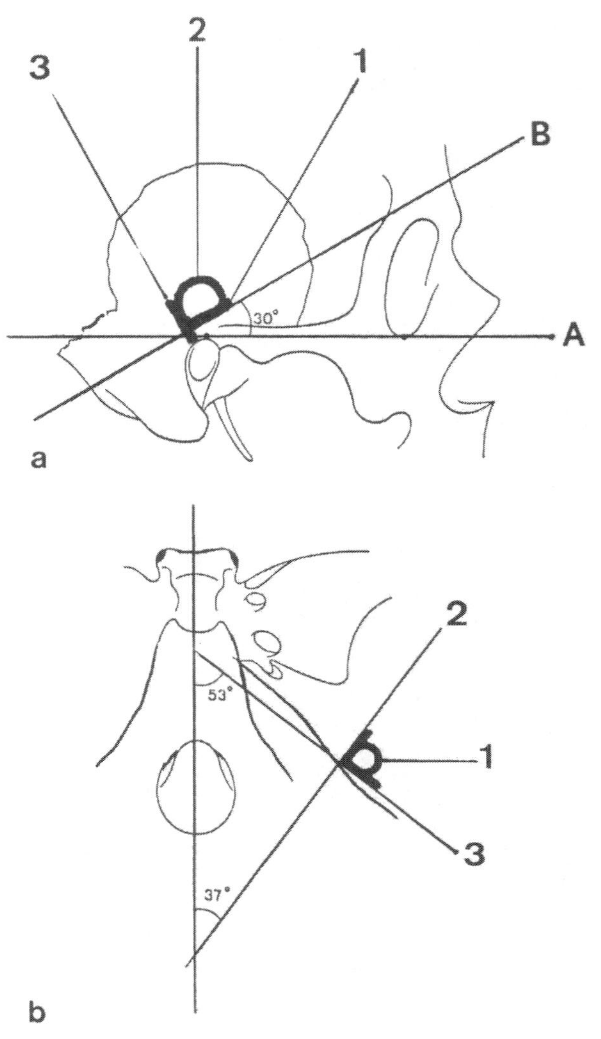

Fig. 11 a, b. Posterior labyrinth. Spatial arrangements. **a** diagram of lateral view. Plane of Francfort *(A)*, plane of lateral semicircular canal *(B)*, lateral semicircular canal *(1)*, anterior SCC *(2)*, posterior SCC *(3)*. **b** Diagram of horizontal section. Plane of lateral SCC *(1)*, anterior SCC *(2)*, posterior SCC *(3)*

Posterior Labyrinth. This is the organ of balance. It is composed of the vestibule on which are implanted the semicircular canals and the endolymphatic duct. It is this posterior labyrinth that is resected during the translabyrinthine approaches in order to gain access to the internal acoustic meatus and its contents.

• The semicircular canals (SCCs) (Figs. 11-14) are eroded initially as they are most lateral. They are 3 small canals, the external or lateral, the superior or anterior, and the posterior, of about 1 mm in diameter, each forming an incomplete loop, the posterior one being more tightly closed. Each loop has a diameter of 7 to 8 mm and is arranged in a plane at right angles to those of the other two. The lateral canal (Fig. 11b), which forms a loop with an external convexity, is located in the horizontal plane of the head. In a subject standing normally and looking ahead, this plane forms an angle of roughly 30° open forwards and upwards (Fig. 11a) with the plane passing through the inferior orbital margins and the upper border of the external acoustic meatuses (plane of Francfort). The anterior and posterior SCCs are located in planes perpendicular to each other and to the plane of Francfort. The anterior canal, forming a loop with a superior convexity, is in a plane practically at right angles to the axis of the petrous. It is not of very great importance during the translabyrinthine approach except that the

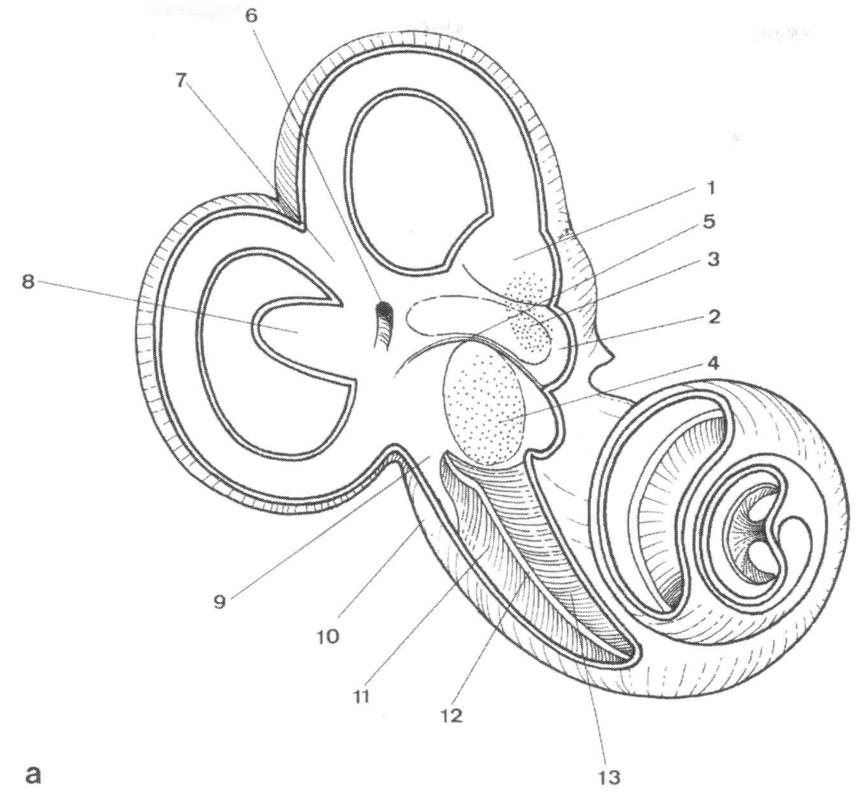

Fig. 12 a–d. Labyrinth. **a** Schematic diagram. Ampulla of anterior semicircular canal *(1)* and of lateral SCC (resected) *(2)*. Utricular fossula *(3)*, saccular fossula *(4)*, vestibular crest *(5)*, orifice of endolymphatic canal *(6)*, crus commune *(7)*, non-ampullary extremity of resected lateral SCC *(8)*, orifice of cochlear canal *(9)*, cochlear window *(10)*, scala tympani *(11)*, spiral lamina *(12)*, scala vestibuli *(13)*, helicotrema *(14)*

internal acoustic meatus is roughly engaged in its concavity and has practically the same diameter. On the other hand, it is an essential feature in the suprapetrous approach, since it represents both one of the major hazards, its accidental opening producing immediate and permanent deafness, and the most useful landmark. The internal acoustic meatus is revealed by drilling on the line which makes an open angle of 60° backwards and inwards (Fig. 39) with the plane of the canal after it has been exposed on the internal slope of the arcuate eminence. The posterior SCC is in a plane parallel to the posterior face of the petrous. Each of the canals is slightly squashed in the plane of its loop, to the extent that on section their diameter is not absolutely circular but rather oval. Each canal also shows an ampullary dilatation (Fig. 12) (2.5 mm by 1.5 mm) at one of its implantations on the vestibule, the other implantation retaining the normal sectional area. The ampulla is anterior for the lateral and anterior canals, and inferior for the posterior canal. Further, the anterior and posterior SCCs have a common posterior limb, the common crus, and therefore a common non-ampullary orifice (Figs. 12 and 13). During the translabyrinthine approach, the burr progressively erodes the labyrinthine massif by following the contours of these canals, which guide it progressively to the vestibule. In the proceedings, the anterior limb of the lateral SCC must be respected, for this safeguards the second or tympanic part of the facial nerve which abuts its anterior border. It is also necessary to respect the ampullae of the lateral and anterior SCCs as the genu of the facial nerve is immediately in front of them (Fig. 35 b).

• *Vestibule* (Fig. 11). This is the central portion of the labyrinth. It is situated between the tympanic cavity laterally, emerging on the internal face of the latter just opposite the external acoustic meatus, and the internal acoustic meatus medially, its medial wall forming the base of this channel. It is a small globular cavity whose long axis runs obliquely forward and outward, making an angle of 45° with the sagittal plane according to Girard [37] and measuring 6–7 mm long, 3 mm wide and 5–6 mm deep. Precise drilling of its lateral convexity, after abrasion of the SCC, and then identification of certain features of its internal wall, are essential stages of the translabyrinthine approach and the operator must be familiar with them.

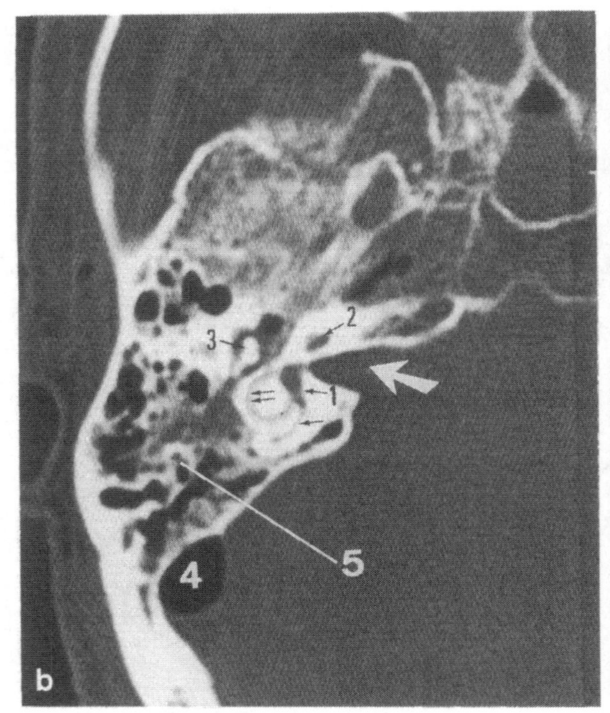

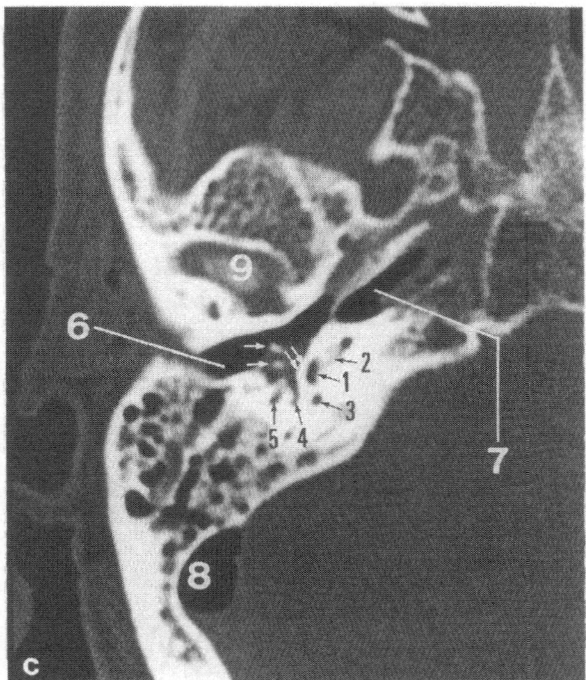

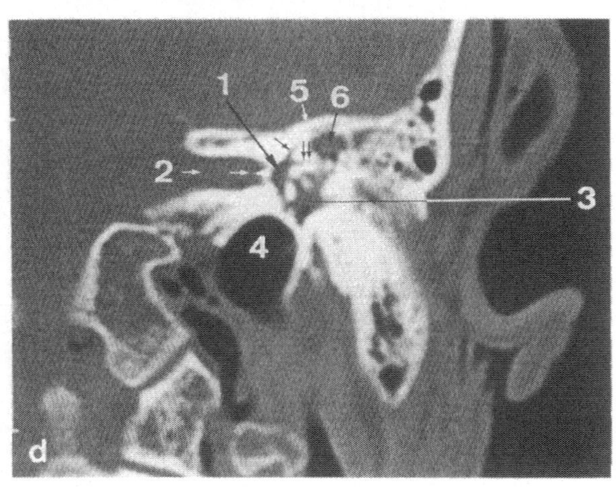

Fig. 12 b, c. Transverse CT section left petrous: IAM *(white arrow)*. Vestibule *(1)*, posterior SCC *(black arrow)*, lateral SCC *(double arrow)*, cochlea *(2)*, malleus and incus *(3)*, sigmoid sinus *(4)*, mastoid cells *(5)*. **c** Transverse CT section of left petrous inferior to section 11 b. Vestibule (lowest level) *(1)*, initial straight portion of cochlea *(2)*, posterior SCC (section of ampullary limb) *(3)*, sinus tympani *(4)*, bend of facial nerve *(5)*, handle of malleus and long limb of incus *(white arrow)*, promontory *(double arrows)*, EAM *(6)*, carotid canal *(7)*, sigmoid sinus *(8)*, glenoid fossa *(9)* **d** Frontal CT section right petrous. Vestibule *(1)*, anterior SCC *(black arrow)*, lateral SCC *(double arrows)*, IAM *(2)*, falciform crest *(white arrow)* tympanum *(3)*, jugular bulb *(4)*, tegmen tympani *(5)*, aditus ad antrum *(6)*

On the convexity there are several orifices:

- The ampullary orifices of the lateral and anterior SCCs at its anterosuperior portion, adjacent, 2.5 mm long by 1.5 mm high, separated only by a slender horizontal ridge
- The non-ampullary orifices of these SCCs, 3 mm behind those just mentioned, 1 mm in diameter, like a double-barrelled gun, the upper orifice for the crus commune, the lower for the lateral SCC, separated only by a small and very narrow ridge
- The vestibular window, which should not be uncovered by drilling except when it is desired also to drill the anterior labyrinth (transcochlear approach). This is an orifice 3 mm long and 1.5 mm high, situated on the antero-inferior part, 2 mm beneath the two ampullary orifices of the lateral and anterior SSC.

There are several landmarks on the internal wall (Fig. 12):

- The most important is the hemispheric or saccular fossula. This is a shallow circular depression occupying the antero-inferior portion of the internal wall, bordered by a poorly-marked ridge, the vestibular ridge, except above and in front where it makes a more marked prominence, the pyramid of the vestibule. To the surgeon it appears mainly as a bluish patch because of the thinness of the wall separating it from the internal acoustic meatus. At this site numerous apertures allow passage of the nerve strands emerging from the saccule (saccular cribriform macula), which join to form the inferior vestibular nerve, revealed by cautious drilling of the base of the fossula (Fig. 12a).

- The semi-oval or utricular fossula is situated above the hemispheric fossula, from which it is separated by the vestibular crest. Above it is adjacent to the ampullary orifice of the anterior SCC. Here there is the utricular cribriform macula, crossed by the nerve strands emerging from the anterior and lateral semicircular ampullae and the utricle to form the superior vestibular nerve. The vestibular crest which separates the utricular (superior) and saccular (inferior) cribriform maculae therefore corresponds to the falciform crest which separates the superior and inferior vestibular fossulae at the base of the internal acoustic meatus (Fig. 12a and d).
- The vestibular orifice of the vestibular aqueduct is situated behind the semi-oval fossula (elliptical recess). It is a punctate orifice (0.2 to 0.3 mm in diameter according to Ogura and Clemis [65], situated 1 mm below and medial to the orifice of the crus commune.
- Aqueduct of the vestibule. This is a fine canaliculus connecting the vestibular cavity with the posterior cranial fossa. Initially, it takes an ascending path, about 1.4 mm long on average according to Ogura and Clemis [65] and parallel to the crus commune: then, at the mid-height of the latter, it curves backwards to cross the crus on its inner border, 1 mm medial to it according to Kartush et al. [53], for whom it marks the lateral limit of abrasion of the posterior wall of the IAM. It then curves again (Fig. 13) to travel outwards and backwards, almost horizontally, to reach the posterior fossa after a second course of 7.3 mm (but varying between 4 and 10 mm according to the pneumatization of the petrous) at the level of the ungual fossula. Its initial very fine diameter (0.5 mm) contracts at its bend and then widens progressively in delta fashion in the plane of the posterior surface of the petrous, to reach an average width of 6.2 mm at its mouth [65]. The canal contains the endolymphatic duct, formed by the confluence of two small canaliculi emerging from the utricle and saccule. This duct widens progressively to form the endolymphatic sac, which lies by the ungual fossula in the posterior cranial fossa where it adheres to the deep surface of the dura.

The Anterior Labyrinth. This is the organ of hearing, consisting of the cochlea and the cochlear duct. Whereas the preservation of the facial nerve and the need to maintain impermeability call for very precise drilling of the SCC and vestibule during translabyrinthine approaches, infraction of the cochlea, performed during transcochlear approaches when the facial nerve is already rerouted and when the problems of impermeability are managed otherwise, dose not call for such profound anatomic knowledge. Therefore the description will be more succinct, as the details can always be read in the fundamental texts [98].

- *The cochlea* (Fig. 12). This is a bony tube, 2 mm in diameter at its origin, progressively narrowing by half, coiled on itself like a snail and performing two and a half spiral turns of decreasing diameter. The coiling occurs round a conical axis, the modiolus, 3 to 4 mm high, the base of which corresponds to the cochlear area at the fundus of the internal acoustic meatus. The first turn of the spiral, slightly oval, measures 9 to 10 mm in its major axis and 7 to 8 mm in its minor axis. The axis of the modiolus is more or less perpendicular to that of the petrosal pyramid. The peripheral cortex is called the contoural lamina while the portions of the wall apposed by the coiling constitute the spiral septum. The lumen of the cochlea is divided in two by a diametric septum, the inner half of which is the bony spiral lamina, attached by winding around the modiolus, while the outer half is a membranous septum, the basilar membrane. These demarcate the scala vestibuli above and the scala tympani below, which communicate at the apex of the cochlea by an orifice, the helicotrema. The cochlea begins under the floor of the vestibule, to which it is attached at the level of the cochlear window. The scala vestibuli opens into the vestibular cavity by an orifice in the floor. The scala tympani below terminates in a blind end from which opens an orifice, the cochlear window, closed by the secondary tympanic membrane. For its first 5 to 6 mm the cochlea still remains straight (Fig. 12c), passing forward and inward in the axis of the pyramid to engage under the lateral extremity of the internal acoustic meatus. It runs tangentially to the inferior border of the cochlear area and then starts to coil. It ascends on the anterior aspect of the internal acoustic meatus without surpassing it in height and then redescends to surround the cochlear area which thus corresponds to the base of the modiolus. This first turn ends in front of the vestibule and the initial portion of the cochlea. The whole of this is apparent on the inner wall of the tympanic cavity, where it constitutes the promontory. The coiling continues for another turn and a half and ends in the cupula. During the suprapetrous approach, the surgeon must locate the position of the cochlea exactly if he wishes to preserve hearing while dealing with a small neurinoma (stage I or II). The cochlea is situated very precisely in the angle formed by the first part of the facial nerve, the geniculate ganglion and the petrosal nerve (Fig. 14). Moreover, it sticks on the anterior wall of the internal acoustic meatus near its base. Drilling of the bone around the meatus to isolate it and expose its contents should not be carried deeply near its base or in a forward direction if the integrity of the cochlea is to be respected during the suspetrous approach.

• *The aqueduct of the cochlea* (Fig. 13) is a narrow canaliculus opening in the scala tympani by a small orifice situated just beside the cochlear window the diameter of which, according to Rask-Andersen et al. [81] does not exceed 0.3 mm. It descends obliquely downward, backward and inward, and after a straight course of an average length of 12.9 mm (ranging from 8.8 to 17.3 mm according to Rask-Andersen et al. [81]) it widens slightly and ends at the summit of the fossula petrosa as an oval orifice with an average maximal diameter of 4.2 mm, though the extremes of range, according to Rask-Andersen et al. [81] are 2.6 to 6.7 mm. This cochlear canaliculus contains connective tissue in continuity

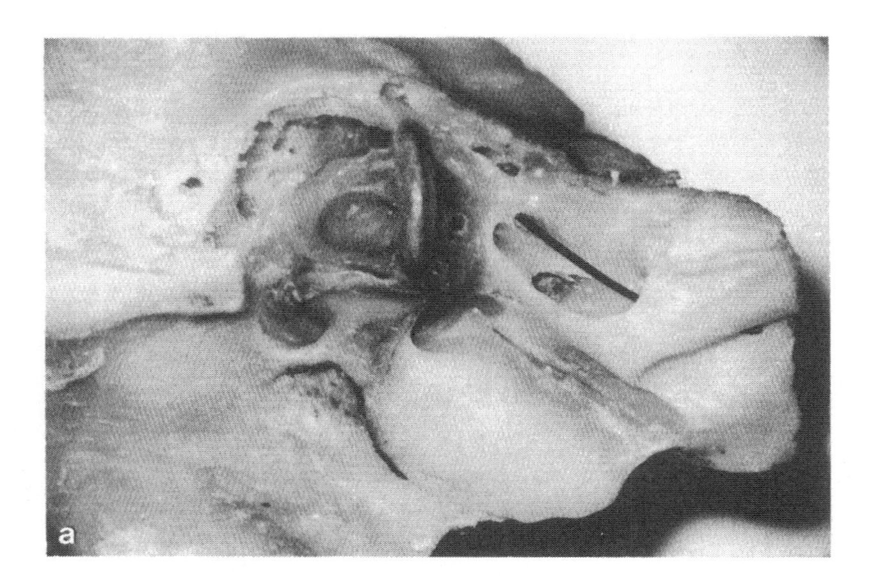

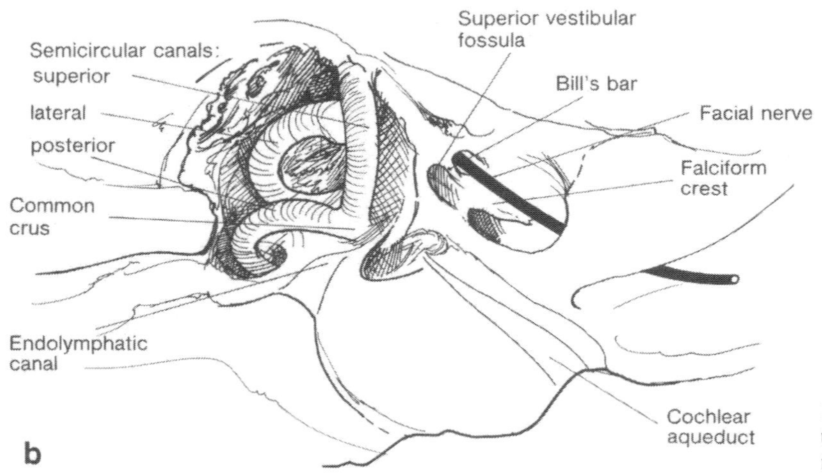

Fig. 13 a, b. Posterior labyrinth: dissection + diagram (Dr. Triglia). Labyrinth injected with methylene blue

with the meninges outside and the periosteum of the scala tympani inside; it is not a true perilymphatic canal comparable to the endolymphatic canal of the aqueduct of the vestibule. Sometimes, however, it is much larger and possibly permeable, which accounts for some cases of deafness developing in normal pressure hydrocephalus or some cases of otorrhea or rhinorrhea, either spontaneous or after operation on the cancellous bone of the ear. It is constantly accompanied by a narrow canaliculus 0.1 mm in diameter, running parallel to it at an interval of 0.3 mm, the canal of Cotugno or first accessory canal, through which travels the vein that drains the cochlea. The cochlear canaliculus forms an important landmark during the translabyrinthine approach, marking the most anterior limit for drilling under the IAM. To go beyond it risks damaging the glossopharyngeal nerve. Therefore one must take pains to identify this small vertical channel, parallel to the lower border of the IAM and halfway between it and the jugular bulb. Kartush et al. [53] state that this is possible in 90% of cases.

The internal Acoustic Meatus (IAM). This is a tubular excavation communicating widely with the posterior cranial fossa and therefore forming an integral part of the latter. It is classically described (Fig. 14) as having two walls, anterior and posterior, a ceiling, a floor and a fundus which corresponds to the inner wall of the vestibule. This fundus is barred by a transverse or falciform crest, which corresponds to the vestibular crest on the other side. This crest extends at length of the anterior wall, where its sickle shape gives it its name. It divides the fundus of the meatus into two levels. The upper level is divided by a vertical crest known by otologists as Bill's bar, so named by the pupils of William (Bill) House who clearly showed that its identification and preservation are the keys to identification and preservation of the facial nerve in the surgery of the IAM, especially in acoustic neurinoma surgery: in front there opens the facial canal, behind is the superior vestibular or utricular area corresponding to the semi-oval fossula. The inferior level shows in front the cochlear area, pierced by numerous orifices located in a spiral (spiral foraminous tract) through which the cochlear nerve strands emerge, while behind is the inferior vestibular or saccular area corresponding to the hemispheric fossula of the vestibule. On the posterior wall, 1 mm in front of the base and 2 mm above the floor, there opens another small orifice 0.3 mm in diameter, the foramen singulare, through which emerge fibers from the ampulla of the posterior SCC forming the posterior ampullary contribution to the vestibular nerve.

There is a surprising contrast between the descriptions found in the neurosurgical literature [66, 76, 80, 83], which are concerned mainly with the content of the meatus and describe its conformation rather vaguely and its length (1 cm) only approximately, and the otologic publications [2, 26, 43], which deal mainly with the proximity of the labyrinth, specify precisely the dimensions, thicknesses, distances, and define the limits not to be exceeded to safeguard the labyrinthine cavities. As long as one is restricted to the problem of total excision of a tumor within the canal while safeguarding the facial nerve, the neurosurgical viewpoint seems altogether adequate; but when it is a matter of preservation of hearing it is impossible not to respect otologic discipline since it is not enough to respect the nerves in the IAM; the integrity of the labyrinth must also be preserved. Logically, therefore, one needs to be thoroughly acquainted with the IAM and its relations with the labyrinth.

Orientation of the IAM:

- The biauricular axis joining the two EAMs passes through both IAMs, but the axis of each of the 2 IAMs makes an 8° angle with this biauricular axis, open backwards and inwards according to Girard [37]. This near-concordance is very useful to locate the IAM during the suprapetrosal approach.
- According to Cornelis [in 41], the axis of the IAM makes an angle of 80 to 100°, open forward, with the sagittal axis of the skull. It makes a 45° angle, open forward, with the axis of the petrous pyramid and, according to Duday [in 41], a 51° angle with the plane of the posterior face of the petrous.
- According to Cohadon et al. [22], the axis of the IAM makes a 35° angle of 35° (ranging from 28 to 45°) with the plane of the anterior SCC, though Parisier [69] puts this at 48° (26 to 62°) and Fisch [34] at 60°.
- Finally, according to Y. and B. Guerrier [41], the line passing through the roof of the IAM and the roof of the vestibule is flush with the upper border of the lateral SCC while the line passing through the floor of the IAM is flush with the roof of the EAM.

Dimensions of the IAM [2]:

- The average length of the anterior wall is 14.9 mm (12 to 19 mm) and there is perfect symmetry in 87% of cases.

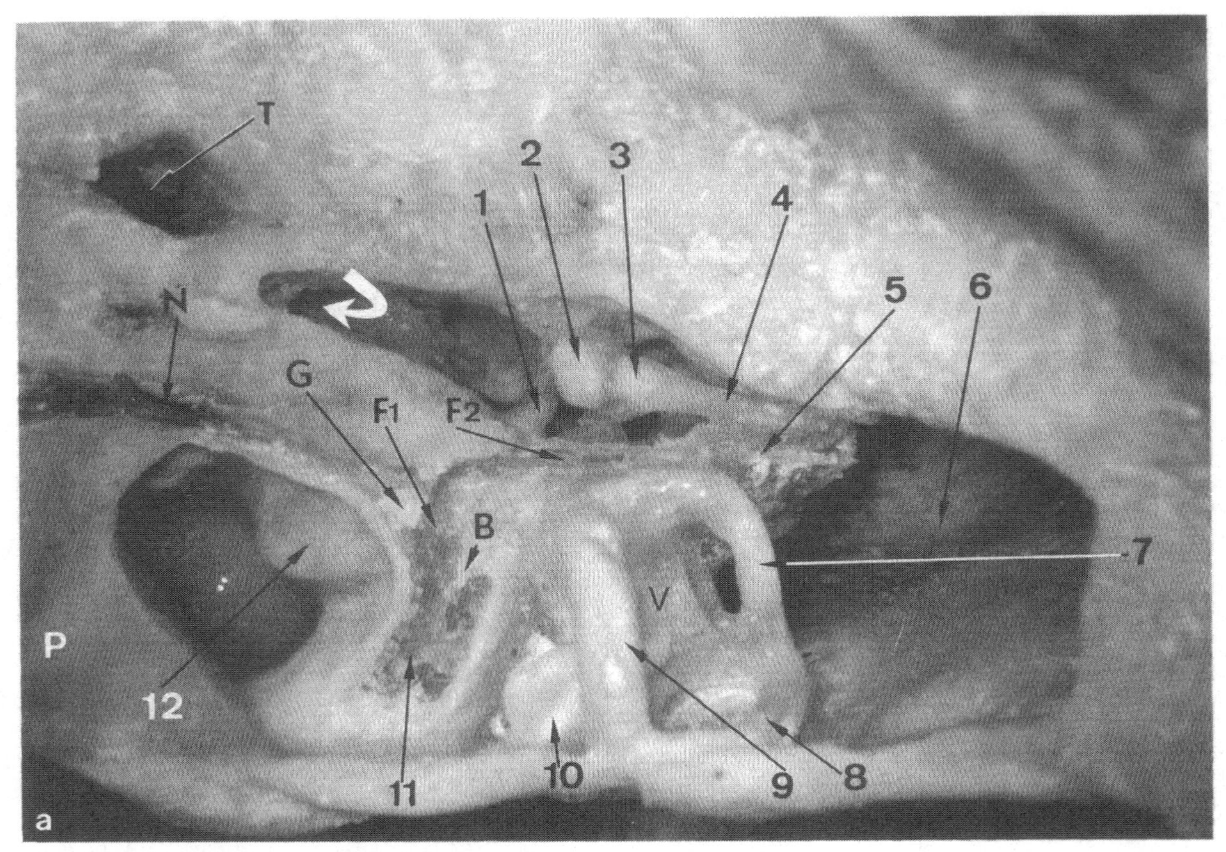
T
2
3
1
4
N
5
6
G
F1
F2
B
7
V
P
12
11
10
9
8
a

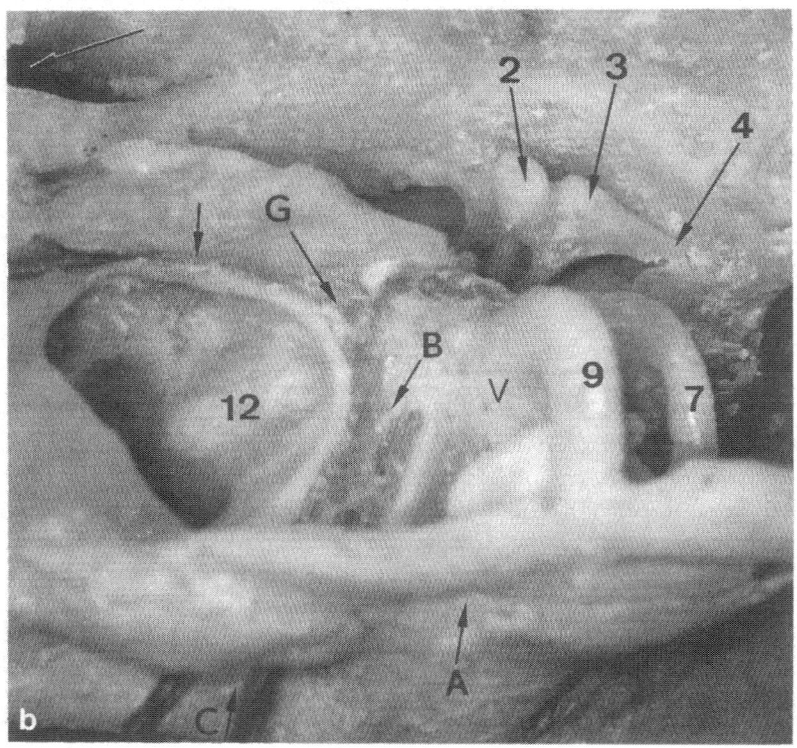
2
3
4
G
B
V
9
7
12
A
C
b

- The average length of the posterior wall is 9.9 mm (8 to 13 mm) with symmetry in 73% of cases.
- The average length of the roof is 10.9 mm (8.5 to 14 mm).
- The average length of the floor is 9.3 mm (7 to 12 mm).
- The mean vertical diameter is 5.9 mm (4 to 7.5 mm).
- The mean horizontal diameter is 5.4 mm (4 to 7 mm).

Measurements [26, 43]:

- On the posterior face of the petrous
  - Starting from the outer lip of the IAM (Fig. 13),
    the common crus is at 7.4 mm (5.5 to 9.5 mm)
    the vestibule is at 7.9 mm (6 to 10 mm)
    the ampulla of the posterior SCC is at 8.2 mm (6.5 to 10 mm)
  - Starting from the petrosal crest (Fig. 13),
    the common crus is at 5.7 mm (3 to 7 mm)
    the vestibule is at 7.9 mm (4.5 to 9 mm)
    the ampulla of the posterior SCC is at 9.2 mm (5.5 to 11 mm)
  - Starting from the surface of the posterior aspect of the petrous (Fig. 14a),
    the foramen singulare is at 6.8 mm depth (3.5 to 12.5 mm)
    the superior vestibular nerve is at 7.7 mm (4.5 to 13 mm)
    Bill's bar is at 9.5 mm (7 to 15 mm)
    the common crus is at 2.2 mm (1 to 3 mm)
    the vestibule is at 4.3 mm (3.5 to 6 mm)
    the ampulla of the posterior SCC is at 5.5 mm (4 to 7 mm)

Overall, these measurements indicate that the labyrinth overlies the last 2 mm of the posterior wall of the IAM and that trephining of the meatus, as recommended by Bucy [12] and Rougerie [87] must in no case extend as far as the bottom of the meatus if hearing is to be preserved. It should also be noted that the obliquity of the IAM requires the neurosurgeon to excavate to an average depth of 7.7 mm to open the meatus and that the facial nerve is to be found at the bottom of a gorge 9.5 mm deep (Fig. 12b).

- On the superior face
  - Starting from the loop of the anterior SCC (Fig. 15). According to Cohadon et al. [23], drilling 10 mm anterior and medial to the "blue line" of the anterior SCC in a direction parallel to the petrous crest leads inevitably to the roof of the IAM
  - Starting from the outer cortex of the temporal squama (Fig. 15). According to Batisse and Clerc [20], the apex of the loop of the anterior SCC is situated 24 mm medial to the outer cortex of the squama on the biauricular line. According to Voisin [99], the distance varies between 18 and 25.5 mm.
    According to Pialoux et al. [75], the inner border of the cochlea does not extend medial to a sagittal plane situated at 28 mm from the outer surface of the squama. Therefore, as advised by Charachon and Accoyer [16], excavation on the biauricular line at more than 28 mm from the outer cortex of the squama, near the petrosal crest, suffices to reveal the IAM
  - Starting from the surface of the superior face of the petrous (Fig. 15). Cohadon et al. [23] locate the loop of the anterior SCC at a depth of 1 to 3.5 mm. Bouche and Frèche [9] assess the average thickness of the roof of the IAM at 5 mm (3 to 7.5 mm).

The multiplicity of landmark systems shows that localization of the IAM is not easy if an approach under the superior face of the pyramid is considered. The anterior SCC is at once the hazard and the essential landmark in the middle cranial fossa (suprapetrous) route.

Facial Canal

This canal carries the facial nerve through the base of the skull and is, as remarked by Y. and B. Guerrier [41], "long, very long". Cawthorne [15] also says that it has "a record length" for the bony passage of a nerve: 28 to 30 mm. Moreover, its course is Z-shaped, threading its way between the posterior labyrinth, the anterior labyrinth, then the tympanic

Fig. 14a, b. The cavities of the petrous. Dissection (Pellet) exposing the cavities through the superior aspect. **a** Superior aspect. Tympanic cavity extending forwards towards the auditory tube *(white arrow)*. Cochleariform process and tensor tympani muscle *(1)*, head of malleus *(2)*, incus *(3)*, short limb of incus applied to fossa incudis *(4)*, Gelle's wall *(5)*, mastoid curettage *(6)*, lateral SCC *(7)*, posterior SCC *(8)*, anterior SCC *(9)*, petromastoid canal *(10)*, IAM *(11)*, cochlea *(12)*, Bill's bar *(B)*, first part of facial canal *(F1)*, geniculate ganglion *(G)*, second part of facial canal *(F2)*, tubercle of Princeteau *(P)*, spinous foramen *(T)*, vestibule *(V)*, petrosal nerve *(N)*, subarcuate fossa *(A)* **b** A slightly more posterior view. Subarcuate fossa *(A)*, aperture of IAM *(C)*

cavity, further complicated by the fact that the three segments do not extend in a single plane. The canal presents:

Three Parts

The first part (Figs. 14–16 and 35) is the labyrinthine portion. With an average length of 4 mm (69: 215 50 5.2), it begins at the fundus of the meatus in the superior compartment, bending forward by 50°

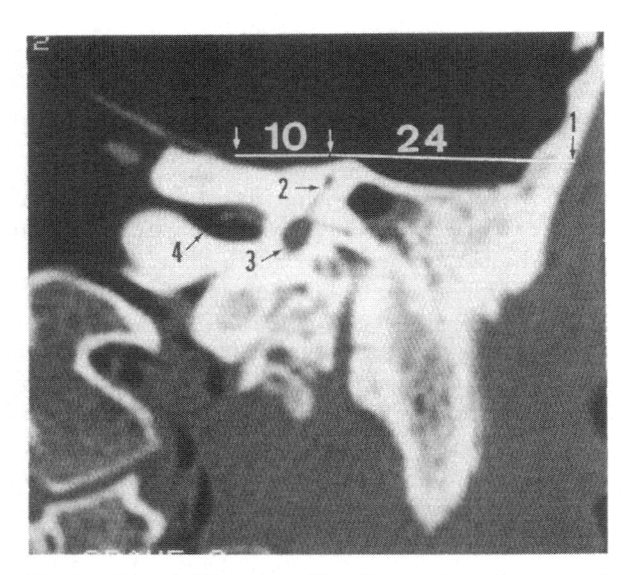

Fig. 15. Frontal CT section. The distance from the external cortex *(1)* to the summit of the anterior SCC *(2)* is 24 mm. This loop elevates the arcuate eminence. Vestibule *(3)*, IAM *(4)*. Excavation at 10 mm medial to the loop of the anterior SCC opens the IAM

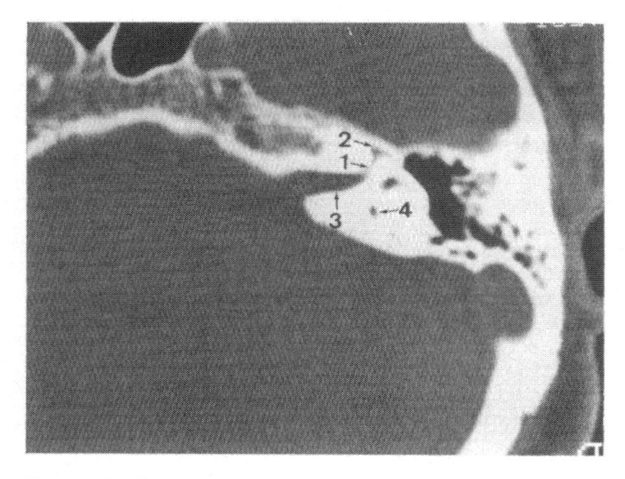

Fig. 16. Facial canal. Horizontal CT section passing through the first part of the canal *(1)* and the compartment for the geniculate ganglion *(2)*. IAM *(3)*, crus commune *(4)*

the general direction of the IAM, which makes an angle of 45° with the axis of the pyramid, so that the canal is practically at right angles to the axis of the latter. This labyrinthine portion describes a curve with a concavity forwards and inwards, skirting the supero-lateral flank of the first turn of the cochlear canal and continuing forward in the direction of the greater petrosal nerve. Moreover, it has a slightly ascending course, especially towards its termination, which brings it closer to the superior face. Its diameter is about 2 mm, with a constriction just at the beginning and at the end of the segment, zones where the facial nerve is particularly endangered during drilling. At its posterolateral border, it skirts the vestibule, the ampulla of the anterior SCC and the canaliculi of the utricular and superior and lateral ampullary fibers, i. e., the origin of the superior vestibular nerve, at the level of the superior cribriform macula, is just lateral to the first part of the facial. At its anteromedial border, it skirts the superolateral flank of the first turn of the cochlear canal and then comes to lie above the second turn of the spiral. According to Parisier [69], the interval between the anterior limb of the anterior SCC from the first turn of the spiral measures 5.8 mm (3 to 8 mm).

The second part (Figs. 14 and 35) is the tympanic portion. It is 10 to 12 mm long. It makes an angle "of 74° according to Claustre, 80° according to Winckler" [42] with the first part and passes outward and backward, making an angle of 35–40° with the sagittal plane of the skull. Like the first part, this segment is not horizontal and descends outward and backward following a slope of 37° according to Guerrier [42]. At its posteromedial border, it skirts the ampullary limb of the lateral SCC but forms an angle of about 10° with it, so that, situated at the same level as the ampulla of the lateral SCC and just in front of it, it gradually comes to lie under the anterior limb. This ampullary limb is thus an essential landmark of the second part of the facial nerve and the best way not to damage this part of the nerve during the translabyrinthine approach is to avoid drilling this ampullary limb of the lateral SCC. The anterolateral border of the canal corresponds to the tympanic cavity, for it is situated on the internal, labyrinthine, wall of this cavity (Fig. 14). Because of its obliquity (35 to 40°), which is greater than that of the tympanic cavity (20°), it projects increasingly on the labyrinthine wall as it approaches the posterior wall; in its first third, it is embedded in the wall just beneath the termination of the tensor tympani canal. In its middle third it increasingly projects into the cavity, taking part in the formation of what is called the neuromuscular lintel, constituted in front by the projection into the tympanic cavity of the tensor tympani canal, terminated by the cochleariform process, and beyond that by the projection of the second part of the facial canal which overhangs the vestibular window. A dehiscence of the wall of the canal at this level can expose the facial nerve in the tympanic cavity. The canal subsequently plunges into the posterior wall of the cavity to the extent that the last third of this segment is embedded in the mastoid behind the fossa of the incus. Thus, exposure of the short limb of the incus indicates the proximity of the facial canal and of the facial nerve (Fig. 14).

The third part (Fig. 17) is the mastoid portion. Its length varies with the development of the mastoid but measures about 13 mm on average. This third portion descends practically vertically, slightly obliquely outwards, to open on to the inferior surface at the level of the stylomastoid foramen. The outward obliquity ensures that the direction of this third part crosses behind the plane of the tympanic membrane, which is oblique inwards (Fig. 17b). The crossing of these two structures in space occurs more or less at the midlevel of the tympanic membrane. The canal descends to the midst of the mastoid air-cells, which separate it from the petrous cortex, in particular from the groove of the sigmoid sinus, with which it bounds the intersinuso-facial space.

Two Angles

The Genu (Figs. 14 and 16). Situated between the first two parts of the facial canal, this consists of a roughly triangular dilatation which forms the compartment for the geniculate ganglion. The two parts have an oblique course, one ascending and the other descending, and the genu is situated at their point of junction, emerging from the plane defined by the labyrinthine and tympanic parts "as the prow of a ship emerges from the surface of the water", to employ the metaphor of Pech-Gourg [73] (Fig. 18). Its floor rests on the second turn of the cochlear canal. Its roof, tangential to the superior aspect of the petrous, is formed of a thin lamella which, according to Hall et al. [44] even has a dehiscence in 15% of cases across which the geniculate ganglion is more or less exposed on the petrous surface, sometimes even completely bare. The inner wall continues the direction of the labyrinthine portion against the superolateral flank of the first turn of the cochlear canal, from which it is separated only by a thin septum no more than a millimeter thick. The outer wall continues into the tympanic portion. It corresponds to the canal of the tensor tympani and the termination of the deep petrosal nerve. This nerve, after having ascended on the promontory, penetrates the wall behind the cochleariform pro-

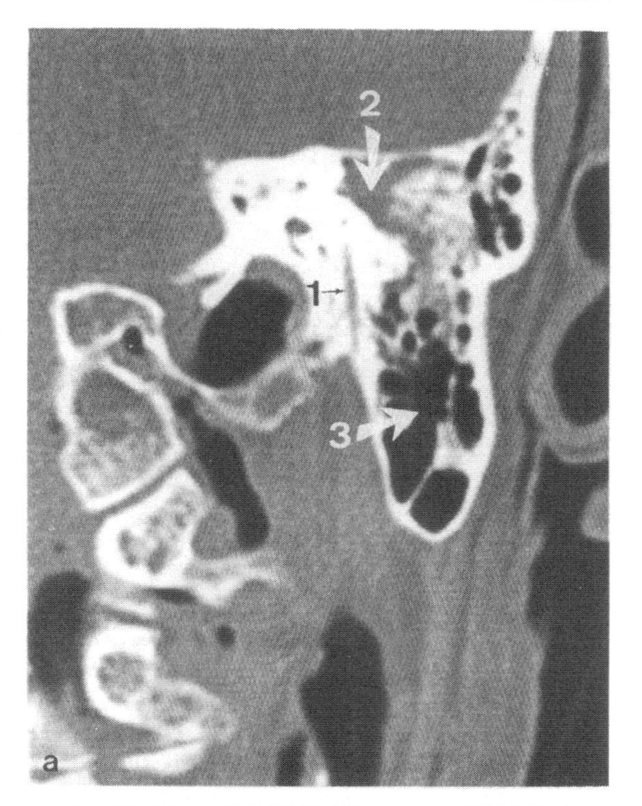

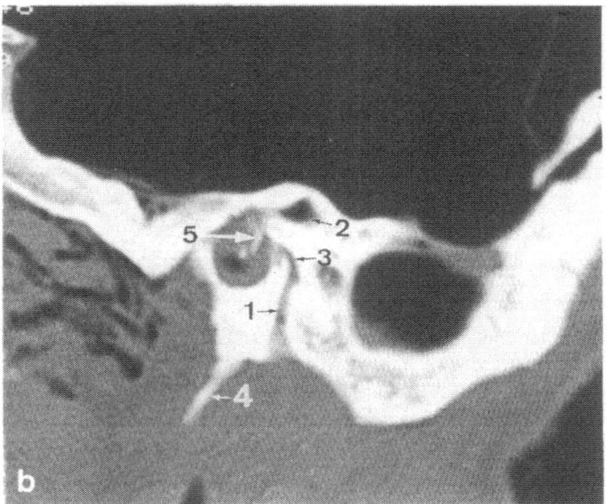

Fig. 17a, b. Facial canal. **a** Frontal CT section. Third part of canal *(1)*, antrum *(2)*, mastoid cells *(3)* **b** sagittal CT section. Third part of canal *(1)*, bend *(3)*, antrum *(2)*, styloid *(4)*, handle of malleus *(5)*

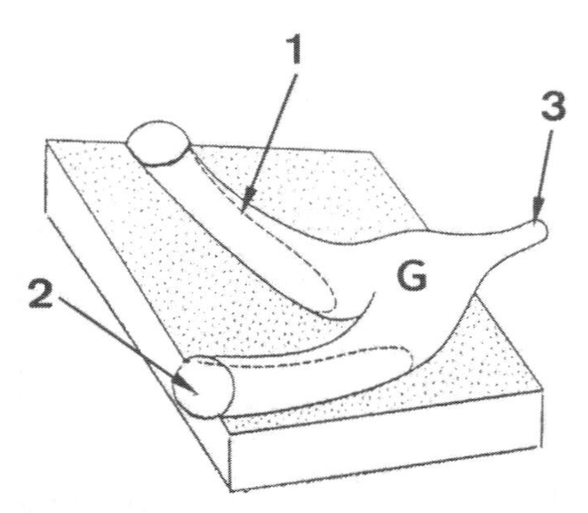

Fig. 18. The genu of the facial nerve: drawn after Dr. Pech-Gourg. First part *(1)*, geniculate ganglion *(G)*, second part *(2)*, petrosal nerve *(3)*

cess, bends to follow the course of the tympanic part of the canal, running up against the canal and the compartment of the geniculate ganglion, and emerges at the level of the accessory hiatus. On the way, it receives an anastomosis from the tympanic portion of the facial nerve, either directly at the level of the compartment of the geniculate ganglion or further forward after having temporarily assumed the course of the petrosal nerve. The posterior wall is very short, 2 mm, and interposed between the orifices of the labyrinthine and tympanic portions. It forms a slender crest marking the angle between these two portions. It corresponds to the anterior pole of the vestibule. The two posterior angles of this compartment correspond to the two orifices. The anterior angle is occupied by the orifice of emergence of the petrosal nerve, a branch of the geniculate ganglion. In 70% of cases [44], this orifice is prolonged by a canal of variable length, up to 6 mm, representing the hiatus of the petrosal canal. In 15% if cases it forms part of the dehiscence of the roof of the compartment. Within the compartment the geniculate ganglion, roughly triangular and of pinkish-gray colour, caps the bend of the facial nerve. At the anterior angle there emerges the greater petrosal nerve, which, after having received some pericarotid sympathetic strands, becomes the nerve of the pterygoid canal which joins the pterygopalatine ganglion. The geniculate ganglion represents the sensory ganglion of the facial nerve, the first relay of the fibers conveying sensation and taste.

The Bend (Figs. 8 and 17). Interposed between the tympanic and mastoid portions of the facial canal, "this is a curve of large radius beginning behind the vestibular window and ending at the upper third of the mastoid portion" [18]. Adjacent to the pyramid of the stapes, it constitutes the pyramidal portion of the English-language literature. The curve describes an angle of 110 to 127° [42]. It passes between the loop of the lateral SCC medially and the summit of the aditus ad antrum and the fossa incudis laterally, from which it is separated by a layer of compact bone some 3 mm thick called the facial beak. In fact, its position is quite variable, sometimes very anterior and then relatively sheltered during the translabyrinthine approach, sometimes very posterior and then exposed to injury by the burr. It is for this reason that we customarily perform a posterior tympanotomy, whose importance is discussed later, for a more accurate safeguard of the facial nerve.

The Canal of the Chorda Tympani

Classically, this is said to consist of a dehiscence in the posterior petro-tympanic fissure. According to Chouard [19] (Fig. 19), its orifice of entry is usually situated (40%) at the level of the inferior third of the facial canal on its anterior wall. The canal travels upward, and slightly forward and outward, to open into the tympanic cavity at the summit of the chordal eminence, just at the margin of the tympanic sulcus, 1 mm below the level of the umbilicus of the tympanic membrane. In 27% of cases the orifice is placed higher, on the middle third of the mastoid portion, and in 20% even higher on the upper third. Finally, in 13% of cases the origin of the chorda tympani is very low, below the level of the stylomastoid foramen. The chorda then ascends in its own canal, whose orifice of entry is situated on the base of the skull just in front of the stylomastoid foramen.

The Carotid Canal

Its diameter is constant throughout its course, about 5.2 mm on average according to Paullus et al. [71]. Its entry orifice is situated on the inferior aspect of the petrous just in front of the middle portion of the posterior foramen lacerum, while its exit orifice is at the summit of the pyramid. This produces a curved course with an initial ascending segment, a bend developed in the axis of the petrous pyramid, and a continuation as a horizontal segment opening against the lateral wall of the body of the sphenoid at the postero-inferior angle of the cavernous sinus.

Ascending Portion (Fig. 20a)

This is about 10.5 mm long on average according to Paullus et al. [71]. It ascends almost vertically from the carotid foramen, slightly inclined in the axis of

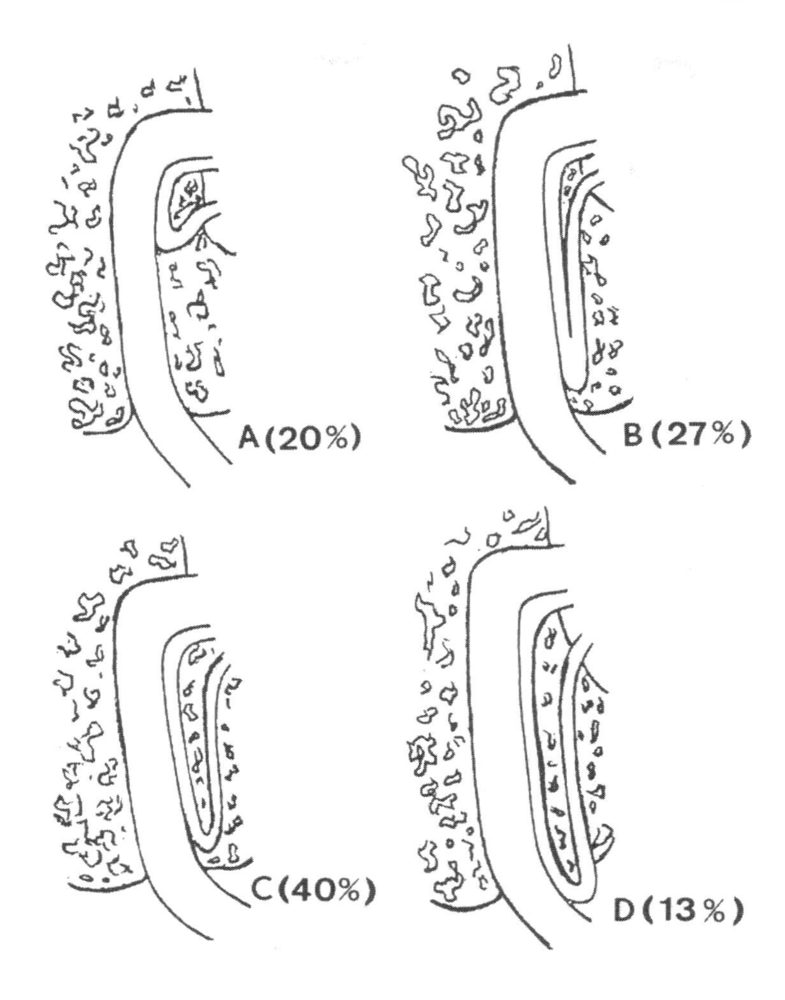

Fig. 19 A–D. Canal of chorda tympani. Drawing after Chouard showing the different types of emergence

the petrous pyramid. Behind and laterally, it corresponds to the tympanic cavity, from which it is separated by a bony wall of an average thickness of 3.2 mm according to Paullus et al. [71], traversed by the carotico-tympanic branch of the internal carotid artery and the strands of the pericarotid sympathetic plexus on their way to anastomose with the tympanic nerve on the promontory. Sometimes (in 2% of cases according to Paullus) this wall is dehiscent, so that the internal carotid is left in contact with the tympanic mucosa or even forms a loop within the tympanic cavity [27]. Behind and medially is the jugular bulb, and medial to that the pars nervosa of the posterior foramen lacerum and, more particularly, the pyramidal fossula. In front is the vagina of the styloid process.

Bend (Fig. 20b)
This is formed beneath the cochlea, particularly the first turn of its spiral, simply coming into contact with the inferior pole of the latter when the petrous is well pneumatized but capable of being molded on the inferior outlines of the entire cochlear cone when the petrous is small. Behind is the summit of the pyramidal fossula and the cochlear aqueduct, exposure of which marks the limit of drilling at the end of the translabyrinthine approach if one is to avoid injuring the glossopharyngeal nerve below and opening the carotid canal higher up. In front of the bend is the entry to the auditory tube.

Horizontal Portion (Fig. 20c)
This has an average length of 20 mm [71] and travels forward and inward in the axis of the pyramid. In front is the auditory tube, from which it is separated by a very thin bony wall, less than 1 mm in 56% of cases according to Paullus et al. [71], and even absent in 6% of cases leaving only a mucosal plane separating the auditory tube and the internal carotid. Above and in front is the tensor tympani canal, separated by a somewhat thicker septum (average 1.3 mm according to Paullus). The upper wall

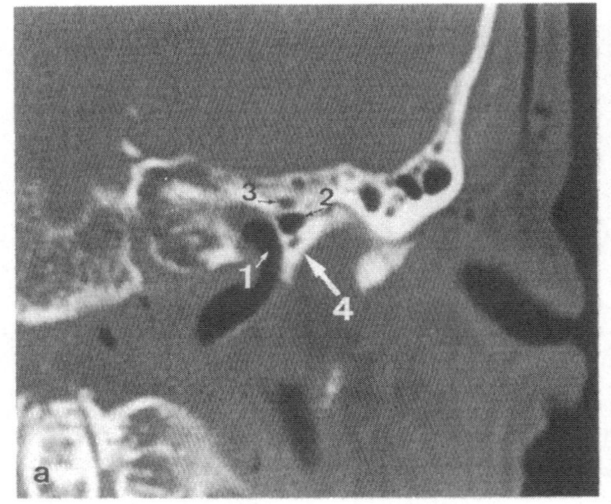

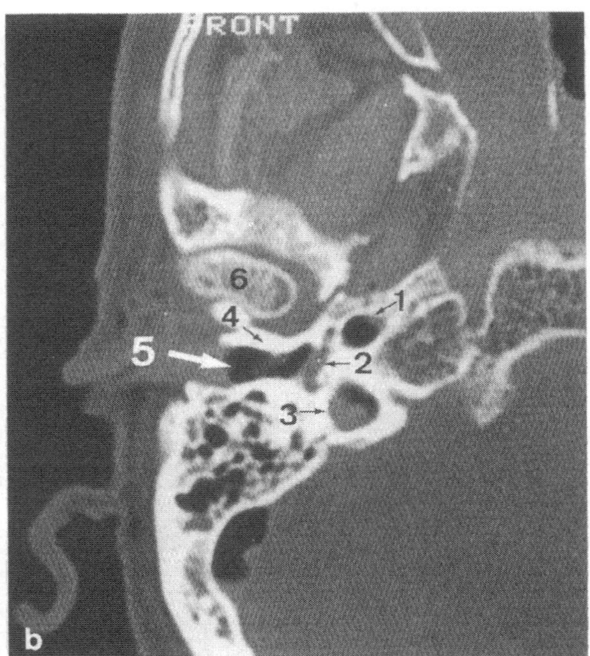

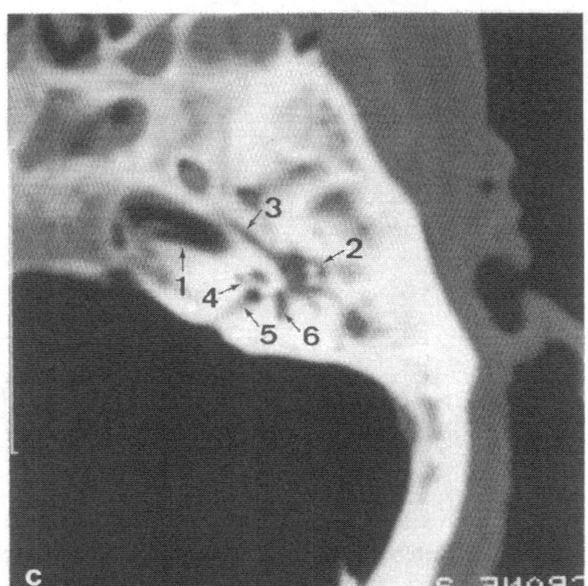

Fig. 20 a–c. Carotid canal. **a** Frontal CT section right petrous. Vertical portion *(1)*, auditory tube *(2)*, canal of tensor tympani *(3)*, vaginal apophysis of tympanal *(4)* **b** Transverse CT section of left petrous. Vertical portion *(1)*, tympanic cavity (lowest level) *(2)*, jugular bulb *(3)*, tympanal *(4)*, EAM *(5)*, glenoid fossa *(6)* **c** Transverse CT section right petrous. Horizontal portion *(1)*, tympanic cavity (attic) *(2)*, canal for tensor tympani *(3)*, cochlea *(4)*, IAM *(5)*, vestibule *(6)*

of the carotid canal becomes thinner as it approaches the apex of the pyramid because of the slightly ascending course of the horizontal portion, to the extent that this wall becomes dehiscent near its apex, under the compartment for the trigeminal ganglion, in the majority of cases (84%) according to a study by Harris and Rhoton [45] and very often (68%) even more laterally, thus leaving the carotid exposed under the dura mater over an extent of about 5.2 mm on average but which may reach 14 mm. On the other hand, the posterior wall is usually thick and, according to Paturet [70], 5 to 8 mm of bone separates the internal carotid from the posterior wall of the petrous at the level of the petrous apex, medial to the aperture of the IAM.

Air Cells

These constitute, as it were, the packing which fills the free spaces between the peripheral shell and the walls of the petrous cavities. While the labyrinthine cavities, in general, have dimensions and therefore volumes that are more or less comparable in all individuals, the air-cells on the contrary may be of very varying extent from one subject to another, so explaining the extreme variability of pneumatization of the petrous. This pneumatization primarily affects the mastoid, of which several types may be distinguished according to its extent. Their incidence is estimated by Zuckerkandl [in 1] as follows:

- Eburnated mastoid (compact bone) or cancellous mastoid (diploëtic bone): 20% (Fig. 21)
- Pneumatized mastoid: 36.8% (Fig. 22)
- Mixed (pneumato-diploëtic) mastoid: 43.2%

The more the petrous is pneumatized, the easier and quicker is the surgical approach.

There is still debate as to the respective roles of hereditary factors and acquired factors (infection, tubal patency, air-pressure) in determining this expansion, which is known to begin during the last weeks of fetal life and to end between the ages of 10 and 15 years [67]. The first component of the cell group is the mastoid antrum, which begins to develop from the tympanic cavity towards the 5th month of fetal life. It opens into the epitympanic recess via the aditus ad antrum. The very great majority of the other air-cells develop from the antrum, with which they remain in communication either directly of via an adjacent cell. Pannier [68] has stressed that the development is radiate, more or less identical in each direction from the antrum. He also states that the size of the cells increases with the distance separating them from the antrum, the largest being the most peripheral. Some cells develop directly on the walls of the tympanic cavity, in particular from the hypotympanum and around the auditory tube. All the cells are invested by a fine mucosa which appears to be an extrusion of the mucosa of the tympanic cavity.

With Hallam [1], we may distinguish three zones:

The Mastoid Cells

Antrum (Figs. 7, 8, 17). This is the most constant and largest of the mastoid cavities. The burr opens into it at about 15 mm depth, at the level of the suprameatal spine. This kidney-shaped cavity, the size of a haricot bean, is described as having the following aspects:

- An internal or labyrinthine aspect corresponding to the outer aspect of the posterior labyrinthine massif. The loop of the lateral SCC is at 2 or 3 mm depth, that of the posterior SCC between 4 and 6 mm. At this site, bleeding from the subarcuate artery, clearly visible under suction-irrigation, marks the external orifice of the petromastoid canal, which has its beginning at the bottom of the subarcuate fossa.
- A superior or tegmental aspect, often very thin or even dehiscent, a posterior continuation of the tegmen tympani. Sometimes it is thicker and distended by some supra-antral cells.

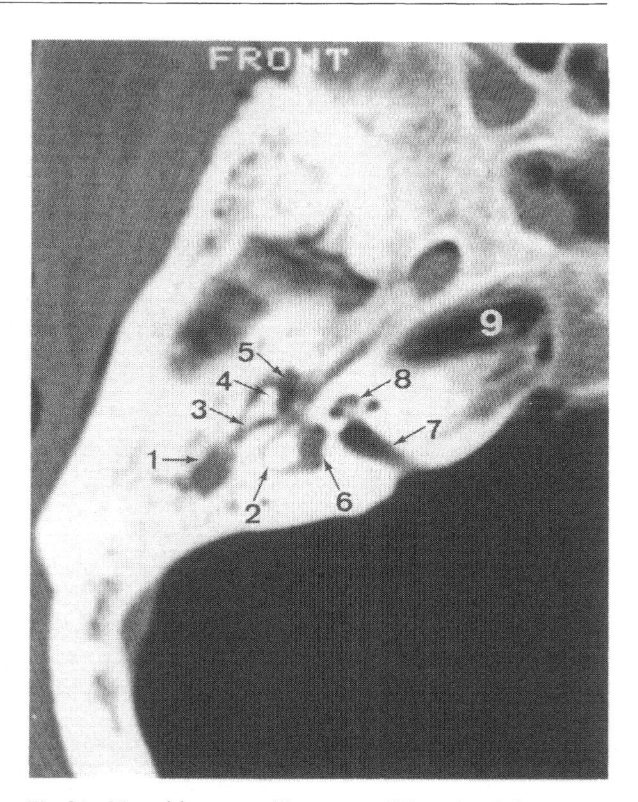

Fig. 21. Mastoid antrum. Transverse CT section left petrous of eburnated type. Antrum *(1)*, lateral SCC *(2)*, short limb of incus *(3)*, head of malleus *(4)*, attic *(5)*, vestibule *(6)*, IAM *(7)*, cochlea *(8)*, carotid canal (horizontal portion) *(9)*

- An inferior aspect generally corresponding to the subantral cells.
- A posterior aspect, corresponding to the bend of the transverse sinus, which is usually at a distance of 6 or 7 mm; but it may be right up against it or even covered by it on its outer face when the sinus is prolapsed.
- An anterior or tympanic aspect opens at its upper segment into the tympanic cavity via the aditus ad antrum. Through this orifice, which is roughly triangular, there can be just seen at its inferior angle the horizontal ramus of the incus and its internal border is elevated by the ampullary limb of the lateral SCC (Fig. 21). These two anatomic details are important landmarks during the translabyrinthine approach. Below the aditus, the anterior aspect is formed by a bony layer which separates the antrum from the tympanic cavity and from the deep extremity of the external auditory meatus: this is the massif of Gellé, within which there travel the third part of the facial canal, the canal of the chorda tympani and the canal of the stapedius muscle which elevates the pyramid within the tympanic cavity.

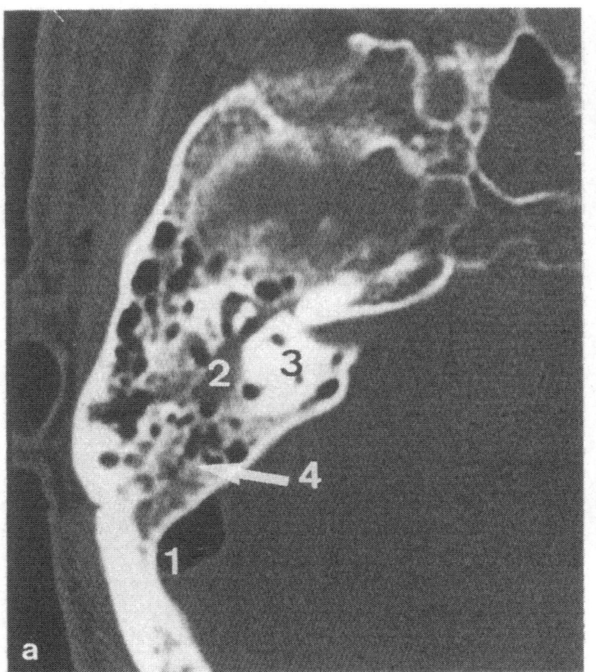

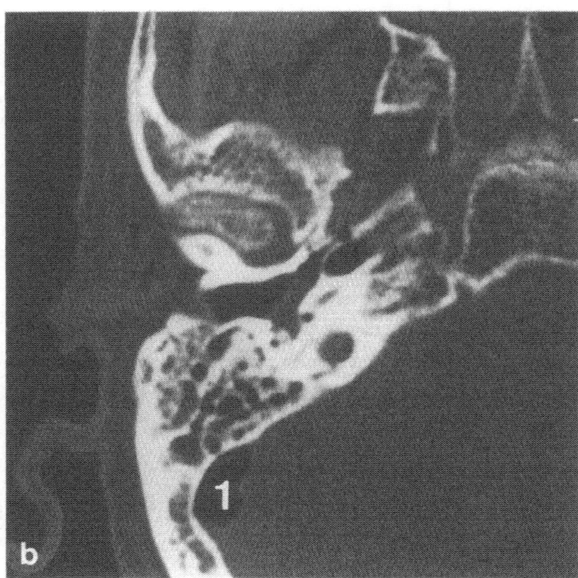

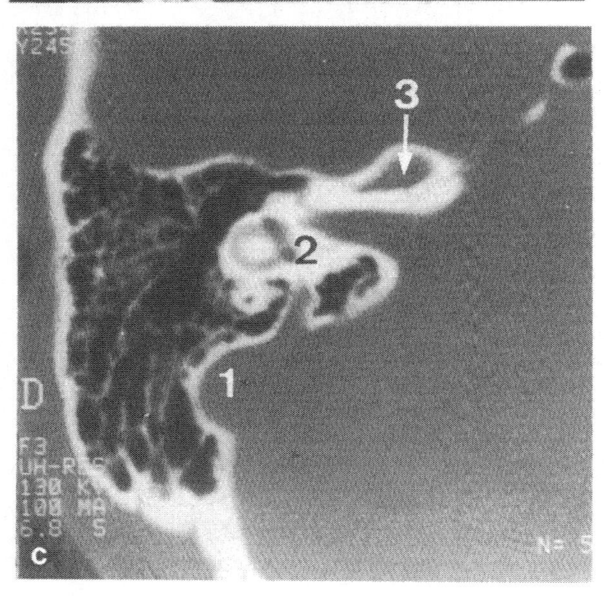

Fig. 22a–c. Mastoid cells. **a** Transverse CT section left petrous. Cells of perisinusoidal type. Groove of sigmoid sinus *(1)*, mastoid antrum *(2)*, labyrinthine massif *(3)*, mastoid cells *(4)* **b** Transverse CT section left petrous. Post-sinusal cells. Groove of sigmoid sinus *(1)*. **c** Transverse CT section left petrous. Groove of sigmoid sinus *(1)*, labyrinthine massif *(2)*, apical cells *(3)*

- Finally, an external aspect which is broken down at the start of the translabyrinthine approach during the extralabyrinthine stage: this is the surgical aspect.

The Mastoid Cells. Since Mouret [in 98], it has been classical to divide the mastoid cells into two groups, anterior and posterior. The anterior group is composed of two columns, external and internal the internal column being formed by the superimposition of the antrum and the deep subantral or intersinuso-facial cells, the superficial column superimposing the cells of the mastoid tip, the superficial subantral cells and the superficial periantral cells. The posterior group is composed of presinusoidal or intersinuso-meningeal cells and retrosinusoidal cells. This "geographic" concept, suited to partial mastoidectomy, takes no account of the dynamics of petrosal pneumatization, with its radial distribution around the antrum, and is not of great importance in terms of total mastoidectomy as it is practised in otoneurosurgery. Here, the concept of Pannier [68] appears more useful. This author suggests distinguishing five types of mastoid, each with its own degree of pneumatization:

- Type I (13%) comprises the cancellous mastoids. The cells are reduced to quite small cavities not exceeding a millimeter and occupying the entire mastoid and also the pyramid (Fig. 21).
- Type II (21%): presinusoidal. The cells are still small (less than 4 to 5 mm) and extend towards the mastoid cortex, but no further back than the anterior margin of the groove of the sigmoid sinus.
- Type III (37%). The lateral wall of the groove of the sigmoid sinus is pneumatized by cells as large a centimeter. The mastoid tip also contains quite large cells (Fig. 22a).
- Type IV (21%): postsinusoidal. Behind, the cells extend beyond the groove of the sigmoid sinus, invading the entire mastoid, the pyramid and even the temporal squama (Fig. 22b).
- Type V (8%): total, is characterized by major pneumatization by large cells over one a centimeter in size, invading entirely the mastoid, pyramid and squama (Fig. 22c).

The advantage of this classification is that it quantifies the degree of pneumatization, presaging the facility or difficulties of mastoidectomy.

The Petrous Cells

These are all cells nested in the petrous pyramid to the exclusion of the mastoid. Some may have developed from the mastoid but others arise from the tympanic cavity or the auditory tube. A vertical plane passing through the axis of the modiolus divides them into two groups.

The *perilabyrinthine* cells are situated behind this plane, some above the labyrinth, others below, without any communication between the two groups [1].

- The supralabyrinthine group is itself divided into three subgroups:
  - The posterosuperior cells, arising from the epitympanic recess of the mastoid, usually from the antrum, extending around the anterior SCC and then along the posterior face of the petrous under the petrosal crest and above the IAM (Fig. 23)

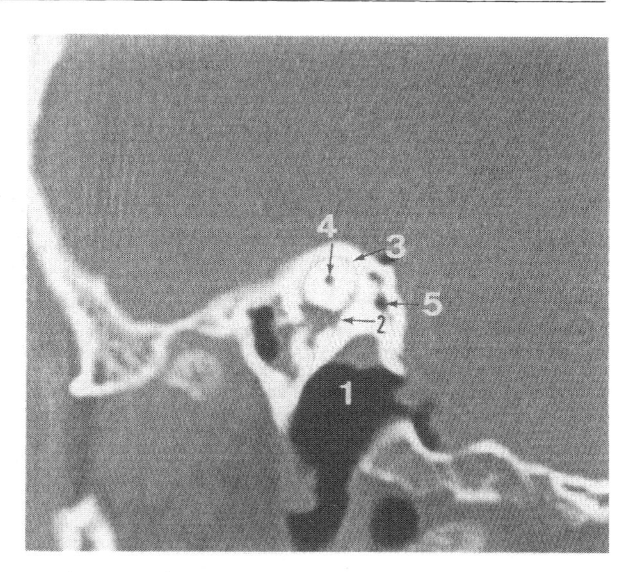

Fig. 23. Posterior lacerate canal. Sagittal CT section. Posterior lacerate canal *(1)*, vestibule *(2)*, anterior SCC *(3)*, petromastoid canal *(4)*, posteromedial cells *(5)*

  - The posteromedial cells, all derived from the mastoid, are subantral or intersinusoidal-facial cells, extending along the posterior face of the petrous beneath the former subgroup. The surgeon may open them when drilling the posterior wall of the IAM (Fig. 23)
  - The translabyrinthine cells, starting from the mastoid antrum, are embedded along the petromastoid canal, and sometimes across the loop of the anterior SCC, in the roof of the IAM.

  This classification is not of great importance. What is certain is that the immediately subcortical cells, under the superior aspect of the petrous, lateral to and behind to anterior SCC, are usually clearly visible by transparency during the approach through the middle cranial fossa. Their opening by drilling of the superior aspect of the petrous may be exploited to uncover the anterior SCC without great risk (see the suprapetrosal approach later).
- The infralabyrinthine group extends under the labyrinthine massif. Some cells arise from the hypotympanum and develop in the roof of the jugular bulb and around the cochlear canal; others arise from the deep subantral or intersinusoidofacial cells and develop under the lateral SCC.

Apical Cells. These are situated in front of the vertical axial plane of the modiolus. Some develop from the epitympanic recess or the auditory tube and extend under the superior aspect of the petrous, around the genu of the facial nerve or along the auditory tube above the carotid canal. Others, arising from the hypotympanum or a prolongation of the preceding cells, also sometimes arising from perilabyrinthine cells, invade the region of the tip medial to the carotid canal and under the cave of Meckel.

Accessory Cells

These develop beyond the limits of the pyramid in the other parts of the petrous, the zygoma, the temporal squama, the styloid or even sometimes in adjacent bones, in particular the occipital bone in the region of the sigmoid sinus groove. It is necessary to be aware of their possible existence, and especially that they may be accidentally opened, in which case they must be properly occluded to prevent the development of a CSF fistula whose aberrant point of origin may be very difficult to locate subsequently.

The Peripetrous Regions

The Skull Base

It is not possible to undertake a detailed description of the base of the skull. But three points must be dealt with:

The Posterior Lacerate Canal

This term, proposed by Aubaniac [5], seems more suitable than the classical names - posterior foramen lacerum or jugular foramen - which do not indicate the precise shape of this wide open dehiscence in the petro-occipital fissure. In fact, it has two orifices, extra- and intracranial, differing in situation, structure and dimensions, situated at the two ends of a "zigzag" canal, a shorter version of the carotid canal (Fig. 23).

Intracranial Orifice
Situated in the plane of the posterior surface of the pyramid, this is almost vertical, facing backward, inward and slightly upward. The sigmoid groove ends at its rather rounded lateral extremity, and the inferior petrosal sinus at its rather tapering internal end. Its upper side is temporal. This is the middle free segment of the inferior border of the petrous, between the anterior and posterior segments of the petro-occipital fissure (Fig. 24a). At about its middle it shows the projection of the jugular spine of the temporal bone, situated between the jugular notch laterally and the pyramidal notch medially. Its inferior aspect is occipital. This is the lateral border of the lateral mass of the occiput. In its anterior half it exhibits a rather blunt projection, the occipital tubercle, more or less corresponding to the pyramidal notch of the temporal and raised to this level by a small elevation, the jugular spine of the occipital. Its posterior half, on the other hand, forms a notch - the jugular notch of the occipital - with a posterior apex corresponding to the jugular spine of the temporal. This notch has as its inner border the posterior aspect of the occipital tubercle. Its outer border is marked by an often well-marked ridge which forms a transverse obstruction in the bed of the sigmoid groove: the sinusoido-jugular ridge. The excavated base forms the jugular fossa of the occipital bone. Thus, the perimeter of this intracranial orifice is composed of two opposed, but staggered, spines corresponding to two notches. This conformation accounts for its typical and well-known radiographic appearance, as well set out by Porot and Aubaniac [77] (Fig. 24b).

Extracranial Orifice
This is somewhat bigger than the former and lies in the plane of the inferior aspect of the pyramid, almost horizontal seen from below. It is situated just medial to the base of the styloid process and behind the inferior orifice of the carotid canal. Its anterior margin is temporal. It is marked by the curved retrotympanic lamina (Fig. 4a) and by the petrotympanic ridge, the internal branch of the bifurcation of the curved retrotympanic lamina, bounding the petrosal fossula in front and medially. The posterior margin of the extracranial orifice is occipital. Its internal portion is elevated by the projection of the jugular spine of the occipital. Laterally lies the anterior margin of the jugular notch of the occipital, concave anteriorly and quite deep. Medially is the internal aspect of the occipital tubercle, which is fairly straight and forming the notch over which pass the mixed nerves. Here also may be evidenced the double staggered opposition of a spine and a notch, giving a very characteristic outline the radiologic contours of which (Fig. 24b) have been well defined by Porot and Aubaniac [77]. The recognition of these two orifices in the films is the essential preliminary to the study of the posterior lacerate canals, an important stage in the diagnosis of tumors developing at this site.

Posterior Lacerate Canal
This is a kind of zigzag, bounded below (Fig. 24a) by the superior aspect of the lateral mass of the oc-

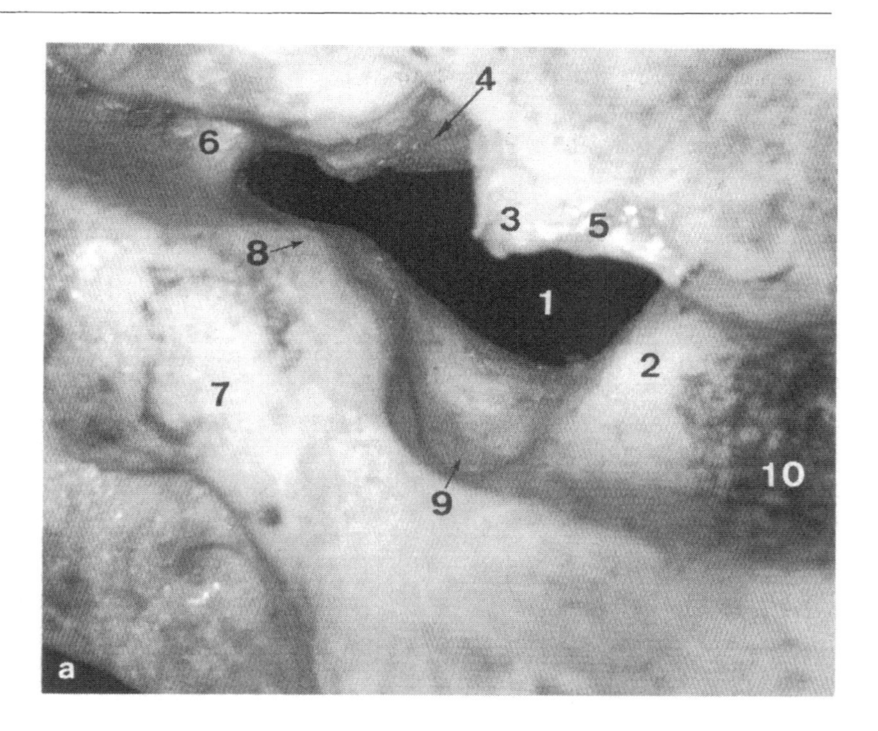

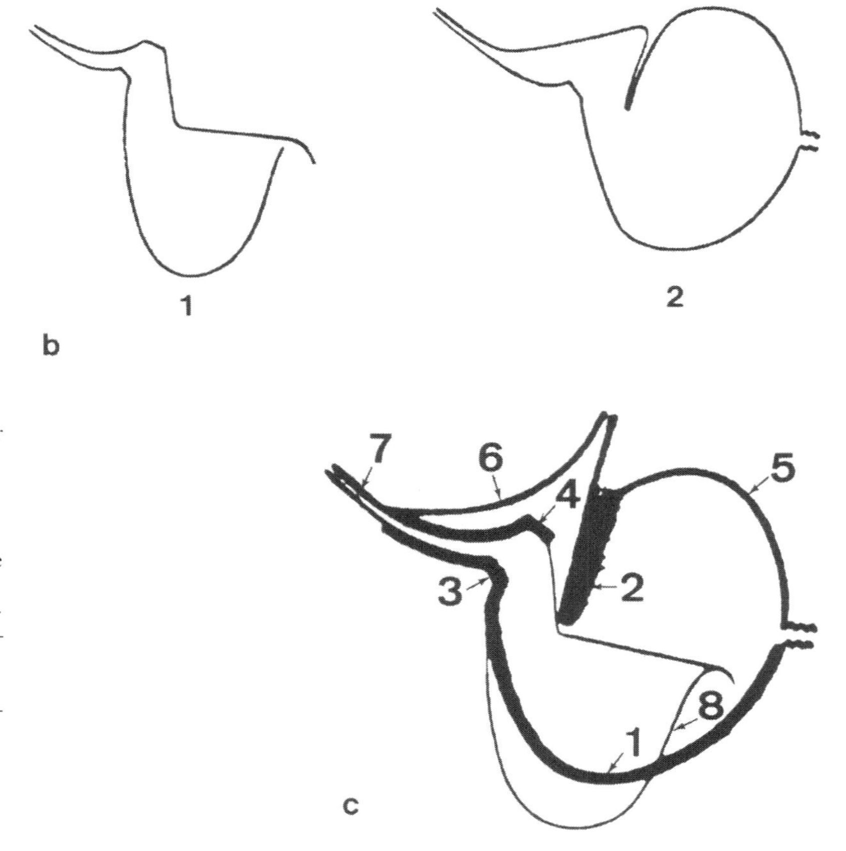

Fig. 24a–c. Posterior lacerate canal. **a** Internal orifice. Lacerate canal *(1)*, groove of sigmoid sinus *(10)*, sinuso-jugular crest *(2)*, jugular spine of temporal *(3)*, pyramidal notch *(4)*, jugular notch of temporal *(5)*, groove of inferior petrosal sinus *(6)*, lateral mass of occiput *(7)*, jugular spine of occipital *(8)*, jugular notch of occipital *(9)*. **b** Drawings after Porot and Aubaniac [77] of outlines of orifices of posterior lacerate canal. Internal orifice (a) external orifice (b). **c** Superimposition of both orifices. Lines always visible. Anterior border of the jugular notch of occipital *(1)*, jugular spine of temporal and petroso-jugular crest *(2)*, jugular spine of occipital *(3)*, pyramidal fossula *(4)*. Lines often visible. Curved retrotympanic crest *(5)*, petroso-carotid crest *(6)*, groove of inferior petrosal sinus *(7)*. Lines rarely visible. Sinusoido-jugular crest *(8)*

cipital bone and excavated at this level by the jugular fossa of the occipital, overhung in front by the occipital tubercle. The inferior aspect of the petrous pyramid partly covers this lateral occipital mass, thus forming the roof of this canal. At this site, this temporal aspect shows the jugular fossa of the temporal bone which overhangs its occipital homologue. It is bordered medially by the petroso-jugular ridge (Fig. 4a) which separates it from the petrosal fossula, the latter being above the occipital tubercle. This anatomic arrangement is sufficient justification for subdividing this canal into two parts: an outer part, wide and rounded, situated between the two jugular fossae, temporal and occipital, and an inner part, narrow and sinuous, between the occipital tubercle and the petrosal fossula. This partition is further accentuated by the habitual presence of a fibrous septum stretching between the jugular spine of the temporal bone and the occipital tubercle, the jugular ligament, a ligament which, according to Rhoton and Buza [84], is replaced by an actual bony bridge in 26% of cases (24% according to Aubaniac [6]). Dichiro et al. [25] place its incidence somewhat higher and stress that it is unilateral in 13.2% of cases and bilateral in only 4.7%. The same authors insist on the asymmetry of the two orifices, the right being the larger in 68% of cases and the left in 20%, perfect symmetry being encountered in only 12% of subjects.

The external portion contains the jugular bulb: this is the vascular portion. Three mixed nerves pass through the internal or neural portion. The dura mater occluding this neural portion seems to produce two orifices, the more internal for the glossopharyngeal nerve, the other - somewhat more lateral and posterior - for the vagus and accessory nerves. In 6% of cases according to Dichiro et al. [25], 4% for Aubaniac [6], the septum between these two orifices is bony and not fibrous, to the extent that the glossopharyngeal nerve leaves the cranium by its own orifice. Shapiro [91] insists on the importance of awareness of these bony septa during interpretation of the images which, because of the disposition of the two orifices in different planes, are the resultant of a superimposition of bony contours which must be properly analyzed if they are to be identified separately. Certain outlines are more readily identifiable and constitute the "lines of force": these are the anterior border of the jugular notch of the occipital bone, the petroso-jugular ridge and the curved retrotympanic lamina, and the petrocarotid ridge (Fig. 24c). The importance of their identification is not just academic but also diagnostic, since, while erosion of the outlines overall suggests only that the lesion is localized at the lacerate canal, erosion of the contours of the neural portion alone rather suggests a neurinoma of the mixed nerves and erosion of the outlines of the vascular portion is primarily suggestive of a tumor of the glomus jugulare. Whether the erosion is clear-cut or irregular in nature obviously also has to be considered in diagnosis.

The Dural Sinuses

These form a venous framework all around the posterior surface of the petrous. While the two petrosal sinuses, superior and inferior, pose no very great problem, the sigmoid sinus, because of its size and situation, has long constituted the main danger in this region. But with optical magnification and good-quality burring equipment, it is now an obstacle easily bypassed.

The Sigmoid Sinus

This is the S-shaped segment of the lateral sinuses, continuing on from the transverse segment. The transverse sinuses, paired and more or less symmetrical, each connects the confluens of the sinuses to an internal jugular vein. As Paturet [70] states, they are the main drainage pathways. The bilateral arrangement suggests that obstruction of one of these two collecting vessels would be devoid of consequences. In fact, Woodhall [104] has stated that in 4% of cases one of the transverse sinuses is absent, the entire drainage being effected by one side. Further, because of variations in caliber, drainage is preferential in 24% of cases (16% by the right side and 8% by the left side). Finally, it sometimes happens that there is no properly so-called confluens [103] and that the superior sagittal sinus in continuous with the right sigmoid sinus, while the straight sinus is continuous with the left sigmoid sinus. As the two flows are only "back-to-back" at the site of what is normally the confluens, the drainage of each flow then depends on the internal jugular with which it is continuous. Apart from angiographic information, this concept makes for safeguarding of these drainage routes. The transverse sinus follows from behind forwards the occipital attachment of the tentorium cerebelli in which it is contained. It is lodged in a transverse groove (Fig. 25). This transverse segment ends at the level of the temporo-occipital suture, where it continues as the sigmoid sinus proper. The latter consists of three parts:

The Bend. At this site, the horizontal course of the transverse sinus abuts against the base of the pyr-

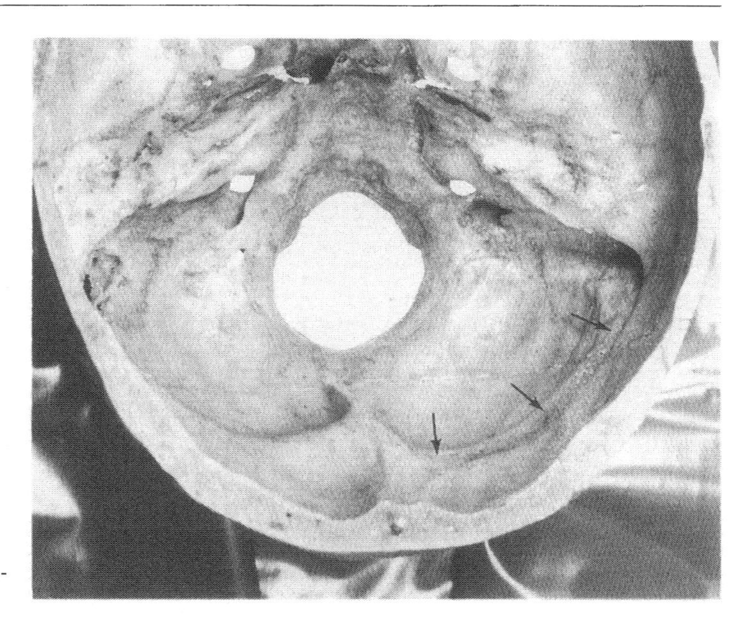

Fig. 25. Groove of transverse sinus. The arrows indicate the groove

amid and bends downwards to become vertical. According to Chatelier [17], the angle so described varies between 70 and 100°. The bony groove in which the sinus travels begins to deepen. It passes under the outer extremity of the petrosal ridge (Fig. 2a) and its superior margin, directly continuous with this petrosal ridge, becomes accentuated and extended behind it. This zone of prolongation of the petrosal ridge on the inner table corresponds to the posterior part of the temporal line on the outer table when it encounters the parietal incisura and the temporo-parietal suture (Fig. 5a). It is by drilling at this level that the surgeon locates the bend of the transverse sinus and the termination of the superior petrosal sinus. The bone is variably thick, 2 to 4 mm according to our experience. For reasons that are still unclear, the position of the bend in relation to the EAM or, more deeply, the mastoid antrum, is variable. In our experience [14], the anterior border of the bend lies at least at 25 mm from the posterior border of the EAM in 75% of cases. In 25% of cases it becomes closer, and in 7% of cases it even comes into virtual contact with it, a source of difficulties during the translabyrinthine approach. It sometimes happens, during exposure of the mastoid, that the surgeon discovers on arriving just above and behind the meatus a fibrous canal which emerges from the bone and continues as a short vein which empties into the superficial temporal vein. This is the very inconstant petrosquamous sinus of Krause-Luschka. The orifice by which it emerges is the temporal foramen of Otto or foramen jugulare spurium of Luschka. It accesses the bend of the transverse sinus via a canal excavated in the diploe in the depth of the temporal line: the petrosquamous canal or aqueduct of Verga. Where it exists, this petrosquamous sinus must be gently freed with the burr, then coagulated and divided in order to allow impaction of the transverse sinus. It will be remembered that certain external cerebral veins – the inferior temporal, 2 in number and large, and the occipital, often single and more slender – empty at the level of this bend or just above it.

The Vertical Portion. This is the mastoid portion. It descends on the intracranial face of the mastoid just in the angle formed by the occipital squama with the posterior face of the petrous. At the inferior border of the latter it bends towards the jugular foramen, still following the line of the temporo-occipital suture. It is often lodged in a deep bony groove (Fig. 2a, 26), excavated astride the petrous and occipital. Like the bend of which it is the continuation, this mastoid portion is in a more or less anterior position, so that the interval separating it from the third part of the facial nerve varies considerably from one case to another. The wider the intersinusoido-facial space, the wider and easier is the translabyrinthine approach. In its upper part, the vertical portion receives the sigmoido-antral vein of Elsworth which collects the venous blood from the antrum and mastoid cells. Lower down, usually in its middle section, the emissary mastoid vein emerges at its posterior border. This vein soon sinks into an orifice that opens just at the posterior margin of the groove and then crosses the mastoid canal from front to back. This last crosses the thickness of the bone obliquely backwards and outwards to open on the outer surface of the cranium near the posterior border of the mastoid at about its middle. Its diameter varies between 3 and 8 mm. This canal contains the emissary vein and an artery to the meninges given off from the occipital artery. The vein, often bulky, bleeds freely during exposure of the sinus but can usually be coagulated without difficulty. It sometimes happens that the sigmoid sinus is atresic or even absent below the emissary vein because of congenital malformation. The entire flow in the lateral sinus may then be diverted towards the occipital veins by a very large emissary vein of considerable caliber [100]. Obstruction of the bulb by a tumor of the glomus jugulare may have the same hemodynamic consequences, though the dilatation of the emissary vein never attains the same size as in the cases of congenital malformation.

The Transverse Portion. This is the terminal portion. The sigmoid sinus, which has followed the petro-occipital suture in the sigmoid groove to arrive at the internal orifice of the posterior lacerate canal, abuts against the outer flank of the occipital tubercle (Fig. 26) and turns practically at 90° forward to engage in the vascular portion of the posterior lacerate canal. It then crosses the entire length of this canal, hugging its outlines closely, and emerges by the external orifice where it continues as the internal jugular vein. The sinusoido-jugular ridge which crosses the sigmoid groove is elevated by this bend of the sinus and marks the boundary between the vertical and transverse portions. This transverse portion lies on the jugular fossa of the occipital bone (Fig. 26). The height of the jugular fossa of the temporal bone dominating this floor explains the domed location of the roof of the sinus at this site, constitutes the jugular bulb. As is known, this bulb is situated under the floor of the middle ear and under the inner ear, at an average depth of 6.5 mm beneath the floor of the IAM according to Gueurkink [43]. It is also known that this height may vary considerably, and that, when excessive, it may produce a prolapse of the jugular bulb. In our experience [14], this is not rare since we noted it in 29% of cases. In virtually one case in two the bulb comes into contact with the floor of the IAM; in one case out of four it extends beyond it, partly masking the posterior aspect of the meatus; and in almost one out of five cases, the prolapse is major, the bulb entirely masking the meatus. Obviously, in such cases of prolapse, the exposure of the IAM by the translabyrinthine route is laborious, but it should be stressed how dangerous it becomes by the suboccipital route. Hence the importance of of preopera-

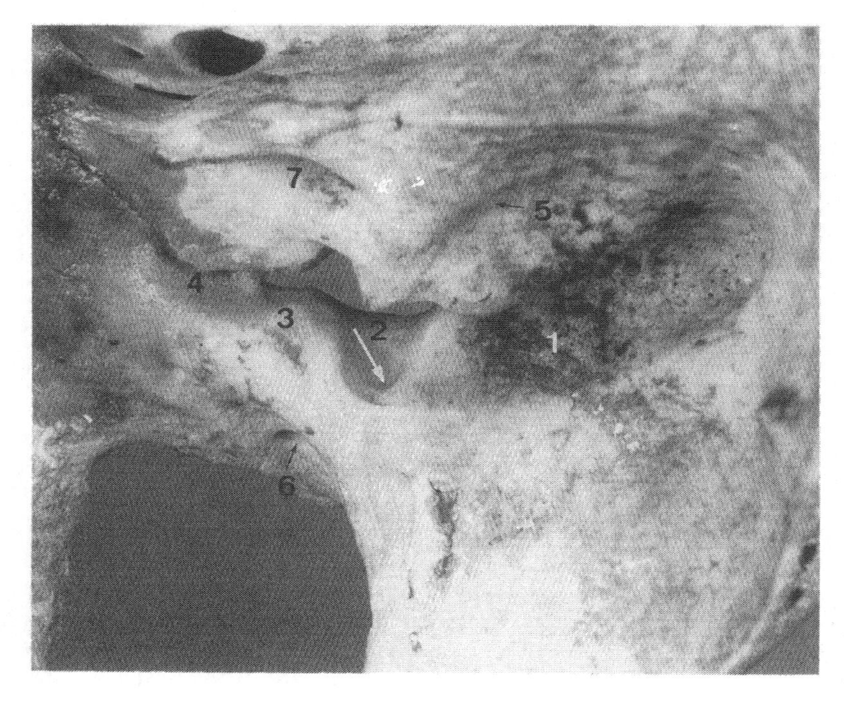

Fig. 26. Groove of sigmoid sinus. Vertical portion *(1)*, transverse portion *(2)*, occipital tubercle *(3)*, groove of inferior petrosal sinus *(4)*, canal of posterior condylar vein *(white arrow)*. Ungual fossula *(5)*, condylar canal *(6)*, aperture of IAM *(7)*

tive tomographic studies to foresee these difficulties and avoid these risks. In the adventitia of the bulb, under the temporal vault, there is a small structure of glomic nature, the glomus jugulare described by Guild [40] in 1941. It is ovoid and measures only 0.5 to 1.5 mm, and is usually symmetrical in both petrous bones. Guild had noted in 1941 that other small bodies might exist, located either along the course of the tympanic nerve in the tympanic canal or in the tympanic cavity on the promontory, or along the course of the nerve of Arnold-Cruveilhier which connects the vagus and facial nerves (the auricular branch of X) by passing over the bulb and then in the lesser innominate canal. Therefore, rather than speaking of the glomus jugulare, it would be better to speak of the tympano-jugular glomic system, known to act as a chemoreceptor and especially as the site of origin of tumors of the glomus jugulare. It is also known that this jugular glomic system is part of a more extensive system from the skull base to the arch of the aorta (glomus of the vagus, carotid, superior and inferior laryngeal, pretracheal and aortopulmonary). This is the non-chromaffin paraganglionic system, the origin of non-chromaffin chemodectomas or paragangliomas, as opposed to the chromaffin paraganglionic system which is the origin of chromaffin pheochromocytomas or paragangliomas. This paraganglionic system is itself incorporated in an even more widespread system, distributed in all or most tissues, consisting of cells of the neural crest which have migrated forwards, while those which migrated backwards have become the spinal ganglia. All these paraganglionic cells have in common histochemical and enzymatic properties that enable them to elaborate, store and sometimes excrete peptide products. Pearse [72] labelled them the APUD cells (Amine Precursor Uptake and Decarboxylation) and the tumors arising from them the apudomas. The tympano-jugular glomic system is supplied by the terminal branches of the ascending pharyngeal artery.

According to Bouchet et al. [10], the occipital sinus opens into this transverse portion. When it is present (65% of cases according to Das and Hasan [24], it descends on the occipital squama from the confluens towards the foramen magnum, more or less at the midline. Before arriving at the foramen magnum, it bifurcates to skirt the margins of this orifice (as the marginal sinus of some authors) and to reach the sigmoid sinuses. Its small caliber (2 mm in 50% of cases according to Das and Hasan [24], explains why it is very often interrupted before reaching the sigmoid sinus and why these authors were able to find the confluence in only 9% of their dissections (6% on the right and 3% on the left). The principal venous affluent on this transverse segment is in fact the inferior petrosal sinus, which usually opens just at the origin of the internal jugular vein, at the extracranial orifice of the posterior lacerate canal. Some of the normal affluents of the inferior petrosal sinus (anterior condylar veins, carotid plexus of Rektorzik, petro-occipital sinus of Englisch) may occasionally empty directly into the bulb. Obstruction of these vessels during excision of tumors of the glomus jugulare which, in principle, only requires occlusion of the sigmoid and inferior petrosal sinuses, then calls for occlusion of the entire bulb, which bleeds like the rose of a watering-can. Finally, the posterior condylar vein connects the transverse segment and the suboccipital venous plexus. It goes through the skull base in a small canal directed posteriorly whose intracranial orifice is situated at the posterior apex of the jugular fossa and whose extracranial orifice is slightly lateral to the foramen magnum and behind the condyle (Fig. 26). Should the bulb be obstructed, the posterior condylar vein can divert the flow in the lateral sinus towards the extracranial venous plexuses.

The Superior Petrosal Sinus

This is paired and symmetrical and connects the cavernous sinus with the bend of the sigmoid sinus. It follows the attachment of the major circumference of the tentorium cerebelli to the petrosal ridge. In its internal portion, it is situated on the posterior margin of the superior surface, bridging above Meckel's cave, but sometimes beneath it or even dividing to pass above and below as described by Coates [21]. It crosses the superior petrosal margin lateral to the arcuate eminence at the level of the notch excavated by the upper part of the subarcuate fossa and runs along the superior border of the inferior surface of the petrous. It encounters the bend of the sigmoid sinus at its internal border. The flow in the superior petrosal sinus takes place from the cavernous sinus towards the transverse sinus and the confluence as a counter-flow. The superior petrosal sinus receives several venules draining the tympanic cavity, the vestibule or the adjacent dura mater. Classically, its main afferent is the superior petrosal vein, or vein of Dandy, a large and constant vein travelling against the outer border of the trigeminal nerve and opening into the superior petrosal sinus just lateral to the cave of Meckel. In fact, this vein seems often to be accompanied by one or two other veins, so that it is better to speak of a "superior petrosal group of veins" as suggested by Matsushima et al. [59]. According to these authors, this venous group consists of three subgroups, distinguished by the site of their junction with the superior petrosal sinus and by their relation to the aperture of the IAM: the intermediate group opens above the porus, the medial group medial to it, and the lateral group externally. These authors also state that the distance of these veins from the center of the cave of Meckel, on average, is 4.4 mm for the medial group, 10.5 mm for the intermediate group and 19.1 mm for the lateral group. The medial group is present in 95% of cases, the intermediate group in 15% and the lateral group in 45%. The superior petrosal sinus is usually easily exposed and safeguarded during the translabyrinthine approach. Its occlusion, on those occasions where its damage has made this necessary, has never had harmful consequences. The superior petrosal margin and its satellite sinus form important landmarks during the approach through the middle cranial fossa (suprapetrous approach): this is the posterior limit for stripping of the dura mater and the posterior limit of burring over the meatal zone. In theory, the superior petrosal sinus bars access to the cerebellopontine angle; however, its section between two ligatures allows considerable enlargement of the approach and sufficient exposure for manipulation in the angle. The petrosal sinus, which passes as a rule above the arcuate eminence, is at least 3 mm removed from the superior border of the porus and the roof of the IAM, and drilling in this interval is usually quite easy. But this space is sometimes more restricted, either because the sinus course is lower, situated perhaps on the posterior slope of the pyramidal crest as a so-called prolapse of the petrosal sinus, or because the tumor has dilated the IAM and its aperture and so elevated its roof. Whatever the exact mechanism, it is certain that in these cases, encountered in 10% of our operations [14], it is more difficult to drill above the meatus and special care must be taken not to damage neither the superior petrosal sinus nor, especially, the facial nerve, known to be situated high in the IAM.

The Inferior Petrosal Sinus

This, too, is paired and symmetrical. It drains the cavernous sinus towards the jugular bulb, descending along the outer border of the basilar lamina in a groove excavated in the petro-occipital fissure (Fig. 26) to reach the extracranial orifice of the posterior lacerate canal at its internal extremity. It then dives into the neural portion of this canal, passing behind the glossopharyngeal nerve but in front of the vagus and accessory, travelling towards the jugular bulb into which it empties just beneath the cranium. According to Bouchet et al. [10], its caliber is sensibly greater than that of its superior homologue. The two inferior petrosal sinuses are connected at their origin by the transverse sphenoidal sinus of Littré. One of the two sinuses, more often the left one, receives the inconstant inferior petrosal vein, found by Duvernoy [30] in 30% of cases. In their extracranial segment, just beneath the skull base, each receives the anterior condylar veins, the petro-occipital sinus of Englisch and the carotid plexus of Rektorzik.

The Dura Mater

This clothes the two surfaces, superior and posterior, of the petrous, while on the superior petrosal margin separating these two faces there is attached the tentorium cerebelli which continues the plane of the superior surface above the posterior cranial fossa. Thus, three fibrous compartments centered on the superior petrosal margin crest and the superior petrosal sinus are demarcated: two intradural compartments, supra- and subtentorial, and the extradural dihedron molded on the petrous pyramid in which the otologic phase of the otoneurosurgical approaches takes place.

Superior Aspect

The dura mater at this level, lateral to the very medially-situated cave of Meckel, has quite an ordinary pattern. It is relatively easy to strip. The surgeon, descending vertically at the level of the external auditory meatus, encounters only some easily ruptured fibrous strands. In the anterior area of the stripping he will reach the spinous foramen (Fig. 2d) and the middle meningeal artery which, as stressed by Brackmann [11], forms the anterior limit of this stripping in the suprapetrosal route. The middle meningeal artery passes forward and outward towards the pterion, but is usually gives off a posterior branch just at its exit from the spinous foramen which travels backward and outward to reach the convexity in the region of the lambda. In addition to this posterior branch of meningeal distribution, Nager [64] reports the origin, just above the spinous foramen, of two other smaller branches (Fig. 27) which both pass backwards and outwards: the superficial petrosal artery follows the course of the petrosal nerve to reach the compartment of the geniculate ganglion via the hiatus of the greater petrosal nerve, while the superior tympanic artery accompanies the termination of the tympanic nerve in the accessory hiatus. Both participate in the vascu-

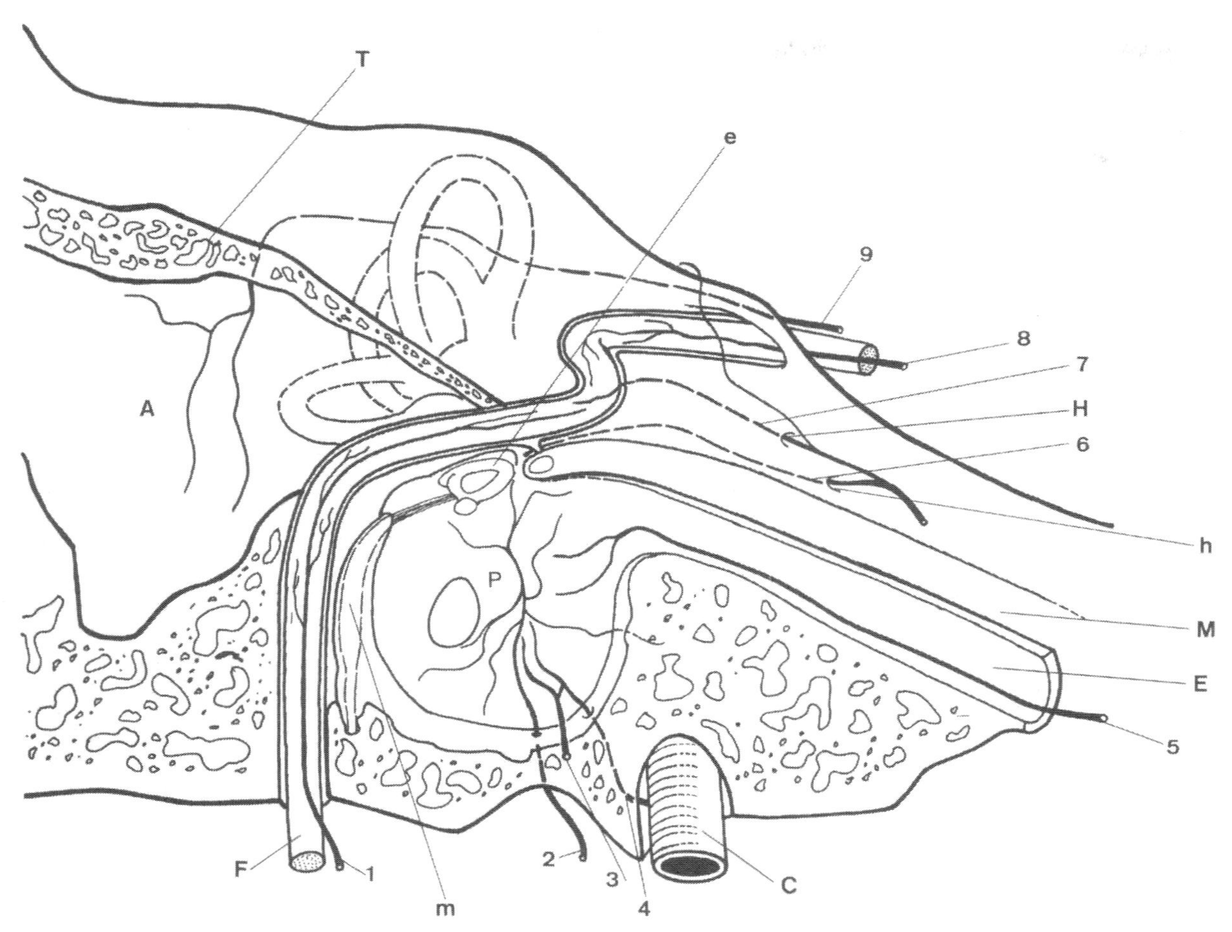

Fig. 27. Arteries of the petrous pyramid. Stylomastoid artery *(1)*, inferior tympanic artery *(2)*, anterior tympanic artery *(3)*, carotico-tympanic artery *(4)*, tubal artery *(5)*, superior tympanic artery *(6)*, superior petrosal artery *(7)*, subarcuate artery *(9)*, facial nerve *(F)*, internal carotid artery *(C)*, promontory *(P)*, stapedius muscle *(m)*, stapes *(e)*, auditory tube *(E)*, canal of the tensor tympani muscle *(M)*, antrum *(A)*, tegmen tympani *(T)*, hiatus of canal of petrous nerve *(H)*, accessory hiatus *(h)*

larization of the middle ear. In 15% of cases [63] they may even form the main supply of the facial nerve in its intrapetrous course. These branches hypertrophy and share in the supply of the superior and anterior sectors [63] of tumors of the glomus which have invaded the protympanum. The superficial petrosal artery, which gives a branch destined for the meninges at the superior petrosal margin crest, carries a superior meningeal vascular component which reaches acoustic neurinomas via the superior margin of the IAM. Once past the arcuate eminence, the retractor places under tension the petrous and tympanic nerves and their two accompanying arteries. Certain postoperative facial palsies have been ascribed to tugging on the geniculate ganglion or to thrombosis of the dominant artery supplying the nerve, but it should not be forgotten that there is always a risk of direct injury to the ganglion and nerve if they are exposed by a dehiscence of the roof at the level of the genu. Right at the back, stripping of the dura exposes the superior petrosal margin and the superior petrosal sinus. And at the very base, starting from the tubercle of Princeteau, the adhesions to the apex of the petrous become the more taut as the mandibular nerve at this level, diving towards the oval foramen, literally tethers the trigeminal ganglion to the base of the skull.

Posterior Aspect

The dura is much more adherent in this region. External to the transverse sinus, dealt with elsewhere, three points of anchorage on the pyramid must be defined.

The Endolymphatic Sac (Fig. 13). According to Bagger-Sjöbäck et al. [7], this sac plays a part in the

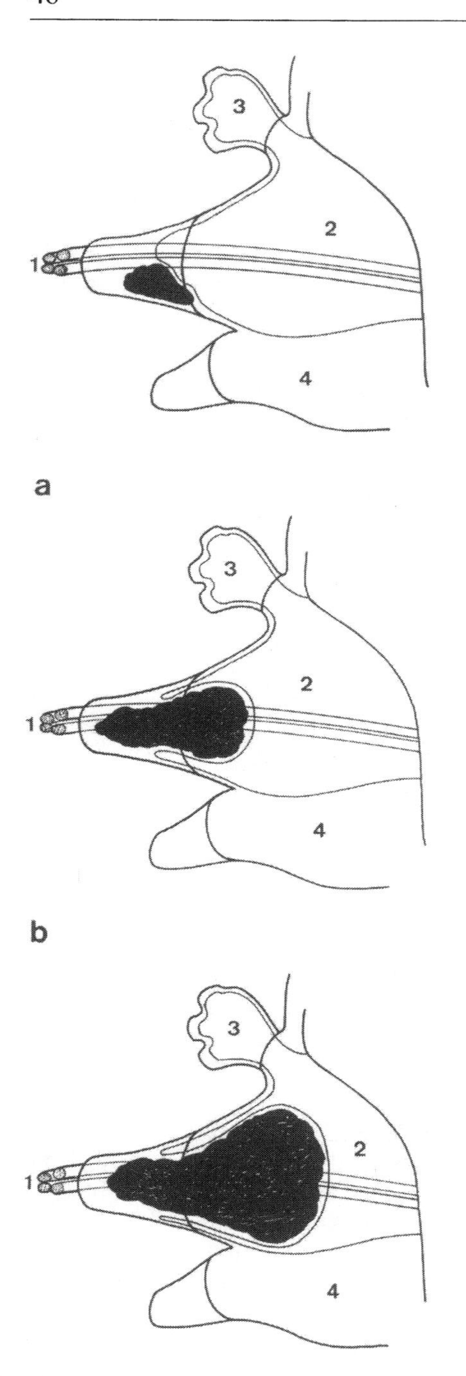

Fig. 28a–c. Relations of neurinoma of VIII and the arachnoid (after Yasargil). **a** Origin of tumor at fundus of IAM. **b** Tumor grows through aperture of IAM. **c** Tumoral expansion in the cerebellopontine angle. Acousticofacial bundle *(1)*, cerebellopontine cistern *(2)*, cave of Meckel *(3)*, cerebellomedullary cistern *(4)*

phenomena of resorption and phagocytosis of the endolymphatic fluid and, it seems, in regulation of the pressure in the inner ear. It has two parts: one intraosseous in the expanded portion of the vestibular aqueduct and one extraosseous in the thickness of the dura, between the periosteal layer and the dura proper according to Anson and Donaldson [3]. The boundary between these two parts fixes the dura firmly to the margins of the ungual fossula. According to Shea et al. [92], this sac measures 8.9 mm high (5.5 to 12.2 mm) and 9 mm wide (5.5 to 11.2 mm). House [47] suggested it be sought on a line extending backwards the direction of the loop of the lateral SCC behind the loop of the posterior SCC as visualized by its blue line. In fact, these landmarks are not always reliable and Shea et al. [92] suggest locating it by reference to the apex of the horizontal limb of the incus, the usually evident prominence of the lateral SCC and the medial border of the sigmoid sinus. From their measurements, there is no risk of opening the posterior SCC beyond a circle of 11.5 mm radius (9.5 to 11.5 mm) centered on the apex of the horizontal limb of the incus, and the base of the sac is found between 13 and 19 mm from this same point. Moreover, the sac is usually situated medial to the line proposed by House.

The Subarcuate Fossa. At this level, the subarcuate artery enters the petromastoid canal to reach the mastoid antrum (Fig. 3b). It supplies the mucosa of the antrum and the mastoid cells but also the semicircular canals and the vestibule, and its obstruction may account for secondary ischemia of these structures. Further, this artery which, according to Mazzoni [60], arises independently from the loop of the anteroinferior cerebellar artery in virtually one case in two, also often arises from a common trunk with the internal auditory artery or with an artery to the cerebellum, so that its coagulation may produce ischemia more prejudicial to the cochlea or cerebellum.

The Acousticofacial Recess. This has been so named by Slivic and Boskovic [94] for the finger-like extrusion of the dura into the IAM, which it closely invests to its base. This fibrous sleeve, though much thinner, continues into the facial canal, ensheathing the facial nerve up to the stylomastoid foramen. At the base, it is adherent to the cochlear and superior and inferior vestibular area and, like these, is perforated by small channels conveying the nerve strands due to converge into the cochlear and vestibular nerve trunks.

This fibrous sleeve is lined with an arachnoidal cuff which extends the cerebellopontine cistern into the IAM. This diverticulum penetrates more or less deeply into the meatus and terminates as a blind ending around the acoustico-facial bundle by adhering to the pial sheath of these nerves. Thus the nerves travel in an intracanalicular cistern whose communication with the cerebellopontine cistern allows the neuroradiologist to perform meatocisternography. Beyond the adhesion of the arachnoid to the pia mater, the nerves, while still intracanalicular, show an extra-arachnoidal segment before traversing the cribriform plate at the fundus of the meatus [in 28]. An acoustic neurinoma arising from the vestibular ganglion of Scarpa in this region will, during its advance towards the aperture of the IAM, push the arachnoidal cuff ahead of it and then, as well described by Yasargil [106] (Fig. 23), will become invested by this layer as it develops in the cerebellopontine angle. This is the origin of the pseudo-capsule of this tumor, the awareness and dissection of which constitute one of the keys to atraumatic excision whatever the route of approach.

In this recess the acoustico-facial bundle travels, the arrangement of which is governed by the proximity of the four orifices of the cribriform plate. The details will be found in the thesis of Mercier [61]. The nerves are accompanied by vessels, and chiefly by the loop which the anteroinferior cerebellar artery describes much more often between the nerves (56%) than under them (38%) [61]. Still according to Mercier [61], this loop usually penetrates into the meatus, even as far as its fundus in 28% of cases. This loop, in which Martin and Rhoton [57] distinguish three segments - premeatal, meatal and postmeatal - gives off branches to the labyrinth according to Mercier [61], in front of the meatus (42%) or at the level of the meatus (50%). Mazzoni [60] and Mercier [61] state that, while the labyrinthine artery is single in 62% of cases as is classically described, it is not uncommon for it to be double (36%) or even triple (2%). While this artery reaches the fundus of the meatus, there are a variable number of others [1 to 10] destined for the walls or contents of the IAM that peter out before reaching the fundus. Marquet [58] has suggested to include these latter in the group of canalicular arteries as opposed to the group of ultracanalicular arteries destined for the labyrinth. We prefer the simpler terminology suggested by Mercier [61], which distinguishes the principal meatal arteries destined for the labyrinth and dividing into 3 branches, cochlear, cochleovestibular and vestibular, and the accessory meatal arteries which are the branches destined for the nerves, the vestibular ganglion and the meninges. Certainly, this complexity quite justifies the term "arterial system of the IAM" proposed by Fisch [33] to replace the term of labyrinthine artery, which suggests a much more elementary concept.

There are also to be added the meningeal arteries derived from the external carotid system, whose importance lies mainly in their hypertrophy when a tumor develops at this site. Some are derived from the middle meningeal artery, particularly via its superficial petrosal branch, others from the ascending pharyngeal via its meningeal branch. Koos [55] has demonstrated their participation in supplying the neurinomas; but it is the meningiomas attached to the posterior aspect of the petrous that provoke the most marked hypertrophy, sometimes with participation of the meningeal branches of the carotid siphon, or the posterior meningeal branch of the vertebral artery where it traverses the dura mater, or of the meningeal branch of the occipital artery which traverses the mastoid canal in contraflow to the emissary vein.

The Sinusoido-Dural Angle

This is the extradural dihedron of dura mater exposed during drilling of the pyramid, bounded above by the dura lining the superior surface of the petrous and behind by that lining its posterior surface. This is the extradural space in which the otoneurosurgical routes of approach operate. The superior plane is simple and uniformly smooth. The posterior plane is traversed diagonally by the sigmoid sinus passing towards its bulb (Fig. 29). The ridge separating these two planes is crossed by the superior petrosal sinus. The anterior plane consists first of the posterior wall of the EAM, then by the wall of Gellé containing the third part of the facial canal, and more deeply by the notional vertical plane separating the anterior from the posterior labyrinth. Is is obvious that the ease of transtemporal progress will greatly depend on the free space left between the two dural planes of the dihedron and the anterior bony wall. The more the sigmoid sinus approaches the anterior wall, which is not to be ruptured, the narrower and more difficult access becomes. Hence the importance of the concept of prolapse of the lateral sinus. Similarly, the closer the temporal meninges approach the roof of the EAM, the more this restricts the space above for use of the burr by the otologist, a situation which leads to the concept of prolapse of the temporal meninges. It can easily be imagined that a well-pneumatized mastoid of good size, with the temporal meninges at a good distance from the EAM and

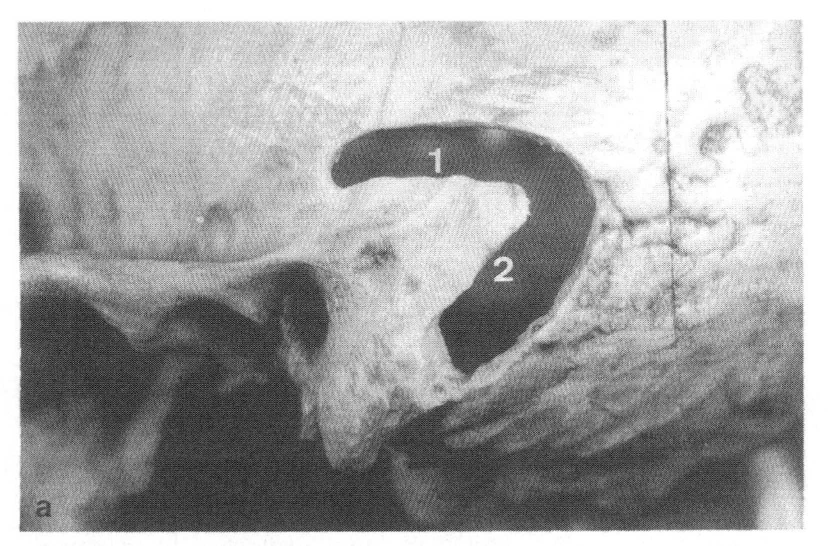

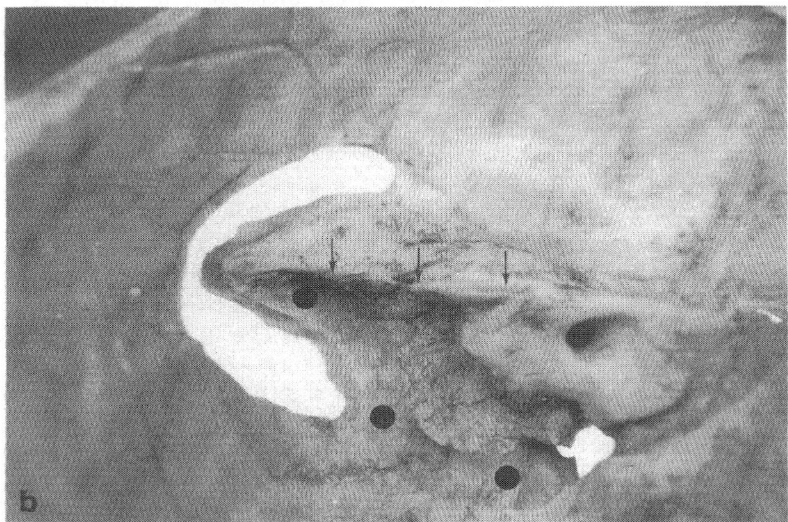

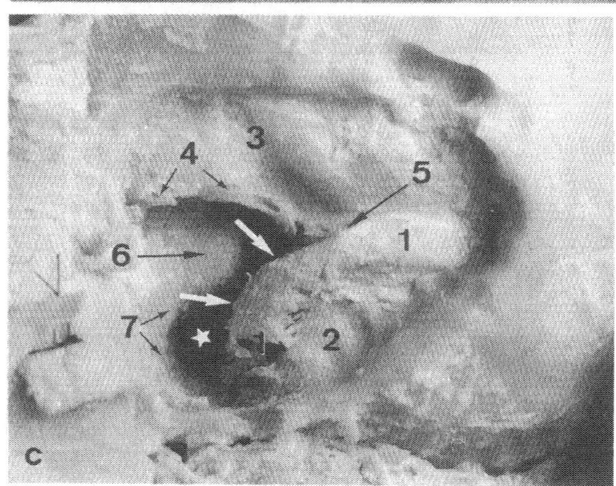

Fig. 29a–c. Sinusoido-dural angle. Trephining along the borders of the superior and posterior faces of the petrous. **a** Lateral view. Outer border of superior face *(1)*, of posterior face *(2)*. **b** medial view. Groove of lateral sinus *(black dots)* groove of superior petrosal sinus *(black arrows)*. **c** Exposure of dura mater. Lateral sinus *(1)*, posterior fossa *(2)*, temporal dura mater *(3)*, superior aspect of petrous *(4)*, posterior aspect *(white arrow)*, sinusoido-dural angle *(5)*, labyrinthine massif *(6)*, 3rd part of facial nerve *(7)*, intersinusoido-facial space *(white star)*

a lateral sinus well posterior to the EAM and the facial nerve, provides very favorable anatomic conditions which, in our experience [14] exist in 60% of cases. Drilling can be carried out with complete safety. Even if the jugular bulb is prolapsed in the depths, very close to the IAM, a condition observed in 15% of cases, it can be freed and depressed without much difficulty. Without being able to call it true prolapse, a limiting situation is found in 10% of cases in which the space between the temporal meninges, the sigmoid sinus and the EAM is relatively less extensive. The approach will be more tedious and will require more extensive retraction of the two dural planes to descend into the depths. Prolapse of the bulb is commoner in these cases (40%), which further complicates the stage of drilling and prolongs the operating time. Finally, in 30% of cases the space is manifestly narrow and the sigmoid sinus and temporal dura markedly approximate to the EAM, either simultaneously, which is commonest, or with isolated prolapse of one or the other and then more often of the sinus. In 6% of cases there is a major prolapse, the sigmoid sinus coming into contact with the cortical bone forming the EAM. Access is really difficult in these cases and calls for sound experience on the part of the surgeon if it is to be carried off well, the more so since there is frequently an associated prolapse of the jugular bulb in the depths (36% in cases of prolapse and 66% when the sigmoid sinus comes in contact with the EAM). In every case, the opening of the extradural dihedron by means of the two blades of a self-retaining retractor is an important maneuver without which the approach cannot succeed. However, this maneuver alone is useless unless it is accompanied by as extensive bony drilling as possible.

The Cerebellopontine Angle

This is the crucial intracranial relationship. It is classically a prism-shaped space bounded medially by the brain-stem, laterally by the cerebellum, in front by the posterior aspect of the petrous and extending from the trigeminal nerve above to the mixed nerves below. But this description is certainly too schematic and the concept of Bebin [8] appears closer to the reality. This author regards it as almost a potential space, a slit lying between the posterior face of the petrous in front and the anterior aspect of the brain-stem and cerebellum behind. He sees this angle as a small space, roughly the shape of a parallelepipedon, lodged between the olive medially and the flocculus laterally, extended by two vertical grooves, above between the pons and the cerebellum and below between the medulla and the cerebellar amygdala. The vestibulocochlear-facial bundle fastened in the center of the space raises an arachnoidal cone compared by Bebin [8] to a tent canvas supported by its pole. This virtually notional space becomes a reality when a tumor developing at this site opens it up widely. Rhoton [86] has described in detail the components of the "neural" aspect of this slit (pontomedullary groove, fissures between the brain-stem and the cerebellum, flocculus, bulbar olive, nerves), which he states may be encountered during operation and then serve to localize and identify the facial nerve, especially its origin from the brain-stem, in the course of surgery for tumors of the angle. In actual fact, as soon as the tumor is of any size, it obliterates all the details to the extent that we find them unrecognizable and devoid of explicit topographic value, apart from the fact that their exposure proves one to be in contact with the nervous parenchyma. The only structures we find it useful to identify are those which are normally contained in the space and which act as a guiding thread in progressing towards the emergence of the nerves constituting the bundle of the vestibulocochlear and facial nerves. We strongly recommend the thesis of Mercier [61] to those desirous of a deeper knowledge of the microsurgical anatomy of the cerebellopontine angle so as to be able to recognize each of the three groups he distinguishes: the superior group, comprising the trigeminal nerve, the abducent nerve and the superior cerebellar artery; the intermediate group, comprising the facial nerve, the nervus intermedius, the vestibulocochlear nerve and the antero-inferior cerebellar artery (the AICA of the English-language literature); and the inferior group, combining the three mixed nerves and the postero-inferior cerebellar artery (PICA).

Apart, of course, from the mixed nerves, the only one of the structures mentioned by Rhoton [86] to have precise localizing value is the choroid plexus when it is identified, as it is situated just laterally to the vestibular fibers at the extremity of the pontomedullary groove.

Nerves

The vestibulocochlear nerve is the most lateral structure in the acoustico-facial bundle. The outermost, vestibular fibers and then the cochlear fibers penetrate the brain-stem just at the level of the pontomedullary groove at its outer end, just medial to the choroid plexus which emerges here through the lateral aperture (foramen of Luschka), and 2 or 3 mm above the first rootlets of the glossopharyngeal nerve (according to Rhoton, [83]). Eyries and Chouard [32] have shown how the nervus intermedius and the facial nerve emerge a little more medially, 1 or 2 mm according to Rhoton [83], the intermedius from the groove itself according to Mercier [61] and the facial from its prominent superior border. The site of this nerve bundle is well-known; as it travels outwards it seems to undergo a twisting movement on its axis, as manifested by the thick cochlear nerve, in such a way that each of the nerves returns to the fixed position it occupied at the fundus of the IAM. It may be less well-known that the fibrous cone formed by the junction of the two portions of the sheath of each of these nerves, the proximal oligodendrocytic and the distal schwannian portions, known by some authors as the zone of Obersteiner-Redlich, is situated quite distally, within the IAM for the vestibulocochlaer nerve, whereas it is always more proximal, in the angle, for the other cranial nerves. This arrangement is explained by the particularly rapid growth of this nerve towards the primitive otic capsule. This rapidity has been invoked to account for the accumulation of schwannoid cells demonstrated by Pirsig [in 8] at this fibrous cone, an accumulation which may be the origin of the schwannomas so common on this nerve. Finally, it should be realized, as pointed out by Sunderland [97], that the facial nerve in the angle is devoid of perineurium and has only very little endoneurium, so that the virtually naked axons are esaily separable on the tumoral convexity. This explains the relative resistance of the facial nerve to compression and also the operative difficulties during dissection of this nerve, giving rise to a doubt as to the true possibility of anastomosis in the angle when the nerve has been divided during operation. The guide-protector suggested by Fisch [36] would seem essential in this context.

Vessels

Of the three cerebellar arteries, only the antero-inferior cerebellar artery (AICA) is really involved in the operative procedure. The postero-inferior cerebellar artery, the longest and largest (1.8 mm in diameter according to Shrontz et al. [93]), arises much lower from the vertebral artery, passes between or above the mixed nerves, but shares with them the shelter of another cistern; similarly, the superior cerebellar artery arises very high up, passes above the trigeminal nerve, is also situated in a different cistern and is only very exceptionally involved in surgery of the angle. The AICA arises from the basilar trunk in its inferior or sometimes middle third. According to Shrontz et al. [93], its mean diameter is 1 mm and remains fairly constant throughout its course. It travels outwards to pass above or below the abducent nerve, and then continues towards the IAM in company with the acoustico-facial bundle. Martin and Rhoton [57] describe it as having three segments (Fig. 30): premeatal, in the direction of the IAM, in an antero-inferior position in relation to the nerves; meatal, where it describes a loop whose variations have already been discussed (see under the acousticofacial recess); and then post- or retromeatal, where it turns back toward the brain-stem following the postero-inferior border of the nerves. Arriving at the flocculus, it divides into branches

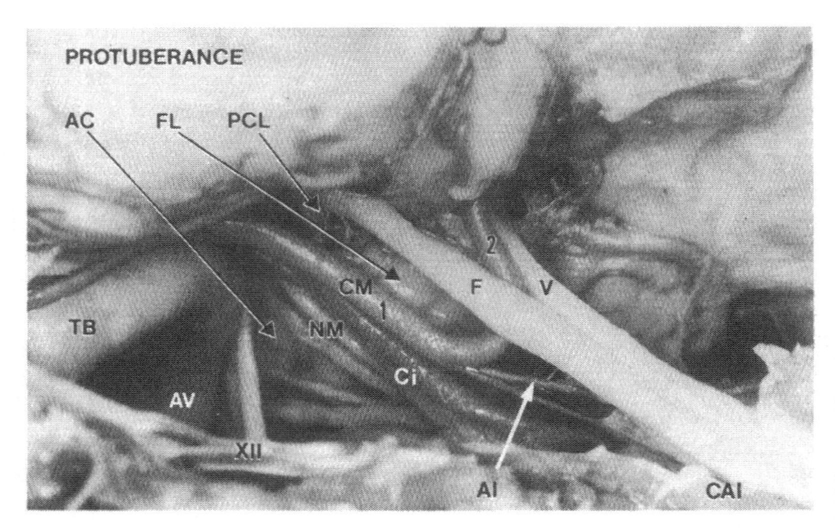

Fig. 30. Cerebellopontine angle. Anterolateral and superior view (dissection by Dr. Zanaret). Facial nerve *(F)*, vestibular nerve *(V)*, vertebral artery (AV), basilar trunk *(TB)*, cerebellar amygdala *(AC)*, flocculus *(FL)*, choroid plexus *(PCL)*, anteroinferior premeatal cerebellar artery *(CMO)*, retromeatal artery *(2)*, posteroinferior cerebellar artery *(CI)*, mixed nerves *(NM)*, hypoglossal nerve *(XII)*, labyrinthine arteries *(AI)*, IAM *(CAI)*

for the brain-stem, for the origin of the facial and vestibulocochlear nerves [61] and, of course, for the IAM and labyrinth. Finally, in almost every case (96% according to Mercier [61]), it gives off the subarcuate artery which enters the petromastoid canal and supplies the mucosa of the antrum and, passing, the vestibule. As emphasized by Rhoton [86], the variable position of the AICA in relation to the nerves accounts for its variable position in relation to the tumor: it is displaced backwards when it travels between the nerves, upwards when it passes above them, and downwards in the rare cases when it passes below (2% according to Mercier [61]). Atkinson [4] first showed that damage to this artery was the source of the pontine softenings observed after surgery in the angle, but it is known that such injury cannot be incriminated exclusively since Popen and many others [74] have interrupted this artery without ill consequences. Relevant factors to be considered are the size of the plexus supplied by the artery and the possible substitution by adjacent arteries, particularly the PICA, especially as Atkinson [4] had initially noted at autopsies after pontine softening the abnormally narrow diameter of the PICA. These discordant results explain why some authors, like Pertuiset [74], have argued for a venous mechanism for this complication.

The veins of the cerebellopontine angle are effectively involved during dissection of the tumor. While they were frequently sacrificed in the past before the advent of the microscope, it should be possible to avoid this now whatever the route of access. Their classification is complex and we refer the reader to the article by Matsushima et al. [59] and the work of Duvernoy [30]. What needs to be borne in mind is that the superior petrosal vein, or vein of Dandy (median group of Matshushima et al. [59], which is the main vein at this site, arises from the convergence on the root of the trigeminal nerve of several veins: the anterior cerebellar vein which emerges from the transverse fissure of the cerebellum slightly lateral to the flocculus and ascends on the brachium pontis, constituting the vein of the cerebellopontine fissure (Matsushima et al. [59]); the lateral pontomedullary vein [30] or vein of the middle cerebellar peduncle [59], arising from the confluence above the bulbar olive of the vein of the pontomedullary sulcus and the lateral medullary or preolivary vein; the transverse pontine vein(s); and the ponto-trigeminal vein, formed by the confluence of the lateral mesencephalic vein and an interpeduncular vein. The arrangement is very variable; the four veins may converge together towards the same confluence that gives rise to the vein of Dandy or may join together in very diverse combinations. Most often, however, the vein of the cerebellopontine fissure constitutes the main trunk to which other afferents are connected, particularly the lateral pontomedullary vein and the pontine veins. Once the dura has been opened during the translabyrinthine approach, the surgeon usually exposes a vein which seems to skirt the posterolateral convexity of the tumor. This is the anterior cerebellar vein, which can usually be preserved provided the fragmentation and excision of the tumor are effected well within the arachnoid layer enveloping the neurinoma. Obviously, at a higher level the depression and excision of the upper pole of the tumor can be effected without risk to the vein of Dandy. On the other hand, very often the dissection of the posterior pole of the tumor over the emergence of the vestibular fibers is accompanied by the exposure and coagulation of a large vein which must be either the vein of the pontomedullary sulcus or already the lateral pontomedullary vein. This is usually the only acceptable venous sacrifice and has never, in our experience, led to any consequences. It is quite exceptional to make a breach in the superior petrosal vein of Dandy and, as recommended by House [51], before resorting to coagulation an attempt should first be made to seal the wound with a tampon of hemostatic material.

The Arachnoid

This is an essential structure for the neurosurgeon. Yasargil [105] has stressed that the cerebellopontine angle, a somewhat reduced space, is not occupied by a single cerebellopontine cistern, but is divided into several closed spaces by the juxtaposition of three cisterns at this site (Fig. 31):

- The cerebellopontine cistern, containing the trigeminal nerve, the acoustico-facial bundle, the AICA and its main and accessory meatal branches, the branches originating from the superior petrosal vein and the trunk of that vein
- The lateral cerebello-medullary cistern, containing the mixed nerves, the hypoglossal nerve, the PICA and the veins of the lateral recess, of the cerebellar amygdala and the latero-medullary veins
- The prepontine cistern, containing the abducent nerve, the basilar trunk and the origin of the two AICAs, the transverse and median pontine veins.

The structures contained in each of these three cisterns remain quite isolated from those of the adjacent cistern by a double-layered septum formed by the adhesion of the arachnoid membrane of each cistern. Theoretically, a tumor originating within one cistern develops within that cavity and is more or less adherent to its contents while remaining separated from those of adjacent cisterns by this double membrane, whose dissection and, if possible, preservation are the keys to atraumatic surgery. Further, Yasargil [106] has described the phenomenon of arachnoidal ensheathing of the neurinoma as it develops at the interior of the angle starting from its intracanalicular point of departure. This layer, initially single, becomes double when the size of the tumor brings it into contact with the arachnoidal membrane of the cerebellopontine cistern, and even triple when this adheres to the double cistern normally separating two cisterns. The neurinoma is directly in contact with the nerves and arteries within the IAM, but is always separated by at least one arachnoid layer from the contents of the cerebellopontine cistern, and by two or even three thicknesses of arachnoidal membrane from the contents of the adjacent cisterns (Fig. 28 c). Thus, the scooping out of the tumoral substance within the arachnoidal envelope without tearing the latter constitutes the surest means of safeguarding throughout the dissection the anatomic structures traversing the angle and adhering to the periphery of the tumor. The neurinoma of the mixed nerves develops within the cerebello-medullary cistern. It may engulf the structures of this cistern but remains separated from the contents of the adjacent cisterns by the arachnoidal layers which demarcate these cisterns. Cautious dissection within the cerebello-medullary cistern allows untroubled stripping of the acoustico-facial bundle. The neurinoma of the trigeminal nerve, however, develops in the cerebellopontine cistern. The small number of cases ob-

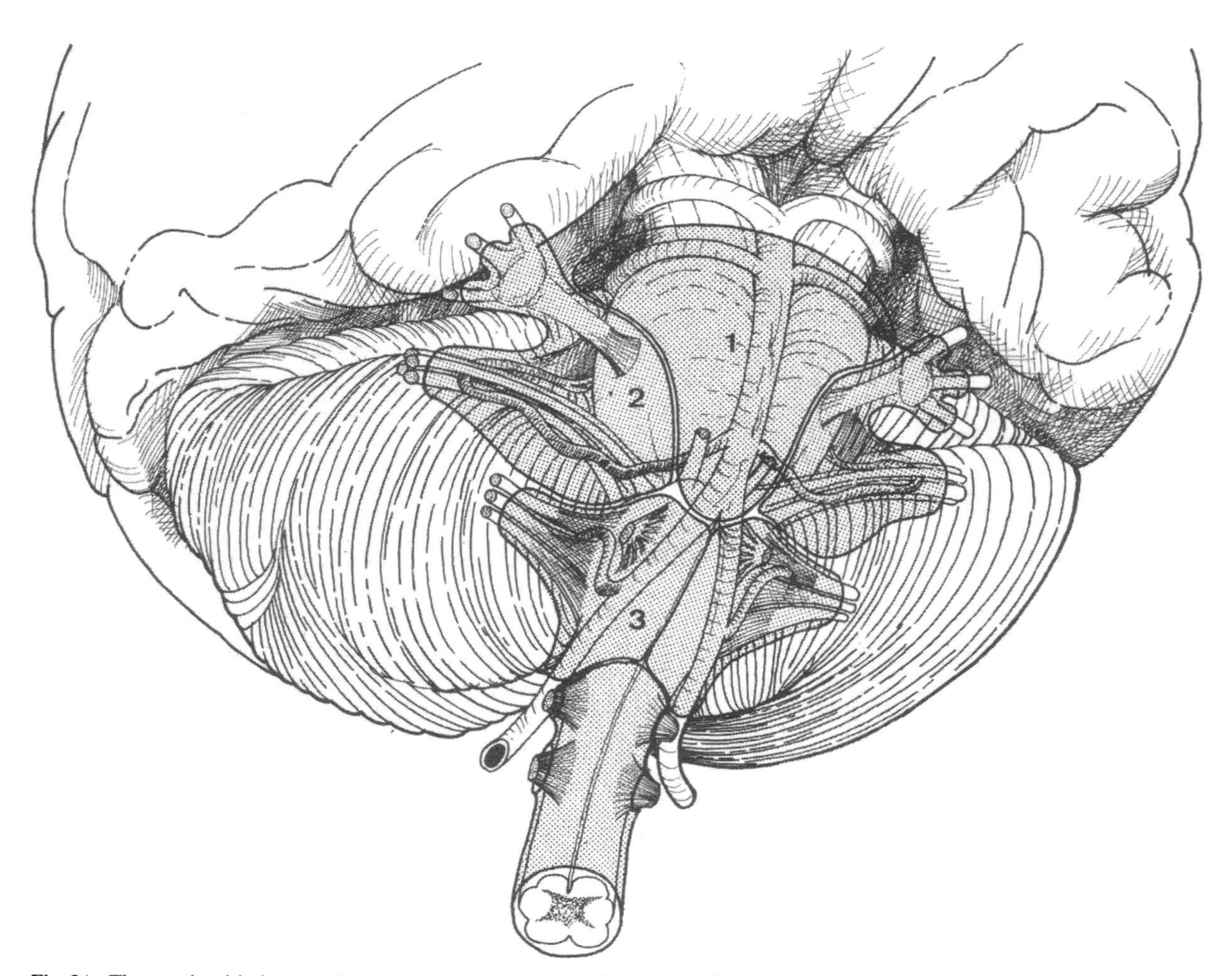

Fig. 31. The arachnoid cisterns. Prepontine cistern *(1)*, cerebellopontine cistern *(2)*, cerebello-medullary cistern *(3)*

served by us is insufficient to allow a precise idea of the relations of the arachnoid to this tumor. In the case of type C trigeminal neurinoma, straddling the supra- and subtentorial fossae, it has sometimes seemed to us that the tumor was ensheathed in an arachnoidal layer which separated it from the other structures of the cerebellopontine angle so effectively that stripping of the acoustico-facial bundle could be performed without any problem, allowing perfect preservation of hearing and facial mobility. This fact seems to suggest that the tumor, doubtless arising in front of the trigeminal ganglion, has, as in the fundus of the IAM, pushed ahead of it the arachnoidal layer which accompanies the nerve up to its exit from the cave of Meckel. On the other hand, in those cases where the neurinoma was of type B, developing only in the cerebellopontine angle, it has seemed that the facial nerve was directly in contact with the tumor substance, and its dissection has always been laborious and accompanied in three out of four cases by a postoperative facial palsy. This fact led us to believe that the tumor filled the entire cerebellopontine cistern. The mixed nerves on the other hand, save for one case, could be stripped without much difficulty because they were protected by an arachnoidal plane. The meningioma attached to the posterior aspect of the petrous develops in the subdural space. A tumor of the glomus jugulare, when it invades the angle, usually does the same. These tumors therefore remain separated from the neuraxis and the structures crossing the cisterns by an arachnoidal layer so effectively that their dissection is relatively easy, provided only that the surgeon respects the arachnoidal layer. However, the problem is sometimes difficult at the perimeter of the orifices of the base of the skull, where stripping is often less simple.

The Infratemporal Region

This term is debatable and some authors, such as Legent et al. [56] reject it because it overlaps several anatomic regions situated beneath the pyramid but also under the greater wing of the sphenoid. However, many accept this term as it corresponds to a defined surgical region approached by specific routes. To some extent, it is the hilum of the hemicranium since it is here that the great vessels and nerves travel as they enter or leave the cranial cavity. It is also the zone of anchorage of the cervical muscles to the base of the skull. Further, it is a region deeply embedded behind an outer bony wall formed by the mastoid process and the ramus of the mandible. Thus everything combines to makes this a complex region, difficult of access. Additional difficulties arise from by the arrangement of the facial nerve, which emerges from the petrous and develops its branches, thus barring access to the region from the outer aspect. Only rerouting the nerve permits such access.

The Extrapetrosal Facial Nerve

After leaving the stylomastoid foramen, the nerve bends to travel forwards and outwards almost horizontally. It passes laterally to the styloid process and medial to the mastoid. It appears in the angle formed by the anterior border of the mastoid and the inferior border of the tympanal portion of the temporal bone, which it virtually bisects (Fig. 32). At this site it crosses the posterior auricular artery, which gives off to the nerve its stylomastoid branch that ascends in the facial canal to supply the nerve and also the tympanic cavity and the mastoid cells. After this very short retroparotid course, the nerve enters the parotid, dividing it into two lobes, superficial and deep, which are reunited above it. It then divides very soon before reaching the posterior border of the ramus of the mandible, equidistant from the gonion and the tragus according to Eyries, Aboulker and Gandon [in 18]. The length of the trunk of the facial nerve from its emergence to its division varies between 5 and 20 mm according to Dargent and Duroux [in 18]. Classically, the nerve divides into two branches: temporofacial and cervicofacial, but actually this mode of division occurs in only 50% of cases. In the other half there is a trifurcation, with a median superior nasobuccal branch as described by Eyries and Biss [31] which arises most often (50%) from a short anastomosis of the two classical branches, sometimes (20%) from a true trifurcation of the facial nerve, and in the remaining 30% of cases either from the temporofacial or cervicofacial branch. Each of these two or three branches is then distributed to the facial muscles. It is conventional to distinguish three zones of distribution: superior or temporo-fronto-palpebral, middle or suborbital and superior buccal, and inferior or inferior buccal, mental and cervical. Chouard et al. [18] stress the variability of the terminal branches, which makes any facial incision risky. Only the cervical branches, destined for the platysma, can be divided without excessive cosmetic defect, since the superficial cervical plexus also innervates this muscle. Section of these lowest branches makes it easier to displace the nerve upwards when it is rerouted and lessens the tension.

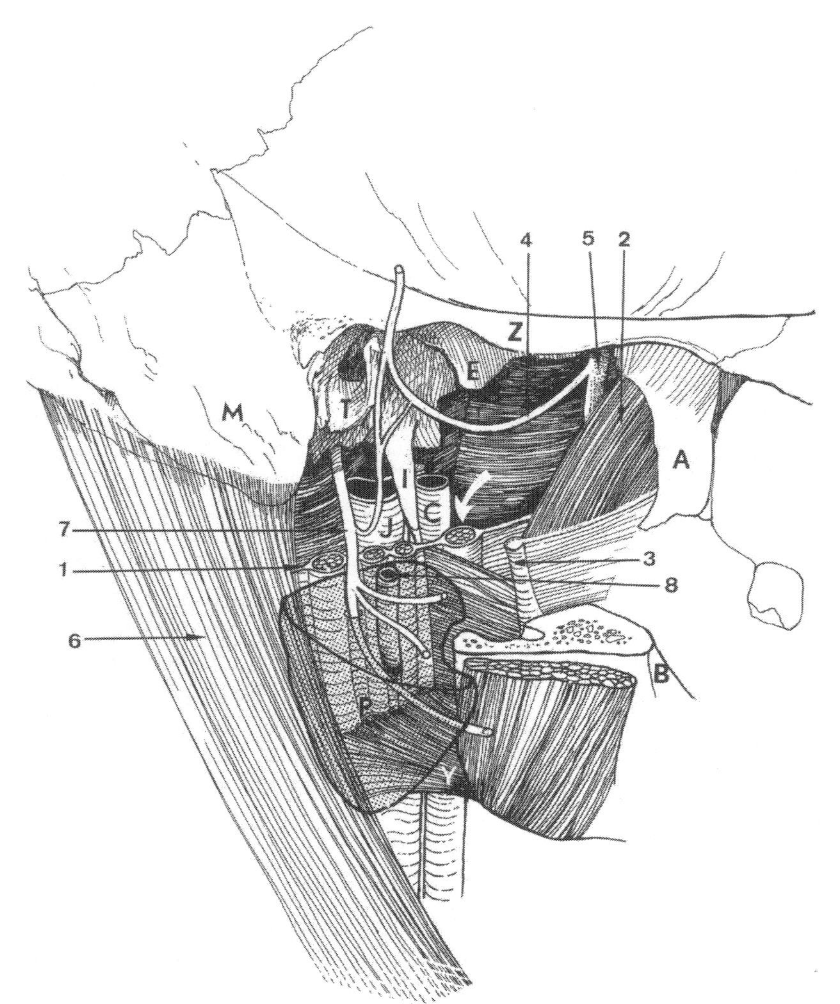

Fig. 32. The infratemporal spaces, schematic drawing. Mastoid *(M)*, tympanal *(T)*, zygoma *(Z)*, styloid process *(I)*, pterygoid apophysis *(A)*, ramus of mandible *(B)*, sphenoidal spine *(E)*, internal carotid artery *(C)*, internal jugular vein *(J)*, styloid process *(I)*, internal pterygoid muscle *(2)*, sphenomandibular ligament *(3)*, auriculotemporal nerve *(4)*, trigeminal nerve *(5)*, facial nerve *(7)*, sternocleidomastoid muscle *(6)*, Charpy's strip *(Y)*, parotid gland *(P)*, external carotid *(8)*, retrostyloid space *(white arrow)*

Overlying Planes

Access to the infratemporal region is limited by a kind of bony arch whose posterior pillar is formed by the tympanomastoid block, the anterior pillar by the ramus of the mandible and the vault by the temporomandibular joint. This arch is prolonged downwards by the sternocleidomastoid and digastric muscles. The vault of this arch is blocked by the parotid gland (Fig. 33).

The Bony Arch

The posterior tympanomastoid pillar is familiar and its complete resection with freeing of the third part of the facial nerve in itself, gives proper exposure of the region. It is also possible, when approach in depth requires it, to resect the anterior pillar, in particular the mandibular condyle and its neck. However, it seems preferable, within the limits of possibilities, to preserve this anterior pillar, especially when the approach makes it possible to safeguard the preglaserian segment of the mandibular fossa. However, in such a case, it is possible to widen the approach a little more forward by dislocating the mandible in front of the temporal articular tubercle after dividing the capsule of the temporomandibular joint, bearing in mind that this joint contains an intra-articular meniscus and that the capsule is reinforced by two lateral ligaments, external and internal, which must be rugined at periosteal level. It should also be remembered that temporomandibular apposition is reinforced by three extrinsic ligaments: sphenomandibular, strengthened by its main tympanomandibular component, stylomandibular and pterygomandibular, and that the sphenomandibular ligament together with the inner aspect of the ramus bounds an osteofibrous slit, the retrocondylar buttonhole of Juvara. Through this slit pass

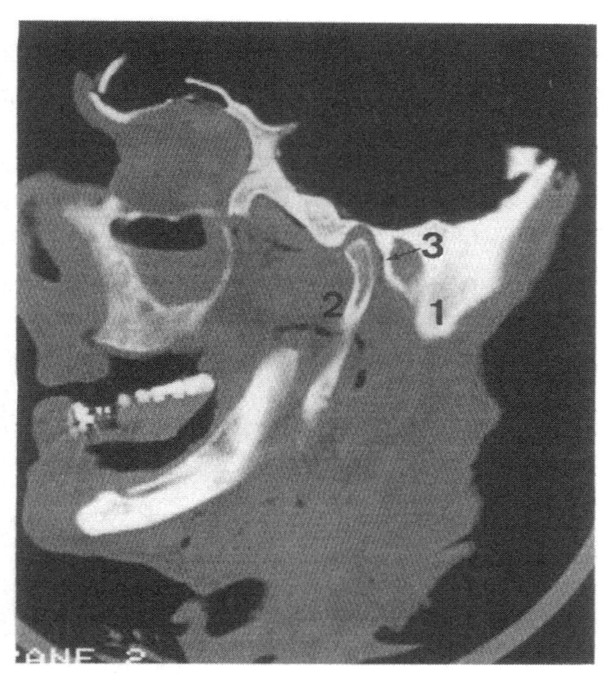

Fig. 33. The bony arch. Sagittal CT section of the mastoid process *(1)*, ramus of mandible *(2)*, tympanal *(3)*

the maxillary artery travelling forward and the auriculotemporal nerve, a branch of the mandibular nerve, travelling backwards. These structures most not be injured during section of the capsule. This maneuver will have been preceded by ligature of the maxillary and superficial temporal arteries, the latter passing over the outer aspect of the zygoma at the joint level of, or just behind it. If necessary, resection of the articular tubercle and its neck is made with the gouge after detaching the medial pterygoid muscle, which has its origin just under the condyle and whose retraction displaces somewhat the maxillary artery as it winds around its lower border. Behind the neck, this artery gives off its first branches, the anterior tympanic and middle meningeal arteries. These two branches are often large in cases of meningioma and especially of tumors of the glomus jugulare and must also be ligatured. At the summit of the arch, against the bone, there passes the auriculotemporal nerve which, after having given two anastomotic branches to the facial nerve, bends upward to accompany the superficial temporal artery and is distributed to the integument of the region (Fig. 32).

Muscles

In the infratemporal or extended transcochlear approaches, the skin incision descends along the anterior border of the sternocleidomastoid muscle as far as the thyroid cartilage. This exposes all of the upper half of this muscle, whose sheath at this level is thick and sends a fibrous strip to the angle of the mandible, the strip of Charpy, which aids in closing the parotid compartment below (Fig. 32). Here, the body of the muscle is usually traversed by the accessory nerve and crossed superficially by the external jugular vein which emerges from the parotid compartment. These two structures run obliquely downwards and outwards. Above and parallel to the vein, the auricular and mastoid branches of the superficial cervical plexus ascend on the surface of the muscle. The attachments of the muscle are divided with the diathermy knife at the tip of the mastoid. Reflection of the muscle bulge downwards and outwards reveals the posterior bulge of the digastric, which is inserted more medially at the bottom of its groove. Before detaching it, it is necessary to identify the occipital artery, which skirts the inferior border of the muscle, and especially the facial nerve, which emerges just at the extremity of the groove and then also follows the lower border of the muscle to run eventually on its outer aspect. Disinsertion of the digastric exposes the styloid process and the stylohyoid muscle, accompanied by the ligament of the same name, descending obliquely downwards and forwards. These structures form the outer border of the styloid curtain. They are crossed over their outer border by the facial nerve, which there enters the parotid compartment (see below, extended transcochlear approach).

The Parotid

This gland is wedged between the ramus of the mandible in front, lined by the pterygoid muscles internally and the masseter externally, and behind by a fibromuscular septum composed from without inwards of the sternocleidomastoid and digastric muscles and then by the outer part of the styloid curtain as far as the stylomandibular ligament (Fig. 32). It is closely moulded and adherent to the contours of its compartment, which is closed laterally by the superficial cervical aponeurosis, medially by an areolar layer stretched between the stylomandibular ligament and the pterygoid muscles, and below by Charpy's strip. The gland is traversed also by various vascular and neural structures which it is customary to describe in three planes according to their depth:

- The arterial plane is formed by the external carotid, which enters the compartment at its inferolateral pole, forming the deep internal pedicle of the gland and passing between the stylohyoid muscle and its satellite ligament laterally and the styloglossus muscle and stylomandibular ligament medially. The artery ascends vertically in the substance of the gland in the direction of the condyle of the mandible and, while still within the gland, divides into its two terminal branches, the maxillary and superficial temporal. These branches emerge from the gland at its upper pole to form two superior pedicles at their emergence. Sometimes the external carotid, while within the gland, gives off its posterior auricular branch which emerges behind at the level of the mastoid tip.

- The venous plane is situated immediately lateral to the external carotid artery. It is formed by the external jugular vein, which arises from the junction of two afferents: an anterior, formed by the maxillary and superficial temporal veins, and a posterior, formed by the posterior auricular and occipital veins. These afferents give rise to a venous trunk which soon divides. One of its branches continues the direction of the trunk downwards and outwards to emerge from the gland against the sternocleidomastoid muscle: this is the external jugular vein. The other branch descends vertically to traverse Charpy's strip: this is the intraparotid communicating vein which opens into the facial vein or directly into the venous trunk of Faraboeuf. Sometimes this venous trunk also receives a vein which accompanies the carotid artery, the vein of Launay.

The Infratemporal Spaces

Spreading out from the styloid process is a bundle of three muscles, the stylohyoid, styloglossus and stylopharyngeus, and two ligaments, stylohyoid and stylomandibular. Like the spokes of a fan, these structures support an aponeurosis which descends from the base of the skull to the lateral wall of the pharynx. This is the styloid curtain, which hangs obliquely forward and inward. It divides into two spaces, pre- and retrostyloid, the infratemporal space, bounded medially by the pharyngeal wall, anteriorly by the pterygoid muscles, laterally by the bony arch and the sternocleidomastoid, and behind by the prevertebral muscles and the scaleni covered by the prevertebral fascia (Fig. 32).

Within the retrostyloid space travel the internal carotid artery ascending towards the carotid canal and, just lateral to it, the internal jugular vein which emerges from the posterior lacerate canal. The three mixed nerves and the hypoglossal (XII) accompany this jugulo-carotid bundle. The glossopharyngeal nerve (IX), which skirts the outer border of the internal carotid, then descends towards the styloglossus muscle and accompanies it as far as the tongue. The vagus (X), swollen above at its plexiform ganglion, descends in the posterior jugulo-carotid dihedral angle. Just as it emerges from the plexiform ganglion it gives off its superior laryngeal branch, which descends a little below the glossopharyngeal towards the lateral wall of the pharynx, where it divides into its two branches, lateral and medial, at the level of the greater cornu of the hyoid bone. The accessory nerve (XI), which very soon gives off its medial branch to the plexiform ganglion, descends obliquely outwards, usually passing in front of, but sometimes behind the internal jugular vein, to cross the bulge of the sternocleidomastoid 4-5 cm beneath the tip of the mastoid. After emerging from the condylar canal, the hypoglossal nerve (XII) descends obliquely downward and outward to pass behind the internal carotid and the vagus, then between the latter and the internal jugular vein, and then travels downward and forward lateral to the external carotid and hooks under the occipital artery medial to the venous trunk of Faraboeuf. In crossing the outer side of the internal carotid artery, it gives off its always well-marked descending branch. Right at the back, plastered against the prevertebral muscles, is the cervical sympathetic and its large superior cervical ganglion. At the base, the ascending pharyngeal artery climbs the wall of the pharynx. All these structures must be located and identified before undertaking excision of tumors of this region (see under the extended transcochlear approach).

Medial to the parotid compartment, the prestyloid space forms the paratonsillar space. The roof of this space consists of the inferior aspect of the greater wing of the sphenoid, pierced by the oval foramen for the passage of the mandibular branch of the trigeminal. This stout nerve divides very rapidly into its terminal branches, the lingual and inferior alveolar. The lingual nerve receives the chorda tympani, which descends obliquely from the tympanomastoid fissure. When the approach has to be very anterior, all these branches and sometimes even the trunk of the mandibular nerve must be identified and safeguarded.

The Pyramid in Operative Position

The surgeon who undertakes to burr a passage across the petrous to approach a tumor developing in this region must be quite familiar with the entirety of the anatomic data rehearsed so far. But even this knowledge, complex as it is, is still inadequate. The surgeon is further required to be able to form a spatial concept of the anatomy of the petrous pyramid in relation to the position at operation and the angle of approach. It is possible to fashion a temporal or occipital flap to gain access to either the superior or inferior aspect of the pyramid. In these cases, despite the existence of a certain degree of obliquity, the view at operation is similar to the orthogonal view of conventional anatomic descriptions, and it is relatively easy to picture the intrapetrous cavities underlying the aspect approached correctly. But it more often happens that the pyramid is attacked from its external, mastoid, base. Then the operative perspective is altogether unaccustomed and calls for special description. Throughout the trans-temporal stage, moreover, the surgeon must meet two obligations: to respect the facial nerve throughout its intrapetrous course and, whether his choice is to preserve or sacrifice hearing, to locate precisely the different structures of the vestibulocochlear apparatus, which he may then preserve or destroy. This underlines the importance throughout the operation of the entity composed by the vestibulocochlear cavities and the facial canal. It is essential to possess a perfect stereoscopic mental vision of this system, and to be able to place it exactly within the pyramid in every position.

Lateral Approaches

These consist of resection of the whole or part of the pyramid. They are in fact "transpetrous" approaches. The one most commonly practiced is the extended translabyrinthine approach. They are performed on a subject in dorsal decubitus, with the head turned to the side opposite the approach. The surgeon positions himself at the side of the table, facing the patient's occiput (Fig. 45) and looking directly at the mastoid.

Operative Perspective

The pyramid approached is seen in three-quarter posterior view (Fig. 34), with the tip below and the posterior aspect directed towards the operator. Its long axis is not vertical but oblique downwards and

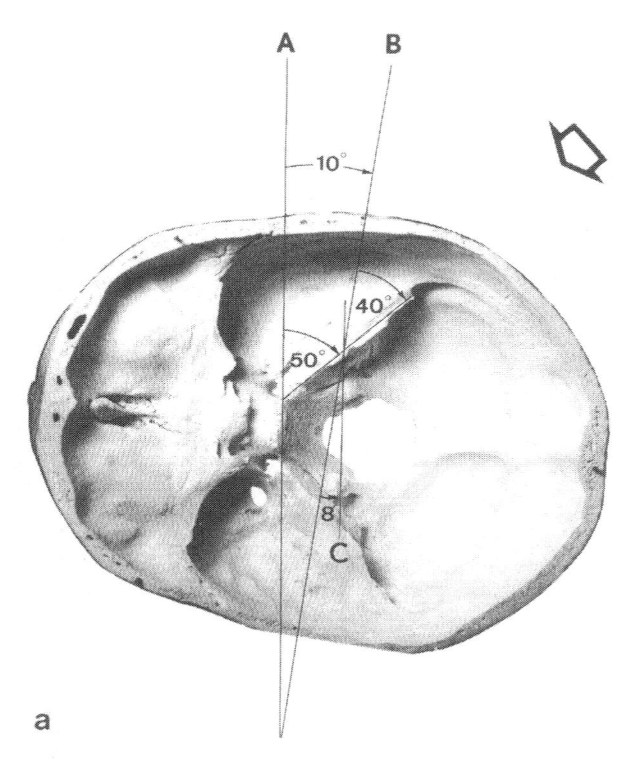

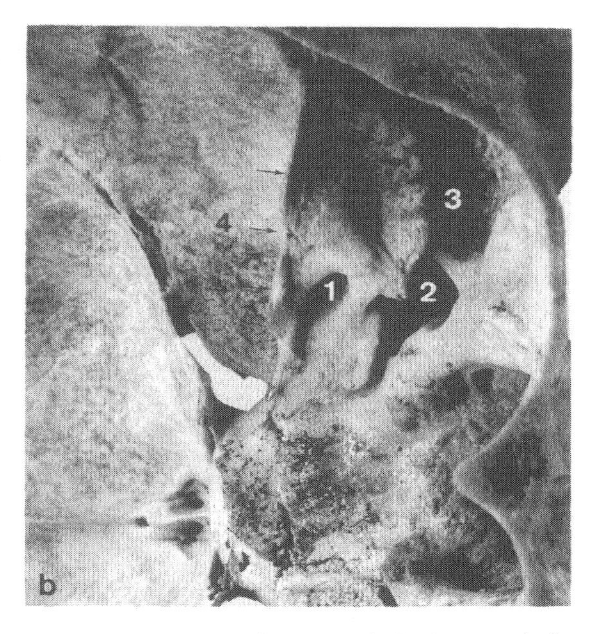

Fig. 34 a, b. External approach. The pyramid in operative position. **a** Placement of right pyramid in operative position. Rotation of the head cannot be complete. The biauricular axis *(B)* is 10° from the vertical *(A)*. The pyramid, 40° oblique to the frontal biauricular plane *(B)* is thus 50° oblique to the vertical *(A)*. The axis of the IAM *(C)* is 8° oblique behind the biauricular plane and thus virtually vertical. The arrow indicates the angle of attack on the mastoid. **b** View of posterior aspect of right pyramid in operative position. Aperture of IAM *(1)*, jugular bulb *(2)*, groove of sigmoid sinus *(3)*, petrous crest *(4)*

towards the opposite side. We know that this long axis normally makes an angle of 40°, open backwards, with the frontal plane of the head and sometimes an angle of 10 to 20° with the horizontal, such that the apex is then higher than the base. Rotation of the head cannot reach quite 90° and, if forced, may be accompanied by a certain degree of anterior flexion which accentuates the obliquity of the pyramid. The outcome of all these mechanisms is that in the operative position the petrous pyramid has its long axis obliquely downward and toward the opposite side (Fig. 34), making an angle of about 50° with the vertical and sometimes accompanied by a lateral obliquity of 10 to 20° to the direction of the patient's head. The axis of the two external auditory meatuses (EAM) is not perfectly vertical and inclines by some 10° towards the surgeon and by a few degrees towards the patient's feet. The axis of the IAM, which is normally inclined by 8° behind the axis of the EAM and 45° behind the axis of the petrous pyramid, is situated practically vertically but, depending on the obliquity of the pyramid upward, there may be an equivalent deviation of its long axis towards the head of the patient. It is important to be aware of this orientation when burring in this region.

The Labyrinth

The posterior SCC, situated in a plane virtually parallel to that of the posterior aspect of the petrous, lies in a plane about 45° oblique downwards and towards the opposite side. The surgeon approaches the canal at its edge (Fig. 35). The plane of the lateral SCC, also perceived at its edge, is perpendicular to the plane of the posterior SCC. Depending on the lateral obliquity of the pyramid in relation to the operator, it may incline by 10 to 20° in the direction of the patient's feet. On the other hand, the anterior SCC, at right angles to the other two, makes its loop in a plane facing the surgeon. The extremity of the IAM is insinuated into the interior of this loop, but since the latter is superior to the meatus in the anatomic position it is the extremity of the roof that is exposed to the operator's sight. The long axis of the vestibule, situated in the plane of the anterior SCC, facing the view of the operator, is practically vertical. It is overhung by the two adjacent ampullae of the anterior and lateral SCC. It is evident that the plane of the anterior SCC and the vestibule will be exposed only after drilling of the two other SCC. During this drilling the surgeon will have perceived in succession the right-angled intersection of the two loops the posterior horizontal and the lateral vertical (Fig. 36), then the sections of the anterior and posterior limbs of the lateral SCC and the superior and inferior limbs of the posterior SCC, which, in the operative position, are situated right and left and aligned horizontally in the case of the posterior SCC and superior and inferior, vertically aligned, for the lateral SCC (Fig. 36). Drilling of this bony plane containing the anterior SCC and the vestibule, leaving in place the two anterior ampullae of the anterior and lateral canals so as not to injure the genu and second part of the facial nerve, opens the fundus of the meatus. Whereas the IAM in this operative position is vertical, the plane of drilling facing the surgeon is at about 45°. Thus, the fundus of the IAM is not opened in the plane at right angles to the long axis of this meatus but obliquely, at the posterior and superior half of its periphery, so that the extremity of the IAM is cut slantwise (Fig. 36). The posterior half of the transverse crest is removed, and its sectioned surface then appears as a practically vertical bony spine pointing in the axis of the IAM. Bill's bar, which in the anatomic position descends perpendicularly on the transverse crest, is situated in a transverse plane in the operative position (Fig. 35 b) and conceals the entry to the facial canal from the surgeon. By inserting a fine hook behind the vestibular nerve it is easy to perceive this transverse ridge, which forms a good landmark in avoiding injury to the facial nerve. After having released the superior vestibular nerve, the entry of the facial nerve into the first part of the facial canal can be uncovered only after drilling this transverse wall. Once this is done, only its upper root is left intact, appearing as a small bony spine which points obliquely in the direction of the patient's feet (Fig. 35). The last stage of the approach consists of drilling of the posterior aspect of the IAM as far as the level of its aperture by wide resection of the posterior half of the periphery. As the exposure of the fundus of the IAM has removed the extremity of the roof and the most upper part of the posterior wall, drilling must extend along this posterior wall so that the resection may be quite symmetrical above and below the meatus and its aperture. To do this, the surgeon must direct his burr slightly obliquely in the direction of the patient's feet, and not vertically as might be suggested by the position of the IAM (Fig. 34 b) unless there is marked obliquity of the pyramid, and therefore of the IAM, towards the patient's head, which would be an indication for keeping the burring vertical.

The Facial Nerve

Throughout its course in the IAM and the facial canal, this follows a tortuous and complex path which

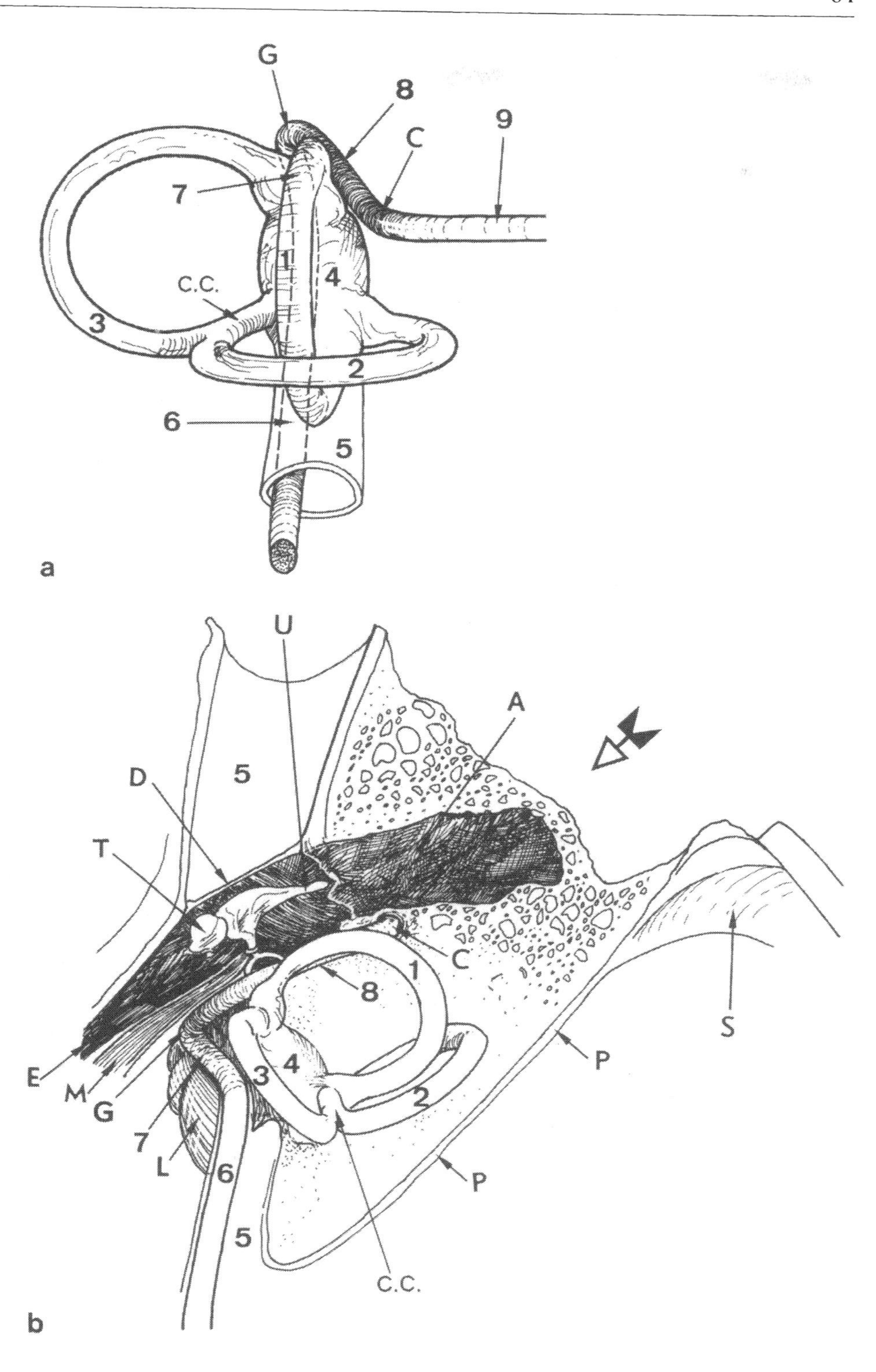

Fig. 35 a, b. Lateral approach. The labyrinth in operative position. **a** surgical view. **b** diagram of lateral view. Lateral SCC *(1)*, posterior SCC *(2)*, anterior SCC *(3)*, vestibule *(4)*, IAM *(5)*, intracanalicular facial nerve *(6)*, first part *(7)*, second part *(8)*, third part *(9)*, genu of facial nerve *(G)*, bend of facial nerve *(C)*, crus commune *(c. c.)*, cochlea *(L)*, tensor tympani muscle *(M)*, auditory tube *(E)*, head of malleus *(T)*, tympanic membrane *(D)*, incus *(U)*, mastoid antrum *(A)*, groove of lateral sinus *(S)*, posterior aspect of petrous *(P)*

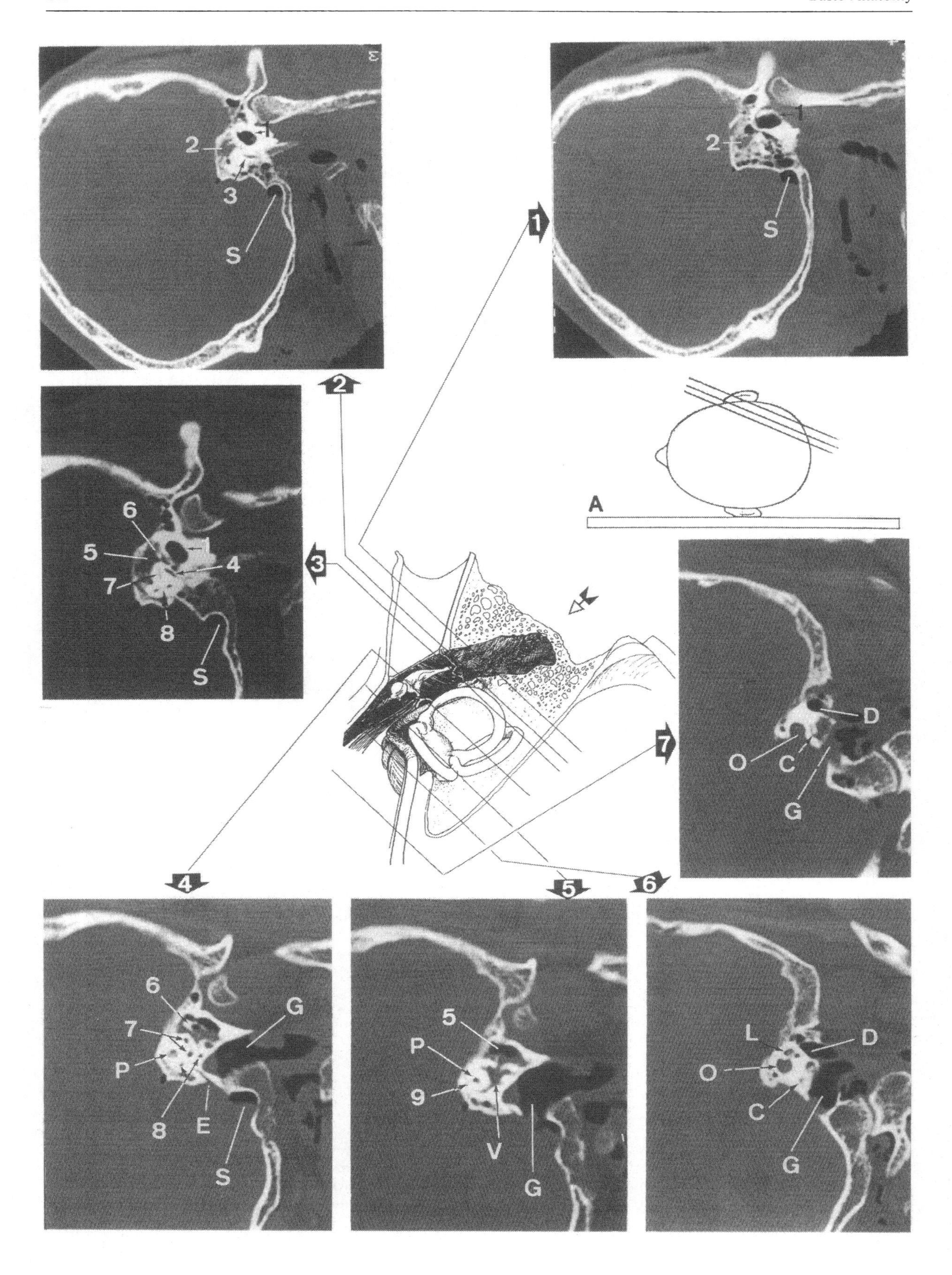
1
2
3
S
1
2
S
2
3
6
5
1
4
7
8
S
A
7
D
O
C
G
4
5
6
6
G
7
P
8
E
S
5
P
9
V
G
L
D
O
C
G

is roughly V-shaped, the acute angle being formed by the genu of the nerve. In the operative field this V is inverted and seen very obliquely, with the deep branch inferior and the superficial branch superior (Fig. 35 a).

The acute angle of this inverted V straddles the two anterior ampullae and the vestibule. Neither limb of the V is straight. The deep limb, comprising the intracanalicular portion of the facial nerve and its first part, twists at 130° as it enters the facial canal. The superficial limb, composed of the second and third parts, bends through 120° at the genu. These two bends are situated in different planes. Moreover, the first portion is itself curved on the plane of torsion of the deep limb (Fig. 37). As is known, in the anatomic position the plane of the roof of the IAM is flush with the superior plane of the mouth of the lateral SCC. In the operative position, this plane is situated vertically and is seen on edge, and it is in this plane that the 130° torsion of the deep limb occurs. Thus, the facial nerve in its intracanalicular portion, situated just under the roof, is found in the operative position to lie vertically and in the plane of the loop of the lateral SCC. As the first portion inclined 50° forward, the direction of the intracanalicular portion becomes obliquely ascending under the plane of the loop of the anterior SCC. At the same time, it bends towards the patient's head. In the anatomic position, it ends internal to the ampulla of the anterior SCC, and therefore in the operative position it is just under this ampulla. The second portion makes an angle of 74 to 80° with the first and therefore emerges, almost at a right angle, from the plane of the loop of the anterior SCC, inclined by some 10° to this plane in the direction of the surgeon. The plane of this loop of the SCC is opposite the surgeon and at an inclination of 55 to 60° (Fig. 37). Thus this portion is an overhanging position above the two anterior ampullae and to expose these the surgeon must burr along this second portion and excavate underneath it. In the anatomic position, this second portion lies obliquely downward at a slope of 35 to 40°, coming to be situated under the anterior limb of the lateral SCC. In the operative position, in addition to its ascending obliquity towards the operator, it inclines by 35 to 40° towards the patient's feet. In depth, it passes under the ampulla of the lateral SCC and then ascends behind the anterior limb of this canal. Emerging from the midst of the labyrinth, the facial nerve then describes its bend. The third portion, vertical in the anatomic position, is situated horizontally in the operative position. It travels transversely in the direction of the patient's feet, to pass 1-2 mm beneath the incudal fossa and the edge of the base of the stapes.

The Tympanic Cavity

This is practically vertical in the anatomic position with a forward obliquity of 20° in relation to the sagittal plane; in the operative position it is horizontal. Drilling of its posterior aspect from the inferior extremity of the aditus, taking care to follow the outer border of the elbow (bend) and then the third part of the facial canal, amounts to the performance of a posterior tympanotomy. This gives a good exposure of the interior of the cavity, the internal aspect of which has become inferior. It is easy to identify the cochleariform process emerging from this aspect and, next to it, the second portion of the facial canal. Below this is the promontory elevated by the first turn of the cochlear spiral. On the posterior slope of the promontory is the vestibular window, in which is embedded the base of the stapes. It will be recalled that the carotid canal ascends under the cochlea and a little in front of it, then curving in the direction of the apex of the petrous. Drilling of the promontory therefore opens into the cochlea and, more deeply, into the carotid canal.

Segmentation of the Petrous

"The internal ear possesses a special skeleton which seems to be separate from the petrous bone in which it is embedded." From this first phrase of his classic work on the surgical anatomy of the labyrinth, Girard [37] thus emphasized the reality of this labyrinthine massif, which forms a compact nucleus in the midst of the cancellous meshwork of the petrous. The heterogeneous structure of this bone is evident in CT section (Fig. 22). In view of this, it is obvious that transpetrous surgery must consist of two stages of drilling, first extralabyrinthine and

Fig. 36. Sections of the petrous in the plane of the anterior SCC according to diagram a. EAM *(1)*, mastoid antrum *(2)*, 3rd part of facial canal *(3)*, bend of facial canal *(4)*, tympanic cavity *(5)*, incus *(6)*, lateral SCC *(7)*, posterior SCC *(8)*, anterior SCC *(9)*, groove of lateral sinus *(S)*, petromastoid canal *(P)*, endolymphatic canal *(E)*, vestibule *(V)*, jugular bulb *(G)*, cochlear canal *(C)*, carotid canal *(D)*, IAM *(O)*

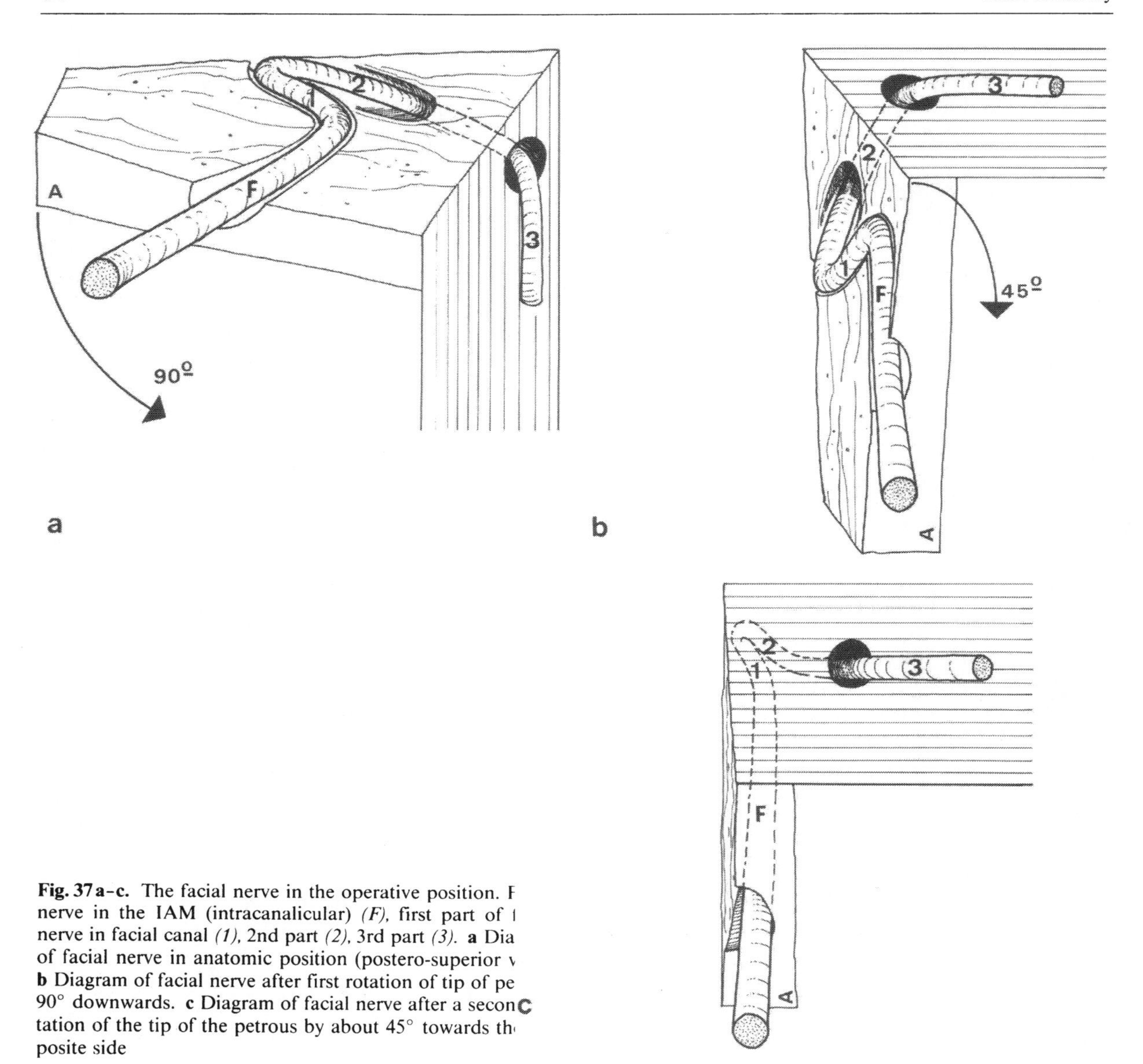

Fig. 37 a–c. The facial nerve in the operative position. F nerve in the IAM (intracanalicular) *(F)*, first part of t nerve in facial canal *(1)*, 2nd part *(2)*, 3rd part *(3)*. **a** Dia of facial nerve in anatomic position (postero-superior v **b** Diagram of facial nerve after first rotation of tip of pe 90° downwards. **c** Diagram of facial nerve after a secon tation of the tip of the petrous by about 45° towards th posite side

then labyrinthine. It is also evident that each of these stages involves resection of a segment of the pyramid whose boundaries are not approximate but, on the contrary, very well defined. In the light of experience, the petrous pyramid is no longer to be regarded as a kind of monolith on which the surgeon must exert himself to carve out a suitable trench. Rather, it should be considered as a sort of assemblage from which each component may be removed, independently or successively. This imposes the concept of segmentation of the petrous and also a unified concept of transpetrosal surgery, which is not made up of the juxtaposition of different operative techniques but, is to be considered as a single operative technique made up of successive operative stages whose deployment can be continued until complete or interrupted at each stage according to the demands of the resection. From this point of view the pyramid may be divided fairly precisely into two parts on either side of a vertical plane (Fig. 38) passing through the posterior wall of the EAM and the anterior wall of the IAM. Thus, the anterior portion comprises the EAM, the tympanic cavity, the cochlea and the carotid canal up to the petrosal apex. This is the cochlear segment, which can itself be subdivided into two other segments situated on either side of the inner aspect of the tympanic cavity: laterally, the anterolateral or tympanic segment; medially, the anteromedial or carotid segment. The posterior portion is composed of the

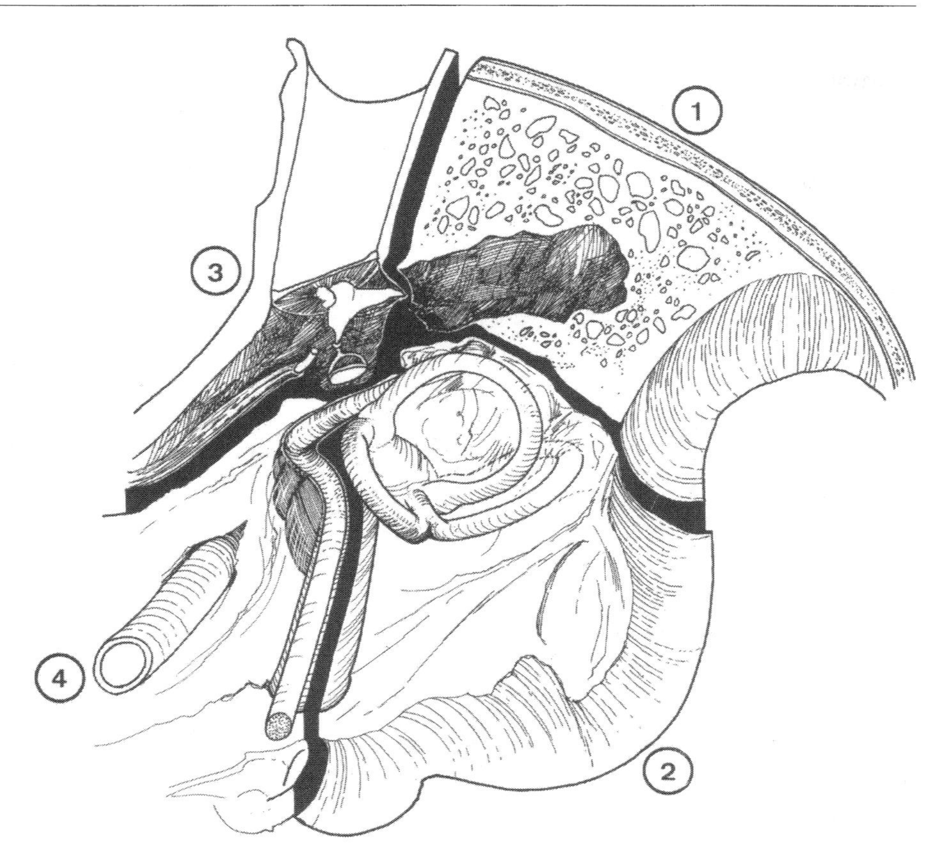

Fig. 38. Petrosal segmentation. Postero-lateral (retrolabyrinthine) segment *(1)*, postero-medial segment (posterior labyrinthine) *(2)*, antero-lateral (tympanic) segment *(3)*, antero-medial (carotid) segment *(4)*

mastoid with its antrum, the posterior or vestibular labyrinth and the IAM. This is the vestibular segment, which can also be divided into two segments by the plane of the outer aspect of the labyrinthine massif: laterally, the posterolateral segment, which may be called either the retrotympanic or retrolabyrinthine segment; medially, the posteromedial segment, conventiionally known as the labyrinthine (more precisely the posterior labyrinthine) segment. The facial nerve follows the demarcation between these different segments very precisely.

The surgeon who drills the mastoid and limits his resection anteriorly at the posterior wall of the EAM and, after having opened the antrum at its anterior wall, restricts his deep penetration by stopping at the outer face of the labyrinthine massif, will have finally resected the posterolateral segment and exposed the dura mater internal to the sigmoid sinus over an extent of more than 1 cm and over the entire height of the posterior face of the petrous from the sigmoid sinus below as far as the superior petrosal sinus above. This is the retrolabyrinthine route of Hitselberger and Pulec [46]. The surgeon can extend his burring, destroy the whole of the posterior labyrinthine massif, skeletonize the entire course of the facial nerve as far as the two anterior ampullaė of the anterior and lateral SCC and then resect the posterior three-fourths of the circumference of the IAM, leaving only its anterior wall. He will then have further resected the whole of the posteromedial segment and widely exposed the dura mater of the anterior aspect of the posterior fossa up to the level of the internal border of the porus of the IAM. This is the translabyrinthine route of House and Hitselberger [49-51], whose first extralabyrinthine stage is simply the retrolabyrinthine route. To progress further, internal to the aperture of the IAM, towards the apex of the petrous, the surgeon must continue drilling the anteromedial segment, destroy the cochlea and denude the intrapetrous internal carotid: this is the transcochlear route of House and Hitselberger [50], which links the resection of three segments: posterolateral, posteromedial and anteromedial. Originally, in order to maintain impermeability, it left in place the skeleton of the EAM and its tympanic membrane, but there is no problem in removing this meatus and sacrificing the external orifice of the EAM. All four segments will then have been removed.

These transpetrous approaches can be extended by enlarging the surface opening. If the superficial drilling exceeds the limits of the base of the pyramid and extends to the temporal squama above and the occipital squama behind, one may speak of an extended translabyrinthine approach. If the upper cervical region is widely opened to gain the maximum exposure offered by the transcochlear route, this is what we call the extended transcochlear route. One may use a somewhat different procedure whose aim is to resect the cancellous portions of the temporal bone while leaving the labyrinthine massif in place. This roughly amounts to resection of the two antero- and posteromedial segments and then drilling the anterior part of the anteromedial segment so as to open the vertical portion of the carotid canal while respecting the cochlea. This is the infratemporal route (type A) of Fisch [35], always accompanied by a wide opening of the cervical region. This route may be extended towards the petrous apex (type B) or forward towards the parasellar regions (type C).

Landmarks

When one of the petrosal cavities has been laid open by the burr, it is relatively easy with a knowledge of the anatomy to continue the drilling by following its structure. On the other hand, if one intends to approach this cavity as closely as possible but without opening it, the maneuver becomes much more difficult and makes precise landmarks essential. Each of the four petrosal segments possesses its landmarks, to transgress which means opening the adjacent segment.

The Retrotympanic Segment

This is the port of entry, the obligatory passage for all the transpetrous routes. On its outer aspect its boundaries are obvious: the posterior border of the EAM and the suprameatal spine in front, the temporal line above, the posterior border of the mastoid behind and below (Fig. 5). Cortical burring within these limits opens the most superficial mastoid cells, exposes the dura mater of the temporal bone and the posterior fossa and the outer limits of the superior and posterior aspects (Fig. 29). Excavating within these limits opens the mastoid antrum, which is a sort of meeting-place where are situated the landmarks essential for completing the resection of this segment and guiding the drilling if it is decided to continue across the petrous. By drilling at the level of the suprameatal spine and the retromeatal cribriform zone, the antrum is usually found at a depth of some 15 mm. Theoretically, its opening poses no problems, but in fact everything depends on the position of the bend of the lateral sinus in relation to the EAM. It is known to come close to the meatus in one out of four cases, and is actually in contact with it in 7% of cases which seriously complicates this route of access.

Once the antrum is open, the surgeon locates the aditus. At its internal border is the smooth and easily identified prominence of the lateral SCC. The slightest drilling at the inferior apex of this orifice reveals the edge of the short limb of the incus, supported against the fossa incudis. In the operative position the bend and third part of the facial nerve pass between 2 and 2.5 mm under this edge. Looking through the aditus into the interior of the tympanic cavity, one sees the cochleariform process, just behind which appears the second part of the nerve which travels backwards to pass under the loop of the lateral SCC. Having fixed these landmarks, one may quite safely burr all this bony corner which forms the retrotympanic segment, expose the first bend of the sigmoid sinus and skeletonize the bend and third part of the facial nerve. Well below, one will have exposed the groove of the digastric, which ends in front at the stylomastoid foramen.

The Labyrinthine Segment

Whereas, in the preceding stage, only the skeletonization of the facial nerve and the exposure of the sigmoid sinus carry any real risks, drilling in this segment must be extremely precise. Abrasion of the SCCs (Figs. 13, 35, 36) leads directly to the vestibular cavity, but one must be careful to stop just at the anterior limb of the loop of the lateral SCC and at the two anterior ampullae of the lateral and anterior SCCs so as not to lay open the second and first portions of the facial nerve. One must be able to identify the internal wall of the vestibule, particularly the hemispheric fossula, behind which the IAM terminates. At this site the strands which form the origin of the superior and inferior vestibular nerves are identifiable and, after opening the fundus of the meatus, allow location of these two nerves. With a fine hook one can then feel and proceed to expose between the two the transverse crest whose posterior border has been drilled away (Fig. 35). It appears as a small vertical spine whose point indicates the direction of the IAM. Behind the superior vestibular nerve is a crest perpendicular to the transverse crest, Bill's bar, and behind this is the entry orifice of the facial canal. With all these structures identified, it becomes possible to drill the posterior periphery of the IAM. One can tell when drilling below this must stop when one notes the narrow groove created by the aqueduct of the cochlea.

The Cochlear Segment

Drilling of this segment can be undertaken only after exposure of the third and second portions of the facial nerve, i. e., after resection of the retrotympanic segment (infratemporal route of Fisch) or of the two posterior segments (translabyrinthine route of House). Opening the facial canal then makes it possible to extract the facial nerve and reroute it upwards, making use of the elongation obtained by mobilization of its two bends. It is then not very difficult to destroy the carotid segment. The promontory indicates the position of the cochlea. In front of it and at 2 or 3 mm depth, sometimes less, the burr opens the vertical portion of the carotid canal. Destruction of the cochlea and then of the cells of the petrous apex exposes more compact bone, the outer border of the basilar layer. In every case, drilling is limited behind and above by the dura mater, which lines the upper and posterior surfaces of the petrous.

The Superior Approach

Properly speaking, the aim of this approach is not to resect part of the pyramid but simply to trephine the roof of the IAM. The patient is still in dorsal decubitus, with the head turned towards the side opposite the approach. The surgeon stands at the head of the patient, looking directly at the temporal and auricular region. After fashioning a temporal flap he approaches the petrous by the extradural route.

Operative Field

The pyramid is approached obliquely but via its superior aspect, so that the classical anatomic data are respected. It is true that the pyramid has its apex directed downwards, but if care is taken to flex the neck, the superior aspect will be displayed more directly to the surgeon, with an inclination of about 30°. The arcuate eminence is an approximate index of the position of the loop of the anterior SCC. Allowing for the obliquity of the pyramid, the plane of this canal in the operative position is situated obliquely downwards and towards the patient's feet. Medial to it is the meatal field, under which is situated the IAM flanked at its anterior extremity by the cochlea. It is orientated obliquely downwards and backwards to open into the posterior fossa. The head being in the maximum possible rotation, the axis of the pyramid is again at an obliquity of about 40° downwards and toward the opposite side, so that the axis of the IAM is almost vertical, slightly inclined by 5 to 10° towards the patient's occiput and obliquely intersecting at about 50° the posterior border of the superior surface. In the anatomic position the IAM is horizontal but the superior surface of the petrous faces obliquely downward and forward, so that the thickness of its roof decreases from the porus to the fundus of the meatus. Because of this, in the operative position the meatus seems slightly oblique in relation to the plane of drilling, and deeper towards its aperture than near the fundus. To reach the latter it is necessary to excavate slightly under the overhang created by the obliquity of the plane of the anterior SCC (Fig. 39).

Landmarks

These are numerous since, according to various authors, it is possible to locate the IAM (Fig. 40) on the biauricular line at 28 mm in relation to the outer table [96], on an axis with 60° obliquity medial to the plane of the anterior SCC [34], or in the continuation of the curve described by the petrous nerve, the geniculate ganglion and the first part of the facial nerve [47]. Actually, as Zanaret has stressed [106], three landmarks are important: the petrous nerve, the exposure of which is simple, the biauricular axis which is known to traverse the lumen of the IAM, and above all the anterior SCC, the identification and preservation of which offers the main technical problems with regard to the preservation of hearing. The principle that used to be generally applied was to display the blue line of this canal, as seen through the bone under irrigation, either spontaneously if the canal were superficial or, more often, after cautious drilling to thin the cortex in which is is embedded. This maneuver called for much experience on the surgeon's part. It seems safer to expose the labyrinthine massif in the midst of the mastoid cells at the level of the attic, as advised by Cannoni [13], without having to burr directly in contact with it (see below in connection with the suprapetrous route).

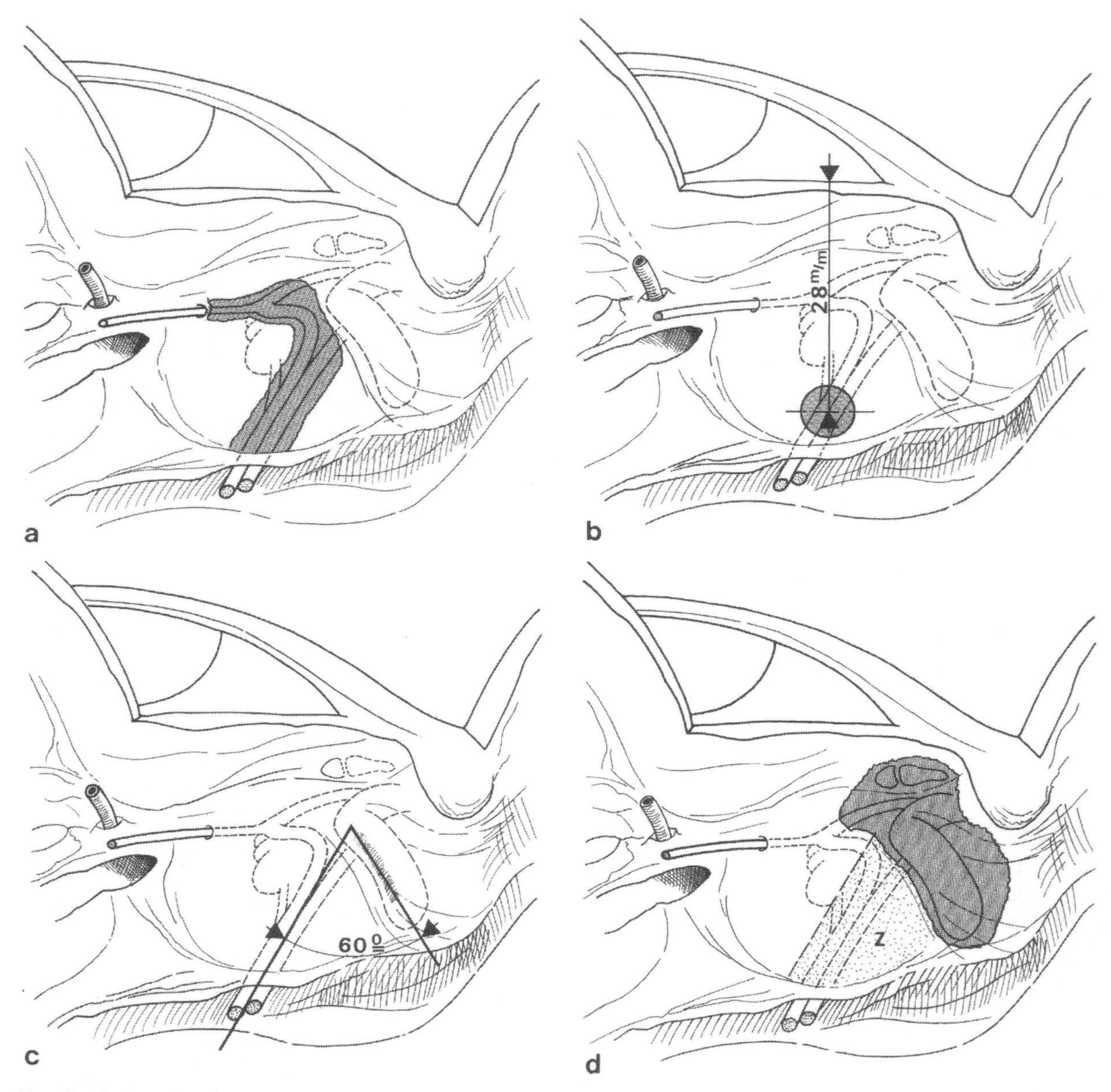

Fig. 39 a–d. Superior approach. The anatomic landmarks of the IAM. **a** Technique of House. Centripetal burring starting from the superior petrosal nerve **b** Technique of Sterkers. Forage at 28 mm of the temporal cortex lateral to the vertical of the EAM. **c** Technique of Fish. Location of the blue line of the anterior SCC, then burring at 60° towards the superior petrosal margin. **d** Technique of Cannoni. Burring of supra-labyrinthine cells outside the arcuate eminence, trephining of attic exposing the anterior and lateral SCCs, then burring at zone Z bounded medially and in front by the line at 60° to the anterior SCC

Variants

Trephining restricted to the roof of the IAM results in cutting a narrow trench which becomes deeper as one approaches the posterior border of the pyramid. In fact, this very limited approach is feasible only for access to a normal IAM within which it is intended simply to divide a nerve (vestibular neurectomy). It is easier in practice to burr progressively the entire meatal area in the limited space, in front via the plane of the anterior SCC and behind via the posterior border of the pyramid. This results, not in a trench, but in an excavation opening more widely at the IAM. This may be exposed from the aperture up to the fundus, where opening the roof reveals the upper end of Bill's bar separating the facial nerve in front from the superior vestibular nerve behind (Fig. 40). If the plane of the anterior SCC may not be crossed laterally, for fear of sacrificing audition, the limits of drilling in front and behind can be kept back and extended towards the petrous apex as far as the tubercle of Princeteau, and even slightly beyond by gliding under the cave of Meckel. This is advised by Wigand [101], who is unique in extending the approach as far as possible, dividing the superior petrosal sinus and then the tentorium cerebelli so as to definitely open the roof of the cerebellopontine angle. This is the extended suspetrous route (Fig. 39).

The Posterior Approach

This is, of course, the trephining of the posterior wall of the IAM well-known to French neurosurgeons since Rougerie [87] advised the method to remove the intracanalicular growth of acoustic neurinomas.

Operative Field

Whatever the position – sitting, half-sitting, in lateral decubitus (Mount's position) or in full decubitus and lateral position – the neurosurgeon can readily picture the direction of the IAM, whether horizontal or vertical. It is embedded in the petrous mass, oblique outwards, making an angle of 50° with the posterior aspect of the petrous, so that the fundus of the meatus is at a mean depth of 7.7 mm and Bill's bar at 9.5 mm. The facial nerve is still behind it, and at a deeper level. The vestibule is oblique in relation to the extremity of the meatus and covers its last 2 mm. It is clear that the lateral obliquity of the suboccipital approach will also limit visibility at the bottom of this trench. According to Stennert and Samii [95], retraction of the cerebellum does not allow a view of more than 52° of lateral obliquity, so that direct viewing of the fundus of the meatus cannot be envisaged without passing well beyond the outer end of the IAM. There is no difficulty in doing this, but it necessarily opens the vestibule and its SCC and thereby sacrifices audition (Fig. 12 b). This approach can only be made by the transdural route. It is then necessary to allow for the position of the jugular bulb, which we know from experience to be as high as to come into contact with the floor of the IAM in 29% of cases, and even to extend above the meatus and extend behind its posterior wall in not a few cases (13%). Clearly, such an arrangement carries a risk of injury to the lateral sinus or the bulb from the moment of exposure of the posterior aspect of the petrous. In this operative position, plugging the wound is often difficult; and even when this has been achieved it is no longer possible to continue this approach.

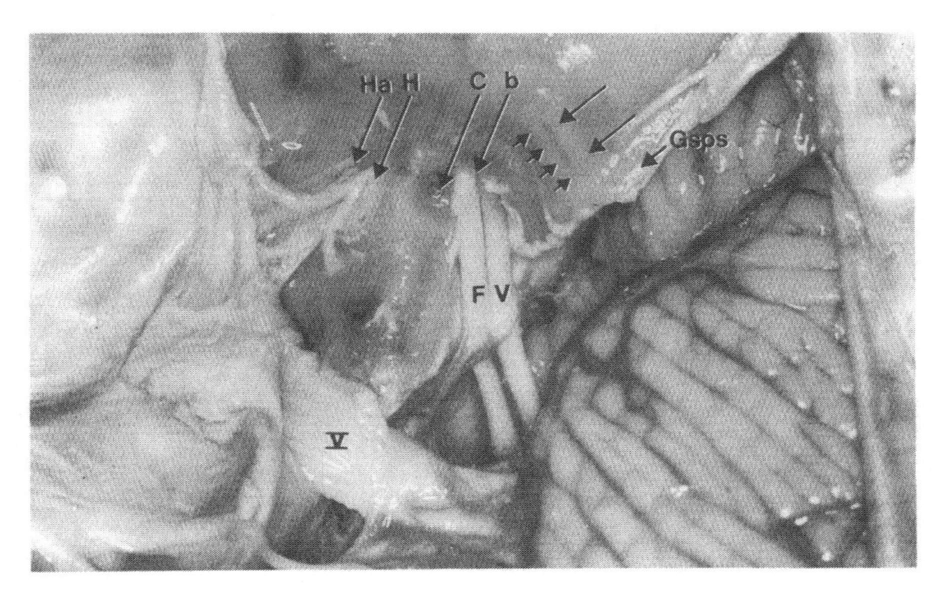

Fig. 40. Superior approach. Dissection (Dr. Zanaret). Suprapetrous approach. Facial nerve *(F)*, superior vestibular nerve *(F)*, trigeminal nerve *(V)*, groove for superior petrosal sinus *(Gsps)*, hiatus for canal of petrosal nerve *(H)*, accessory hiatus *(Ha)*, cochlea *(C)*, Bill's bar *(b)*. Between the two rows of arrows the blue line of the anterior SCC. The tentorium cerebelli has been incised and the superior petrosal sinus divided

Landmarks

These will naturally differ, depending on whether it is desired or not to preserve audition. In the former case, it is essential not to open the vestibule or the crus commune, which are estimated by Domb and Chole [26] to be situated at an average of 7.9 and 7.4 mm respectively lateral to the outer border of the aperture of the IAM. Kartush et al. [53], in order to avoid destruction of the crus commune, suggest that drilling be stopped as soon as one sees the aqueduct of the vestibule, whose bend crosses the crus medially. As these authors locate this bend at 1 mm medial to the crus, this further limits the possibilities of exposing the IAM and in particular its fundus, which is situated at an average distance of 9.9 mm from the outer border of the aperture. It should also be borne in mind that the crus commune is the most superficial component of the posterior labyrinth, and that it is embedded at an average depth of only 2.2 mm under the surface of the posterior aspect of the petrous, whereas the vestibule is at 4.3 mm depth. Of course, when the problem of hearing can be ignored, this drilling problem is simplified. It can then be much more extensive, limited only by the jugular bulb. In these cases of more extensive drilling, the risk of cerebrospinal rhinorrhea arises. This is an ever-present risk in view of the existence of perilabyrinthine cells, but it is greatly increased whenever the burr goes beyond the outer side of the labyrinth, when it may open the innermost mastoid cells.

References

1. Allam AF (1969) Pneumatisation of the temporal bone. Annals Otol Rhinol Laryngol 78: 49-64
2. Amjad AH, Scheer AA, Rosenthal J (1969) Human internal auditory canal. Arch Otolaryng 89: 709-714
3. Anson BJ, Donaldson JA (1981) Surgical anatomy of the temporal bone. Third edition. WB Saunders Company. Publisher. Philadelphia
4. Atkinson WJ (1949) Anterior inferior cerebellar artery: its variations, pontine distribution and significance in cerebello-pontine angle tumors. J Neurol Neurosurg Psychiatry 12: 137-151
5. Aubaniac R (1951) Le canal déchiré postérieur. Travaux du Laboratoire d'Anatomie de la Faculté de médecine d'Alger. Imprimerie Imbert: 5-22
6. Aubaniac R (1951) Les variations du canal déchiré postérieur. Travaux du laboratoire d'anatomie de la Faculté de médecine d'Alger. Imprimerie Imbert: 23-33
7. Bagger-Sjöbäck D, Friberg U, Rask Andersen H (1986) The human endolymphatic sac. Arch Otolaryng Head Neck Surg 112: 398-409
8. Bebin J (1979) Pathophysiology of acoustic tumors. In: Acoustic tumors – Volume I Diagnosis. Edit House WF, Luetge CH University Park Press. Baltimore Chap 5: 45-83
9. Bouche J, Frèche CH (1968) Problèmes méningés dans la chirurgie de l'oreille. Problèmes actuels d'ORL Maloine Edit
10. Bouchet A, Lapras C, Goutelle A (1963) Les voies de drainage veineux de l'encéphale. CR Ass Anatom 138: 58-70
11. Brackmann DE (1979) Middle cranial fossa approach. In: Acoustic tumors. Volume II Management. Edit House WF, Luetge CH University Park Press. Baltimore: 15-41
12. Bucy PC (1951) Surgical treatment of acoustic tumors. J Neurosurg 8: 547-555
13. Cannoni M, Pellet W, Zanaret M, Scavennec C, Collignon G (1985) Les voies d'abord du rocher. EMC (Paris). ORL 10: 2052 A 10 28 pp
14. Cannoni M, Pech A, Pellet W, Zanaret M, Triglia JM (1986) Les procidences veineuses et méningées au cours des voies trans-labyrinthiques élargies. Ann Oto Laryng (Paris) 103: 303-308
15. Cawthorne T (1963) Bell's palsies. Ann Otol (St Louis) 72: 774-779
16. Charachon R, Accoyer B (1976) Note technique sur le repérage du conduit auditif interne. Acta otolaryng. Belgica 30/1: 76-83
17. Chatelier (1923) Le sinus latéral, Arnette Edit Paris
18. Chouard CH, Charachon R, Morgon A, Cathala HP (1972) Anatomie, Pathologie et chirurgie du nerf facial. Masson et Cie Edit 377 p
19. Chouard CH (1977) Wrisberg intermediary nerve. In Facial Nerve Surgery. Edit U Fisch-Kugler Med Pub Amstelveen. Aesculapius Pub Comp Birmingham (Alab): 24-39
20. Clerc P, Batisse R (1954) Abord des organes intra-pétreux par voie endo-cranienne. Ann Oto-Laryng (Paris) 71: 20-38
21. Coates AE (1934) A note on the superior petrosal sinus and its relation to the sensory root of the trigeminal nerve. J Anat 68: 428-430
22. Cohadon F, Castel JP (1968) Incidence de quelques constatations anatomiques sur les voies d'abord chirurgicales du conduit auditif interne. Rev Laryng (Bordeaux). 89/11.12: 643-658
23. Cohadon F, Castel JP, Leman P, Portmann M (1969) Situation et rapports du conduit auditif interne. CR An Anat 144: 1665-1672
24. Das AC, Hasan M (1970) The occipital sinus. J Neuro-Surg 33: 307-311
25. Di Chiro G, Fisher RL, Nelson KB (1964) The jugular foramen. J Neurosurg 21: 447-460
26. Domb GH, Chole RA (1980) Anatomical studies of the posterior petrous apex with regard to hearing preservation in acoustic neuroma removal. Laryngoscope 90: 1769-1776
27. Droissart A (1982) Les anomalies vasculaires à incidence otologique. J Head neck Pathol 1: 39-44
28. Dutillo MV, Malkasian D, Rand DW (1978) A critical comparison of neurosurgical and otolaryngological approaches to acoustic neuromas. J neurosurg 48: 1-12
29. Duvernoy HM (1975) The superficial veins of the human brain. Springer-Verlag (New York) Edit 11 Opp
30. Duvernoy HM (1977) Human Brainstem Vessels, Springer-Verlag (New York) Edit 188 pp

31. Eyries CH, Biss R (1953) Signification de la Rhombomerie chez l'embryon humain. CR Soc Anat
32. Eyries CH, Chouard CH (1970) Les origines réelles du nerf facial. Ann Otolaryng 87/6: 321-326
33. Fisch U (1968) L'anatomie chirurgicale du système artériel du conduit auditif interne chez l'homme. Rev Laryngol (Bordeaux) 89/11 12: 659-671
34. Fisch U (1971) Neurectomy of the vestibular nerve. Surgical technique, indication and results in 70 cases. Acta oto Rhino Laryngol Bel 25: 729-752
35. Fisch U (1978) Infratemporal fossa approach to tumors of the temporal bone and base of the skull. J Laryng Otol 92: 949-967
36. Fisch U (1986) Reconstruction intra-cranienne du nerf facial après exérèse du neurinome de l'acoustique. Communication colloque "Neurinomes de l'acoustique acquisitions et controverses." Toulouse 7 et 8. 11. 1986 (à paraitre)
37. Girard L (1914) Atlas d'anatomie chirurgicale et de technique opératoire du labyrinthe. 2è Edit. Maloine Edit Paris 69 pp
38. Gitenet P (1973) Les voies d'abord de l'oreille interne. Thèse médecine Marseille 158 pp
39. Graham MD (1977) The jugular bulb: Its anatomic and clinical considerations in contemporary otology. Laryngoscope 87/1: 105-125
40. Guild SR (1941) A hitherto unrecognised structure, the glomus jugularis in man. Anat Rec 79: Suppl 2-28
41. Guerrier Y, Guerrier B (1976) Anatomie topographique et chirurgicale du rocher. Acta oto-rhino-Laryngol. Belgica 30/1: 22-50
42. Guerrier Y (1977) Surgical anatomy, particularly vascular supply of the facial nerve. In Facial Nerve Surgery. Edit U Fisch. Kugler Med Pub (Amstelveen). Aesculapius Pub Comp (Birmingham) Alab: 13-23
43. Gueurkink NA (1977) Surgical anatomy of the temporal bone posterior to the internal auditory canal: an operative approach. Laryngoscope 87: 975-986
44. Hall GM, Pulec JL, Rhoton AL (1969) Geniculate ganglion: anatomy for the otologist. Arch Otolaryng 90: 568-571
45. Harris FS, Rhoton AL (1976) Anatomy of the cavernous sinus. A microsurgical study. J Neurosurg 45: 169-180
46. Hitselberger WE, Pulec JL (1967) Trigeminal nerve posterior root retrolabyrinthine selective section. Arch Otolaryngol 96: 412-415
47. House WF (1961) Surgical exposure of the internal auditory canal and its contents, through the middle crania fossa. Laryngoscope 71: 1363-1385
48. House WF (1964) Monograph of trans-temporal removal of acoustic neuromas. Arch Oto-Laryngol 80: 587-756
49. House WF (1968) Monograph II: Acoustic neuroma. Arch Otolaryngol 88/6: 575-715
50. House WF, Hitselberger WE (1976) The trans-cochlear approach to the skull base. Arch Otolaryngol 102: 334-342
51. House WF, Luethje ChM (1979) Acoustic tumors. Univ Park press (Baltimore) Vol I Diagnosis 289 pp. Vol II management 280 pp
52. Kapur TR, Bangash W (1986) Tegmental and petro-mastoïd defects in the temporal bone. J Laryngol Otol 100: 1129-1132
53. Kartush JM, Telian SA, Graham MD, Kemink JL (1986) Anatomic basis for labyrinthine preservation during posterior fossa acoustic tumor surgery. Laryngoscope 96: 1024-1028
54. Körner O (1890) Über die fossa jugularis und die Knochenlücken im Boden der Paukenhöhle. Arch Ohr Nas U Keblkheilk 30: 236-239
55. Koos WTh, Perneczky A (1983) Pathomorphologie et pathophysiologie des neurinomes de l'acoustique. In: Diagnostic et traitement des neurinomes de l'acoustique. Edit Lazorthes Y, Clanet M, Fraysse B, Fournie impr: 37-59
56. Legent F, Perlemuter L, Vandenbrouck C (1979) Cahiers d'Anatomie ORL Masson Edit 125 pp
57. Martin RG, Grant JL, Peace D, Theiss C, Rhoton AL (1980) Microsurgical relationships of the anterior cerebellar artery and the facial vestibulo-cochlear nerve complex. Neurosurgery 6: 483-507
58. Marquet JM (1973) Contribution à l'angiologie du conduit auditif interne. In: Progresos en Radiologia ORL, M Trujillo. Editorial Marban. Hilarion Eslava, 55 Madrid 15: 374-379
59. Matsushima T, Rhoton AL, De Oliveira E, Peace D (1983) Microsurgical anatomy of the veins of the posterior fossa. J Neurosurg 59: 63-105
60. Mazzoni A (1969) Internal auditory canal. Arterial relations at the porus acousticus. Ann Otol (St Louis) 78: 797-814
61. Mercier Ph (1980) Anatomie microchirurgicale de l'angle ponto-cerebelleux. Thèse Médecine. Lyon 256 pp
62. Mercier Ph, Fardoun R, Cadou B, Le Patezour A, Guy G (1983) Etude microchirurgicale du nerf facial intra-cranien. In diagnostic et traitement des neurinomes de l'acoustique. Lazorthes Y, Clanet M, Fraysse B Edit. Fournie Imp Toulouse: 61-70
63. Moret J (1982) La vascularisation de l'appareil auditif. Normal - Variantes - tumeurs glomiques. J Neuroradiol (Paris) 9: 213-260
64. Nager GT, Nager M (1953) The arteries of the human middle ear with particular regard to the blood supply of the auditory ossicles. Ann Otol Rhinol Laryngol 62/4: 923-949
65. Ogura Y, Clemis JD (1971) A study of the gross anatomy of the human vestibular aqueduct. Ann Otol Rhinol Laryngol 80/6: 813-825
66. Pait TG, Zeal A, Harris FS, Paullus WS, Rhoton AL (1977) Microdissection anatomy and dissection of the temporal bone. Sur Neurol 8: 363-391
67. Palva T, Palva A (1966) Size of the human mastoid air cell system. Acta Oto-laryng 62: 237-251
68. Pannier M (1970) La pneumatisation de la mastoïde. Etude radio-anatomique. CR Soc Anat Fasc II. 481-486
69. Parisier SC (1974) The middle cranial fossa approach to the internal auditory canal. An anatomical study stressing critical distances between surgical landmarks. Laryngoscope 84: 1-20
70. Paturet G (1958) Traité d'anatomie humaine. Tome III. Fasc II. Masson et Cie Edit: 665-1305
71. Paullus WC, Pait TG, Rhoton AL (1977) Microsurgical exposure of the petrous portion of the carotid artery. J Neurosurg 47: 713-726
72. Pearse AGE (1968) Common cytochemical properties of cells producing polypeptide hormones (the APUD series) and their relevance to thyroid and ultimobranchial C cells and calcitonin. Proc Roy Soc (Biol) 170: 71-81
73. Pech-Gourg F (1976) Anatomie appliquée du ganglion géniculé. Thèse médecine Marseille 276 pp
74. Pertuiset B (1970) Les neurinomes de l'acoustique. Neuro-chir (Paris) 16: Suppl I 147 pp
75. Pialoux P (1973) Contribution à l'anatomie stéréotaxique

du conduit auditif interne. Ann oto-laryng (Paris) 90/7-8: 409-422
76. Pool JL, Pava AA, Greenfield EC (1970) Acoustic nerve tumors. Early diagnosis and treatment. Second edition. Ch C Thomas. Pub Springfield (ill) 161 pp
77. Porot J, Aubaniac R (1953) Le canal déchiré postérieur (constitution et types radiologiques chez le sujet normal). J radiol Electrol 34: 18-27
78. Portmann M (1955) Quelques observations sur la systématisation des nerfs transpétreux. Revue laryngol 76: 281-298
79. Proctor B (1969) Surgical anatomy of the posterior tympanum. Ann Otol-Rhinol-laryngol 78: 1026-1040
80. Rand RW (1969) Microneurosurgery. CW Mosby comp. St Louis
81. Rask-Andersen M, Stahle J, Wilbrand H (1977) Human cochlear aqueduct and its accesory canal. Ann-Otol-Rhinol-Laryngol 86/5: part 2: 2-16
82. Rhoton AL, Pulec JL, Hall GM, Boyd AS (1968) Absence of bone over the geniculate ganglion. J neurosurg 28/1: 48-53
83. Rhoton AL (1974) Microsurgery of the internal auditory meatus. Surg Neurol 2: 311-318
84. Rhoton AL, Buza R (1975) Microsurgical anatomy of the jugular foramen. J neurosurg 42: 541-550
85. Rhoton AL (1979) The suboccipital approach to removal of acoustic neuromas. Otolaryngol. Head-neck-Surg 1: 313-333
86. Rhoton AL (1986) Microsurgical anatomy of the brainstem surface facing an acoustic neuroma. Surg Neurol 25: 326-339
87. Rougerie J, Guyot JF (1964) Essai de conservation du nerf facial dans l'ablation des neurinomes de l'angle ponto-cérébelleux. Neuro-chir (Paris) 10/1: 13-21
88. Santini JJ (1980) Atlas d'ostéologie du crâne. Masson Edit Paris 129 pp
89. Sauvage JP, Vergnolles Ph (1976) Anatomie de l'oreille moyenne. EMC (Paris) ORL 20015 A 10 18 p
90. Savié D, Djerié D (1986) Morphological variations and relations of the epitympanum. Rev Laryngol 107/1: 61-64
91. Shapiro R (1972) Compartmentation of the jugular foramen. J Neurosurg 36/3: 340-343
92. Shea DA, Chole RA, Paparella MM (1979) The endolymphatic sac: anatomical considerations. Laryngoscope 89: 88-93
93. Shrontz C, Dujouny M, Ausman JI, Diaz FG, Pearce JE, Berman SK, Hirsch E, Mirchandani HG (1986) Surgical anatomy of the arteries of the posterior fossa. J Neurosurg 65/4: 540-544
94. Sljivic B, Boskovic M (1955) Recessus acoustico-facial de la dure-mère. CR Soc Anat 42: 1246-1253
95. Stennert E, Samii M (1986) Anatomical studies related to the different approaches to the internal auditory canal. Communication colloque "Neurinomes de l'acoustique. Acquisitions et controverses". Toulouse 7-8/11 (in press)
96. Sterkers JM, Batisse R, Gandon J, Cannoni M, Vaneecloo JM (1984) Les voies d'abord du rocher. Lib arnette Edit 195 pp
97. Sunderland S (1977) Some anatomical and pathophysiological data relevant to facial nerve surgery and repair. In Facial Nerve Surgery. U Fisch Edit Kugler Med Pub Amstelveen. Aesculapius Pub Comp Birmingham (Alab): 47-61
98. Tran Ba Huy P, Bastian D, Ohresser M (1984) Anatomie de l'oreille interne EMC (Paris) ORL 20020 A 10 18 pp
99. Voisin R (1973) Intérêt du polytome dans le repérage millimétrique des structures de l'oreille interne. Thèse Médecine Marseille
100. Waltner JG (1944) Anatomic variations of the lateral and sigmoid sinuses. Arch Otol 39: 307
101. Wigand ME, Rettinger G, Haid T, Berg M (1985) Die Ausräumung von Oktavus Neurinomen des Kleinhirnbrückenwinkels mit transtemporalem Zugang über die mittlere Schädelgrube. HNO 33: 11-16
102. Winkler G (1965) Le nerf facial. Morphologie, topographie, structure et systématisation fonctionnelle. CR Ass Anat 125 b: 11-50
103. Woodhall B (1936) Variations of the cranial venous sinuses in the region of the torcular Herophili. Arch Surg 33: 297-314
104. Woodhall B (1939) Anatomy of the cranial blood sinuses with particular reference to the lateral. Laryngoscope 49: 966-1009
105. Yasargil MG, Kasdaglis K, Jain KK, Weber HP (1976) Anatomical observations of the sub-arachnoïd cisterns of the brain during surgery. J Neurosurg 44/3: 298-302
106. Yazargil MG, Smith RD, Gasser JC (1977) Microsurgical approach to acoustic neurinomas. Adv Tech standards neurosurg 4: 93-129
107. Zanaret M (1979) La voie sus-pétreuse. Problèmes techniques. Possibilités actuelles. Thèse médecine Marseille 156 pp

Surgical Approaches

W. Pellet, M. Cannoni, and A. Pech

In Collaboration with
C. Lacroix and P. Querruel

It is by doing that one discovers what one wanted to do.
(Alain, after Balzac)

Otoneurosurgery is not a field for improvisation. On the contrary, it can only be considered in the context of a strict framework without which it would be unable to prove its efficacy and fully succeed in its aims. It is a team enterprise and we have already stressed, in the introduction, that each member of the group, whether otologist or neurosurgeon, must be animated by the team spirit. Obviously, the staff is not limited to these specialists alone; these often protracted procedures call for the special competence of the anesthetist and of all the medical auxiliaries. It is also a matter of qualification, as demonstrated by the range of anatomic detail discussed in the previous chapter. However, this theoretic basis would be inadequate if not coupled with the practical aptitudes that are acquired only after long apprenticeship. The inadequate drilling by an inexperienced otologist will afford only a restricted approach, unsuitable for the proper exposure of the tumor to be removed. Similarly, a rough and ready neurosurgical dissection carries a great risk of hampering the attainment of the essential aims of surgery of tumors of this region, namely safeguarding of the brainstem, complete excision of the tumor and, as far as possible, preservation of the different cranial nerves of the region. In fact, it is a matter of technical competence.

Instrumentation

We find it cardinal to stress the importance of acquiring adequate equipment: first, for the drilling of the petrous pyramid, next for fragmentation of the tumor and dissection in the cerebellopontine angle, and finally for peroperative monitoring of the facial nerve and sometimes, in certain cases, of the cochlear nerve.

Drilling

Motors

The duration of the drilling stage renders the motors normally used for conventional otology unsuitable for this type of operation: they are not fast enough, they lack power and they become overheated. It is essential that the surgeon should not have to press. He should be able to handle the burr "as if making strokes with a pencil" (Sterkers [55]). This gentleness is the best way to both avoid slipping and perform precise level drilling, plane by plane, and not by making excavations to be subsequently connected. Only motors which are rapid, powerful and well-cooled can simultaneously ensure rapidity and security during the otologic stage. Unlike Sterkers [55], our preference goes to air-driven and not electric motors, especially now that they are equipped with reverse systems allowing a choice of the direction of rotation so that the burr can always be driven in the opposite direction to the vulnerable structures. Experience has shown that supply from wall-sockets is inadvisable as the gas so delivered is at the temperature of the operating-room. It is necessary to have nitrogen cylinders available in the theater: the proximity of the pressure-reducing valve and the shortness of the connecting tubing ensure the supply of gas at a low temperature which effectively cools the motor. The mastoid can be disintegrated without excessive precautions while the drilling of the labyrinthine massif has to be very precise. Because of this, we usually use two different motors in succession, each suited to a particular stage of the operation.

- The C100 3M craniotome (Fig. 41) is used for the extralabyrinthine stage. This powerful and robust air-driven neurosurgical motor, fitted with a "pineapple" cutting burr (C470), so-called because of its shape and striations, is exactly right for rapid erosion of the mastoid cortex and destruction of the air-cells.

– The Zimmer osteon (Fig. 41) is used for the labyrinthine stage and for all precise procedures (such as the suprapetrous route). It is an air-driven otologic motor developed by the Los Angeles team. It combines the following advantages: it is a light tool perfectly suited to delicate work, the powerful motor delivers a regular smooth rotation in clockwise and anti-clockwise directions which markedly reduces the risk of slipping or skidding, and there is a very wide range of highly effective tungsten carbide burrs and of diamond burrs. This is the most reliable current equipment, provided the maintenance instructions are observed and the gas supply does not exceed the pressure at the point of use laid down by the manufacturer. It is important also that it should be reserved exclusively for otoneurosurgery and, in the special case of the transpetrous approaches, only for the labyrinthine stages.

Burrs

Quality is more important than quantity. Of course, one must have at one's disposal burrs of different sizes, from 1 to 7 mm, and also of different qualities, some cutting, others diamonded. It is best to have available the range of burrs provided for the Zimmer ototome. We use 12 burrs, 5 diamonded from 1 to 5 mm, and 7 cutting tungsten burrs from 1 to 7 mm. These last are very markedly superior in quality, resistance and safety to the conventional cutting metal burrs. With these tungsten burrs it is possible to burr right up to the dura mater and even the walls of the venous sinuses without risking excessive bony infraction. In addition to the quality of cut, it adds greatly to the safety of the procedure which thus becomes much faster. The diamond burrs are reserved for the more delicate stages and the finishing touches. The burr used should always be the largest possible, allowing for the working space available. The two types, tungsten und diamond, suffice; there is no need for intermediate

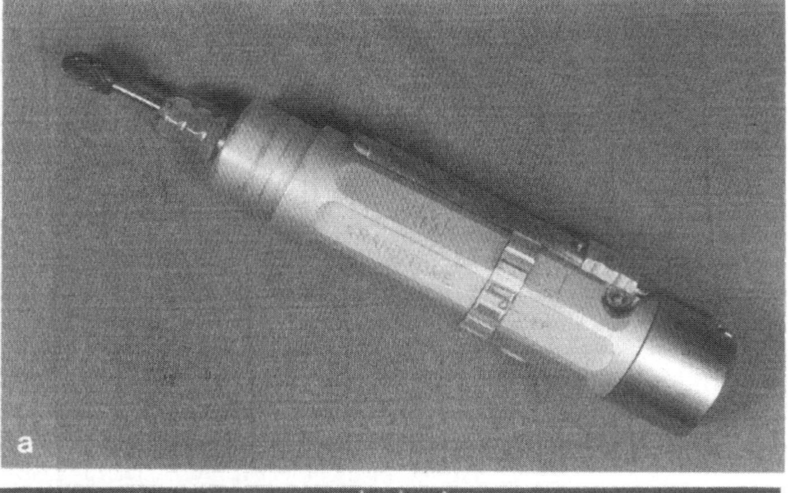

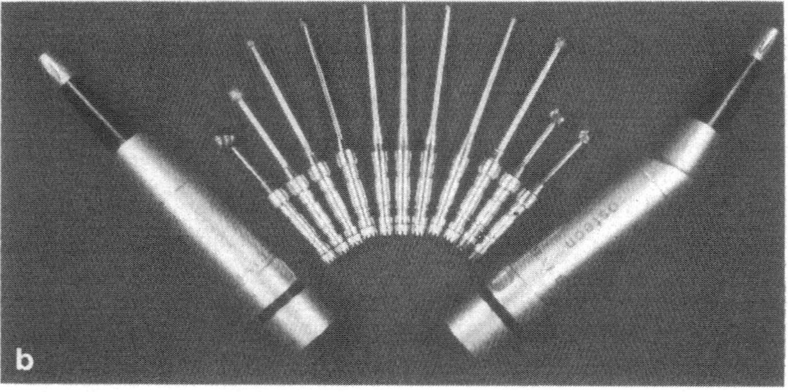

Fig. 41 a, b. Motors required. **a** Craniotome fitted with a stout polishing burr to abrade the mastoid. **b** 'Osteon' for more precise burring

types of burrs. It is obvious that they must be kept in good condition, for when their cutting or abrasive capacity is lost the surgeon is tempted to press, the worst way of rough working with risk of slipping. There should be no hesitation in changing burrs, ideally every 3 or 4 operations.

Irrigation

This is an important aid, combined of course with suction. Continuous irrigation ensures the lavage and continued removal of bone-dust without which the operative field would rapidly become unmanageable. It also ensures cooling of the burr and stops it from becoming choked. It also cools the abraded bone and particularly its contained anatomic structures, especially the facial nerve. Finally, the moistening of the bone, when fully adequate, renders visible by transparency the different cavities and the organs they contain. Without irrigation it is impossible to visualize the famous "blue line" which heralds the proximity of the anterior SCC for instance. We agree with Sterkers [55]: "Irrigation of the bone during drilling is the basis of all transtemporal surgery."

Continuous irrigation of the operative field can of course be ensured by the assistant; but the best arrangement is to use double-flow cannulas fed by a sterile perfusion system (one must beware of tubing that is only sterile internally) provided with a stop-cock and a system of variable flow. In view of the length of the operation, large liter flasks of distilled water or normal sline are more practical than smaller bottles. A range of double-flow cannulas of different caliber (2 to 6 mm MICRO-FRANCE) should be available. As these are used at the same time as the burrs, held in the other hand, they are rapidly made sharp and pointed; they should therefore be handled cautiously in contact with soft or vulnerable parts and their use suspended during the neurosurgical stage. However, irrigation remains very important during this second stage, removing all trace of bleeding and ensuring good visibility of the dissected tissues. It also facilitates hemostasis by demonstrating the small bleeding vessels. For this second stage a second set of cannulas should be abailable, with multiple orifices and blunt ends, such as those of Brackmann (Storz). As these are difficult to obtain in France, the best course is to procure the double-irrigation blunt-ended cannulas supplied by MICRO-FRANCE (Fig. 42) and to have two or three side-orifices made at about 3 mm from the tip, either by the biomedical department of the hospital or by an artisan. The existence of these multiple orifices prevents the soft tissues from being subjected to forcible suction and allows perfectly safe suction even in contact with nerves or small vessels, and this with no need to reduce the suction pressure. Though their construction is simple, we consider these cannulas a decisive advance in microsurgery. They should be available in different diameters (every ½ mm from 2 to 4 mm).

Microsurgery

The Microscope

What is required here is a high-quality visual field and great maneuverability allowing easy orientation at every angle, so very necessary for the neurosurgical stage. The commercial choice is not very extensive but our unhesitating preference goes to the op-

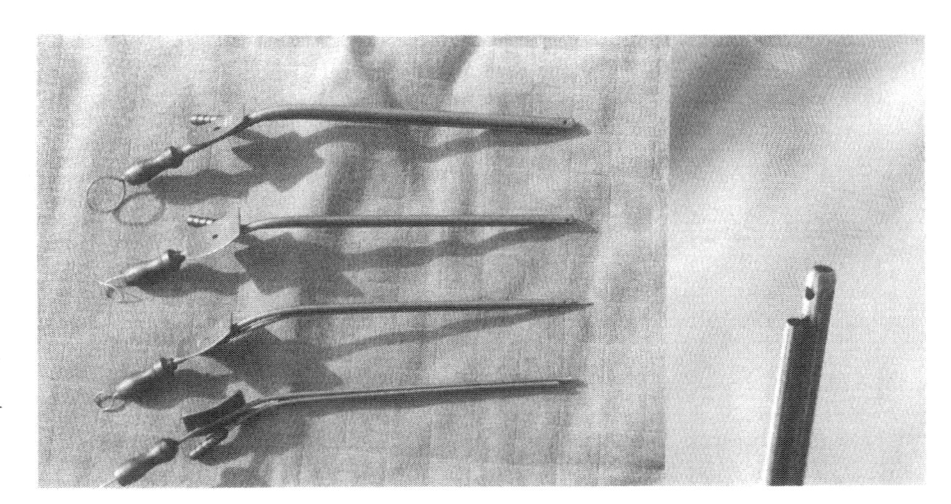

Fig. 42. Double-flow cannulae. Note in B the blunt end and the lateral discharge holes

erating microscope of Wild, which ensures the best illumination and the best depth of field; this is particularly valuable at somewhat greater magnifications, above ×10, where the focus is sharp not only at the working plane but also above and below. Dissection is greatly facilitated by this, just like the "shooting" of photographic or filmed documents. The telescopic control arm is also highly maneuverable. Several focal lengths can be used, but the most "comfortable" seems to be a 250 mm one as this leaves sufficient space under the lens for managing even the longest instruments without difficulty. From this point of view, a focal length of 200 mm seems inadvisable except for small tumors, while a focal length of 300 mm causes much loss of illumination and depth of field. The magnifications most often used are ×6, ×10 and ×16. The greatest magnifications are reserved for the more delicate stages in the IAM and in the course of the facial nerve. Sterilization can be achieved either by a sterile plastic or cloth envelope, or by enclosing the whole microscope in a bag containing 7 or 8 containers of formalin. We prefer the latter, but it requires at least 24 hours out of action.

Micro-Instruments

Otologic Stage

The entire micro-instrumentation of Fisch (Fischer) must be available during the period of dissection in the fundus of the IAM, in the facial canal and in the middle ear. It comprises:

- An elevator for stripping the dura mater
- A microdissector angled to the right and another to the left
- A micro-hook angled at 90°
- A curved micro-hook
- A fine needle

To this may be added Yasargil's self-retaining retractor, fitted with its connecting head allowing the attachment of several arms and with a support arm attached to the operating table.

Neurosurgical Stage

Fisch's micro-instruments are retained for this stage, but one instrument seems particularly important and in constant use: Fisch's fine stapedectomy hook. This is the ideal instrument throughout the neurosurgical stage for dissecting the arachnoid as well as the vessels and nerves.

The micro-scissors are certainly necessary. The long scissors of Malis (Codmann), made in titanium with straight or curved blades, seem particularly well suited for this surgery thanks to their bayonet shape, length and lightness.

A set of small malleable curettes, from 2 to 9 mm, is very useful when starting to scoop out the tumor and sometimes even longer when the tumor is soft enough. They can be used to remove some material for histologic examination before using the equipment for fragmenting the tumor.

Bipolar Coagulation

This is considered here as its employment is essential in microsurgery. Various generators are available to surgeons. As our teacher, J. E. Paillas [37] wrote: "its choice is very important indeed". It must be constantly reliable, both for coagulation and fine section and for coarser usage. We particularly value the Malis CMC II system of bipolar coagulator and bipolar knife. Thanks to its very low exit impedance, this system ensures a stable exit voltage at the tips of the forceps that guarantees regular and reliable coagulation whether the tissue is irrigated, dry or bloodsoaked. It minimizes charring, sticking at the end of the forceps and especially the risk of vascular perforation. We have found it particularly well-suited to microsurgery in general and neurosurgery in particular. Finally, the bipolar incision it delivers is particularly valuable at the time of tumoral evacuation, especially when the texture of the tumor (neurinoma type A of Antoni or meningioma) is so firm as to resist the method of fragmentation available. The bipolar forceps must also be selected with care. We use those supplied by Codmann, adaptable to the Malis CMC II system. They are 22 cm long, bayonet-shaped, and their tip is either straight (4 sizes: 0.25, 0.5, 1 and 1.5 mm) or angled (1.5 mm) upward or downward. The entire set must be available if one is to be able always to grasp in the forceps, without any difficulty, whatever it is desired to coagulate. In theory, these forceps can be readily passed under a microscope fitted with a 25 lens. However, when the tumor is very bulky and one has to work at great depth, it may sometimes be awkward to introduce these forceps. This is why we also use the shorter Mathys & Son bayonet forceps: either straight (20 cm long, 0.7 mm and 1 mm at the tip) or angled (21 cm long, 1 mm at the tip) upward or downward.

Fragmentation

Whatever the technique of approach, the excision of an extracerebral tumor, unless it is too vascular like a jugular glomus tumor for example, is always performed by scooping out the central portion followed by dissection and fragmentation of the peripheral shell. Recently, sophisticated techniques have been introduced, based on vaporization of the diseased tissue by a beam of energy applied to the tumor. Whether this is a laser beam absorbed by the tumoral tissue or a beam of ultrasound applied to the tumor, the outcome is comparable: pulverization of the tissue subjected to the energy beam. We have had no real experience with these systems. As far as the laser is concerned (neodyme - Yag and CO_2), the otologists of our team have acquired over a number of years an experience which has gradually induced them to limit its use to well-defined cases, amounting in practice to treatment by the endoscopic route. It is true that, as stressed by Roux et al. [45, 46], this system has the advantage of feeble penetration and precision, but its hemostatic power is effective only on infra-millimetric vessels and then on condition of reduction of power and defocussing, which markedly reduces the power density of the beam, reduces the power of vaporization and thus notably prolongs the operating time. On the few occasions when we have used the ultrasound knife, it was our impression that vaporization of the tumoral pulp of a neurinoma was quite limited, even at maximum power, and that the time of evacuation was rather long. The only meningioma we tried to reduce with this apparatus proved entirely resistant to the energy beam. It also seemed to us that the borderline between the tumoral tissue and the brain tissue could be transgressed involuntarily quite easily. For all these reasons we have not adopted these new systems and have remained faithful to the vacuum rotary dissector of Hurban House (Fig. 43). This apparatus consists of a rotary motor, contained in the sleeve, which drives a long mandrel provided at its tip with a side opening bristling with sharp teeth. This mandrel rotates within a chuck that also has a large lateral window near its tip. A sucker is connected to this system and provides continuous suction in the mandrel. Whenever the window of the latter passes opposite the window of the sheath a fragment of tumor is aspirated and sectioned. The window in the chuck itself can be directed at will by means of a steering mechanism mounted on the sleeve of the apparatus. The speed of rotation of the mandrel can be regulated from a panel independent of the motor, controlled by the theater nurse. The rotation of the motor is controlled by a pedal. Finally, a safety valve, ounted on the chuck and manipulated by the thumb allows the surgeon to reduce the suction pressure at will. With a little practice, the handling of this apparatus is reliable and devoid of risk. The brief intermittent suction brought about by the rotary movement of the mandrel produces a brief period of suction in the evacuation cavity at each rotation. When the tumor shell is sufficiently thinned, this depression has the advantage of jerky retraction of the periphery of the tumor, which is thus stripped progressively from its parenchymatous bed. This apparatus would seem to offer great advantages. Resection is much more rapid than with a laser or an ultrasound knife and the jerky suction facilitates gradual stripping of the tumor. The ease of handling of this very light apparatus, which is held in one hand and whose orientation is very simply ensured by rotation of the guide mechanism held between thumb and index, is very considerable. The suction ensures the evacuation of both tumoral debris and blood, keeping the operative field clear. Its extremely simple principle is a guarantee of robustness and easy maintenance. Finally, it must be stressed how small is the investment required, since its cost is about 20 times less than that of an ultrasound knife. We may hope that its commercial supply in Europe will soon be assured. We consider it essential for suction-resection of tumors of the cerebellopontine angle, but its use is not restricted to this pathology alone and we habitually use it for all extracerebral tumors, especially at the base of the skull, and also for intraventricular tumors, even in the third or fourth ventricles. It is the more effective when the tumor is soft, but it can also be used for firmer tumors provided these are slashed open with a bipolar diathermy incision, the tumoral fragmentation so produced facilitating suction of the debris.

Peroperative Monitoring

Recent advances in methods of neurophysiological detection and recording now permit peroperative monitoring of spontaneous or evoked electrophysiological activities crossing the facial nerve, the auditory nerve or even the proprioceptive pathways in the brain-stem. Normal activities reassure the surgeon as to the correctness of his procedure, while the appearance of modifications warns him that his dissection is becoming aggressive. These methods, much more refined than the traditional clinical methods of detection (a hand on the patient's cheek, the anesthetist's direct surveillance for facial twitching), constitute a real advance contributing to improved operative results, not only at the functional level but also at the vital level, since they con-

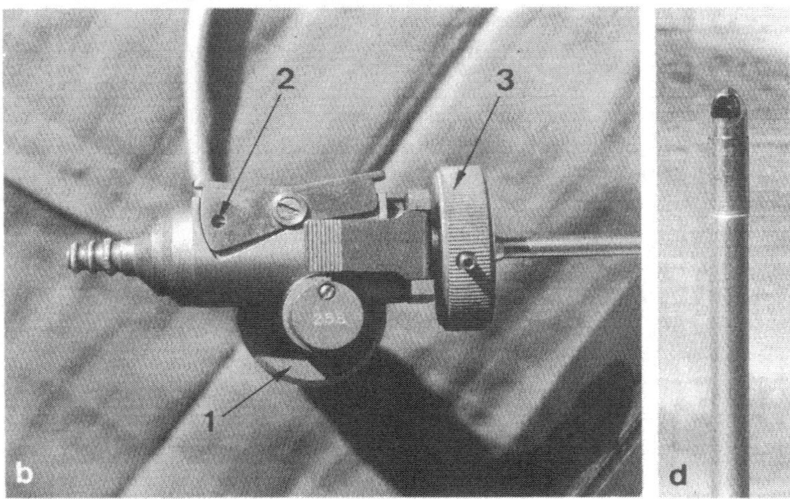

Fig. 43a–d. The vacuum rotary dissector of Hurban-House. **a** complete outfit. **b** the working tool. Motor housing *(1)*, safety-valve *(2)*, guide-wheel for sleeve *(3)*. **c** sleeve with guide-wheel *(1)*, mandrel *(2)*. **d** side-opening of mandrel

strain the surgeon to even greater gentleness in procedure. We shall not discuss the techniques of neurostimulation of the facial nerve employed to locate it when hidden as the preliminary exposure of the nerve in the fundus of the IAM ensured by otoneurosurgical approaches renders this technique unnecessary.

The Facial Electromyogram

In 1979, Delgado et al. [11] suggested using modifications of the facial electromyogram to detect traumas to the nerve during dissection. Møller and Jannetta [35] improved this technique by converting the EMG signal into a sound signal, relayed through a loudspeaker in the operating-room. Recently, Prass and Lüders [41] have reported their experience with a comparable technique, simultaneously visualizing the recording on a video screen, recording it on paper and hearing it on a loudspeaker. They have made a special study of the EMG disturbances that may arise during operation and distinguish three types:

- "Explosive" disturbances, consisting of several synchronous discharges of unit potentials over some hundreds of milliseconds only. These activities are evoked by direct manipulation of the facial nerve during dissection, sometimes by irrigation with cold water or the pressure of a swab. The coincidence between the procedure and the signal usually establishes their relationship and allows correction of the dissection. It is the "mechanical" effect which is responsible for stimulation of the nerve, and it is the sudden rise in compression that seems to be effective since the duration of the disturbance is unrelated to the duration of the pressure, but rather proportional to the aggressivity of the maneuver. According to Prass and Lüders [41], these acute disturbances are not affected by the type or depth of anesthesia, nor by the patient's temperature or respiratory condition. They consider that the anatomic state of the nerve seems to play a part and that, the less the nerve is invaded by the tumor, the more excitable it is.
- "Prolonged" disturbances, composed of prolonged polyphasic discharges often lasting several minutes. The sound-track of these disturbances depends on the frequency and amplitude of the potentials, ranging from the sound of a motor when the frequency is high to a crackling when it is lower. These disturbances often seem to be evoked by stretching of the nerve but the appearance of the abnormal activity is often delayed (up to 2 or 3 minutes). They disappear when the nerve is slackened. Other provocative factors, such as a jet of cold water, compression by a swab or diathermy coagulation in the vicinity of the nerve may be responsible. According to Prass and Lüders [41], it is nerves most damaged by the tumor that give this type of response.
- Finally, there are "periodic" disturbances, characterized by their regular sound (like that of a machine-gun) and elicited by application of the electrical stimulator.

In this context, we have found it very practical to use an anesthetic monitor designed not only for gas monitoring but also for continuous recording of the patient's EEG, EMG and neuromuscular tonus (Datex ABM 100 system). This makes it much easier to detect the smallest variation in the basic electromyographic activity. This apparatus gives a continuous display of facial electromyographic activity in graphic form, on a video screen, as a print-out and in numerical data. Further, a sound alarm gives warning whenever the chosen threshold is exceeded. We are convinced that this system is a great contribution to the improvement operative procedure and plays a large part in the results obtained at both functional and vital levels. It is possible to be continuously informed of the electrophysiological activity in the facial nerve throughout the operation. Like Prass and Lüders [41], we have observed both brief and prolonged disturbances. As we do not use a stimulator, we have naturally never recorded any periodic activity. Not using a loud-

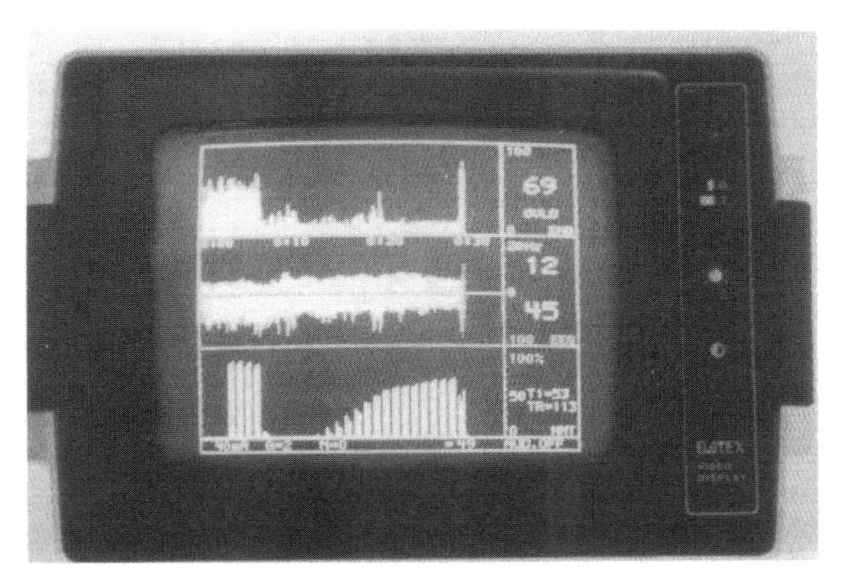

Fig. 44. Anesthetic monitor (ABM 100-Datex system). The upper trace records facial EMG activity. Note the peaks corresponding to direct stimulations of the facial nerve

speaker, we are unable to distinguish the various sound characteristics of the disturbances recorded. On the other hand, the alarm provides a warning whenever the intensity reaches the critical threshold. In practice, we depend largely on indications from the anesthetist who advises us of variations in the numerical values displayed, especially when these are elevated. Of course, at critical moments we do not hesitate to ask the nurse to keep a hand on the patient's cheek and warn us of the least twitch.

Evoked Auditory Potentials

The persistence of evoked auditory potentials despite general anesthesia allows evaluation throughout the operation of the functional condition of the auditory pathways from the cochlea to the upper brainstem. This method was recommended in 1980 by Levine [32] and is now used by numerous teams [18, 23, 36, 42, 49] who unite in stressing its importance in the early detection of aggressive operative proceedings so as to safeguard both the auditory nerve and the brain-stem as much as possible. It is obvious that such a method can be employed only if the integrity of the auditory apparatus is respected, and that it cannot be applied during transpetrous approaches which necessarily involve sacrifice of the cochlea. On the other hand, it does seem to have some value during suprapetrosal approaches intended to involve the IAM while respecting the cochlea and its nerve. As stressed by Fischer et al. [18], the method is only practicable in the operating-room by overcoming major technical difficulties related to the electrical environment of the operating theater unit and the limitations imposed by the relevant operation, but it does provide very useful information in guiding the surgical proceedings. While the main value of the method relates to the functional outcome, it is not without importance at the vital level as it provides the surgeon with additional security and allows him to improve some of his maneuvers. This method is likely to be more generally employed as the inevitable improvement in the apparatus facilitates its application. Nevertheless, it has to be stressed that the recorder cannot do everything and that the value of the method depends largely on whoever interprets the recording and also on his availability throughout the operation. The prolongation of latent periods or disappearance of the recording indicates damage to the auditory pathways. These anomalies may be transient, especially if the provocative maneuver (placing of a retractor, tugging, mobilization of an artery, especially the anteroinferior cerebellar artery) can be checked or modified. Regression, even if somewhat delayed, is the index of the benign nature of the trauma and, as stated by both Fischer et al. [18] and Soulier et al. [52], the operative outcome in these cases is free from neurologic complications and in particular from auditory disorder. On the other hand, persistence of the anomalies and especially complete extinction of the recording is always followed by cophosis and must arouse the fear for severe neurologic complication. Any impairment of electrophysiologic conduction along the auditory pathway indicates damage to the neuronal chain, though without specifying the severity or extent of the lesion. This sets the limitations of the method, since, it must be stressed, the technical imperatives inherent to acquisition of the necessary data mean that the information is not given immediately but subject to a delay of several minutes, thereby correspondingly delaying correction of the causal factor.

Other Methods

These include, for example, study of evoked proprioceptive potentials or of antidromic conduction in the facial nerve as recommended by Richmond and Mahla [43]. This latter technique consists of measuring the speed of conduction between a stimulatory electrode implanted at the stylomastoid foramen and a detector electrode positioned during operation at the emergence of the facial nerve from the brain-stem. It has the advantage of eliminating the muscular factor, for the electromyographic response may be blocked by the intensive use of muscle relaxants.

Anesthesia

As with any surgical procedure, the concerns of the anesthetist relate to:

- Preoperative assessment and preparation
- The actual anesthesia and monitoring techniques
- Recovery and the postoperative period.

Preoperative Care

The patients are hospitalized 10 days before operation for the conduct of otoneurosurgical and medical assessment. On admission, at the time of the pre-anesthetic visit, the anamnesis and complete clinical examination will indicate any need for specialist consultations which may sometimes lead to particular preparation. An Allen's test is made on each arm and, if revascularization by the brachial artery should prove of poor quality or nonexistent, puncture of the radial artery is ruled out [34]. If the circumstances so demand, puncture of an artery in the foot may be preferable, but in all other cases automatic pressure recording by a cuff is used.

Because thromboembolic complications are favored by the duration of the operation and the immobilization of the first 24 or 48 hourse following operation, we routinely employ elastic stockings from the eve of operation and sometimes, in high-risk cases, even from the day of admission. This method is less risky than preoperative calcitherapy or the prescription of platelet antiaggregants and is of comparable efficacy [31].

This initial contact provides an opportunity to discuss with the patient the anesthetic procedure, the necessary monitoring methods and the postoperative sequence, including the recovery period in the intensive care department. This discussion is aimed at lessening the patient's anxieties and may possibly lead to the prescription of anxiolytics, usually Lorazepan (Temesta). Whatever the patient's age, the paraclinical preanesthetic review includes an ECG, a chest Xray, laboratory studies and a study of respiratory function. The laboratory studies comprise: ionography and blood chemistry; blood-grouping with search for any irregular antigens; blood-picture, blood and platelet count; and an assessment of clotting status combining prothrombin levels, activated cephalin time and fibrinemia.

The tests of respiratory function should allow detection of patients who may need delayed weaning from artificial ventilation (TV (Tidal Volume) less than 35 ml kg^{-1}; MMEF (Maximal Mid-Expiratory Flow) (75-25) less than 50% of theoretic value; FEV1 (First Second Forced Expiratory Volume) less than 1 liter associated with capnia above 45 mm Hg; MVV (Maximum Voluntary Ventilation) less than 50% of theoretic value; lowered PaO_2) [21, 22]. In these cases weaning will necessarily be conducted in the intensive care department. Detection of these patients is the more important in view of the known postoperative respiratory complications described after this type of surgery [33].

Peroperative Care

Anesthesia for acoustic neurinoma must have regard for certain imperatives common to all anesthesia (security of the airway, adequate perfusion of vital organs) and for others which are more specific (loss of access to the head, long duration, neurosurgery, recovery).

The translabyrinthine route of approach, performed in strict dorsal decubitus, eliminates the disadvantages inherent in more upright positions and especially the seated position. It lessens or even suppresses venous pooling in the lower half of the body, limits the risk of displacement of the intratracheal tube and other tubing systems, and eliminates the risk of serious gas embolism.

There is no ideal anesthetic agent for the surgery of neurinoma. However, one must consider the effects of different agents on intracranial pressure and cerebral oxygen consumption [51]. Our own choice goes to the volatile halogen anesthetics, particularly isoflurane [2].

Some authors advise light anesthesia, sometimes with spontaneous ventilation so as not to mask any autonomic discharges due to manipulation of a tumor close to the brain-stem [4]. We prefer deep anesthesia, now that the development of such autonomic disorders has been markedly lessened since we have been able to continuously monitor the stimuli transmitted by the facial nerve by means of the Anesthesia Brain Monitor (the ABM-Datex system discussed elsewhere). This monitoring also minimizes lesions of the facial nerve and postoperative sequelae.

Anesthesia

Premedication. The evening before operation the patient is sedated with flunitrazepam (Rohypnol) 1 mg or diazepam (Valium) 10 mg. This premedication is repeated the following morning in the same oral dosage, one hour before operation, when it is combined with any medication required for associated disease (beta-blocker, nitrate derivative, anticalcic or anti-ulcer agent).

Induction and Installation. The patient is then transferred to the operating suite and installed on a heated mattress. A short venous catheter of 18 gauge is inserted to begin the anesthesia. After 3 mcg/kg^{-1} of Fentanyl the patient receives a loading-dose of pancuronium bromide (Pavulon): 0.015 mg/kg^{-1} and is then denitrogenated for 3 minutes. Induction is then made with a barbiturate: thiopental (Nesdonal) 5 mg/kg^{-1} or methohexital (Brietal)

1.5 mg/kg^{-1}. After loss of consciousness, the muscle relaxant is completed to a dose of 0.1 mg/kg^{-1}. Intubation is performed under direct laryngoscopy after local anesthesia with 5% lidocaine, using a disposable reinforced tube and a low-pressure balloon cuff. The position of the tube is checked by auscultation, as well as after each manipulation. The patient is placed under controlled intermittent positive pressure ventilation, without expiratory assistance, using warmed humidified gases. The respiratory parameters (frequency = 10, flow volume = 10 ml/kg^{-1}) ensure a moderate hypocapnia of between 30 and 35 mm Hg. These are modified as indicated by capnimetry and by the initial and subsequent gas analyses (every 3 hours). Packing is inserted, all the connections are fixed by adhesive strips and the circuit is checked to be airtight. The position is carefully checked to avoid stretching or compression of nerve or vascular paths. The heels are rested on pillows to prevent pressure sores. The eyes are protected by a vitaminized ointment and carefully kept closed by an adhesive occlusive bandage. The head is arranged in lateral rotation to the side opposite to the operation on a foam-filled head-piece, taking care not to obstruct jugular venous return.

Maintenance. This is ensured by combining a halogen gas and a morphinomimetic drug. Isoflurane is delivered in low dosage (0.8%) in an equimolecular gas mixture of O_2/N_2O ($FiO_2 = 0.5$). Fentanyl is injected by continuous flow from an automated syringe, from an initial basal level of 500 mcg/hr^{-1} (1 ampoule an hour) to reach ½ or ¼ of an ampulla an hour towards the end of the operation. The fentanyl injection is discontinued an hour before the presumptive termination of the procedure. This technique provides excellent hemodynamic stability. The muscle relaxant, which is not essential, is renewed as required by monitoring of the neuromuscular block, using Pavulon 0.5 to 1 mg. Even though autonomic disturbances have become very rare, their development requires stopping surgical manipulation of the tumor. This is usually adequate, but if they persist the administration of atropine 1 mg IV is necessary. The effect of this agent lasts for 30 to 45 minutes, during which period any possible further autonomic discharges will be masked, and the surgeon must be warned of this.

Monitoring

It is determined by this particular type of surgery and by its duration. Certain equipment is required.

- An electrocardioscope with leads routinely arranged in CM5.
- The arterial blood-pressure is measured by radial puncture, allowing gas analysis and biochemical review at repeated intervals.
- A unit for centralized monitoring, the ABM (Anesthesia Brain Monitor – Datex system) visualizes on a video screen with graphic recording the five channels grouped on an amplifier:

- Arterial pressure
- Frontal electromyogram (EGM)
- Surface electroencephalogram (EEG)
- Neuromuscular function (curarization)
- Capnia.

The electromyographic and electroencephalographic activities and the assessment of curarization are collected by means of a system of electrodes distributed in two groups as follows:

- 5 electrodes arranged over the course of the ulnar nerve and in its muscular territory send a train of 4 stimuli every 20 seconds and receive the corresponding muscular response. The curarization is thereby precisely monitored. The apparatus visualizes the train of the 4 waves and gives the relation (T4/T1 (T4 ratio: ratio of amplitude of 4th stimulus to amplitude of 1st stimulus). The module must be calibrated before every curarization.
- There are 3 cephalic electrodes, 2 frontally on the face on the side of operation and 1 reference electrode on the contralateral mastoid. These collect the EEG and EMG signals. The EEG (frequency and amplitude) only give a tendency, providing information on the depth of anesthesia. The EMG deals with signals from the frontal platysma muscles and therefore with the stimuli transmitted by the facial nerve. These signals can only be collected if curarization is incomplete ($T4/T1 \geqslant 10\%$). In our experience, however, direct stimulation of the facial nerve elicits a muscular response even when curarization is total.

All the electrodes are preimpregnated with gel. The quality of the signals recorded depends on their good fixation, and this calls for careful skin degreasing and the application of benzoin to secure perfect adhesion. The electrodes are then stuck on and held by an adhesive bandage.

- Continuous measurement of blood CO_2 and O_2
- Central venous pressure. The value of this measurement is modified by the impossibility of monitoring the position of the catheter introduced by the brachial route (basilic or cephalic vein). To avoid false passages, we therefore endeavour to place the tip of the catheter in the subclavian vein and insert

it to a maximum of 40 cm. This gives a reliable measurement of CVP [44]. In any case it is routine to insert a second short venous catheter, siliconized and of wide caliber (16 or 14 gauge).

– Instrumentation of the patient is completed by the insertion of a stomach tube which is verified and fixed, of a bladder catheter for hourly measurement of diuresis, and of a rectal thermometric probe.

Fluid Supply and Antibiotic Treatment

The fluid supply should initially compensate for the losses due to preoperative fasting according to a conventional schema, and then maintain a nil fluid balance. This means that the supply, in the form of Ringer-lactate (⅔) and 5% glucose solution (⅓) must ensure an hourly diuresis of the order of 0.5 to 1 $mg/kg^{-1}/hr^{-1}$, with a CVP maintained at 6 cm H_2O [8]. We never use an osmotic diuretic. On the other hand, if the duiresis remains below the fixed level for two hours consecutively we restore it with boluses of 5 mg of furosemide.

Transfusion is not routine. However, packed red cells (4 units) must be rapidly available to compensate for a wound of a venous sinus or the jugular bulb. In practice, this operation gives rise to little hemorrhage. In a general way, in view of its duration which increases the thromboembolic risks, a slight hemodilution is desirable. A hematocrit of 32–35% has no effects on cerebral oxygenation and metabolism [10].

Laboratory monitoring is performed every 3 hours (hematocrit, blood ionogram, gas analysis) and this may lead to corrections, especially as regards potassemia.

The prophylactic antibiotic treatment is based on wide-spectrum antibiotics, capable of traversing the barrier of even the "inflamed" meninges [1, 12]. It combines amoxicillin (Clamoxyl) 1 g and trimethoprim-sulfamethoxazol (Bactrim) 400 mg. The injection must be completed before the incision. It is repeated every 8 hours.

Postoperative Care

At the end of the operation the pupils are examined to give a point of reference. The patient is then transferred, still asleep, to the intensive care department where he comes to. Weaning from the mechanical ventilation is progressive while the patient is warmed. A period of autonomous spontaneous ventilation precedes extubation, performed on objective criteria (respiratory frequency, flow volume, gas analysis) [21].

The patient is quickly arranged in the half-sitting position, which promotes cerebral venous drainage and improves ventilation.

We prefer rapid reawakening, no later than two hours after the end of operation, an interval sufficient for our surgical team to confirm the peroperative hemostasis. During this period, it is best to prevent any manifestation that might produce a rise in intracranial pressure (coughing, attempts to vomit). This early wakening has the advantage of making it possible to assess the early postoperative neurologic status with the shortest delay. At this stage, if there is any doubt at all (elevation of arterial pressure, confusion, anisocoria, neurologic deficit), a cerebral CT scan is done and a rapid decision made as to any necessary treatment. We feel that a return to consciousness delayed for several hours is not justified, and may even be dangerous save in cases where the patient's condition calls for delayed weaning from assisted ventilation (whether planned or after respiratory complications). Indeed, if signs of neurologic complications are detectable in such cases they are indicative of advanced neurologic disturbances (mydriasis, severe autonomic disorders) and evidence of delayed diagnosis.

The monitoring system (central venous catheter, radial artery catheter) is left in place till the next day, when the patient returns to his original ward. His fluid and sodium intake is kept restricted (1500 ml of fluid/24 hrs) for 48 hours, after which all the catheters and tubes are removed and oral feeding begun if there is no risk of false alimentary passages. The patient is then allowed to get up and starts active and passive remedial exercises.

The peroperative antibiotic treatment is continued. Anti-ulcer anti-H2 agents, designed to combat gastritis and stress ulcers, favored by the stomach-tube, are prescribed routinely. Medication is sometimes required to promote the mental state. We then use, for short periods (7 to 10 days) minaprine (Cantor) and/or sulpirid (Dogmatil) in the usual doses, bearing in mind the epileptogenic risk associated with the use of minaprine.

Very special care is given to the prevention of thromboembolic complications. Calciparin is begun at the 48th hour to produce moderate hypocoagulability (TCK: control-patient ration 1.5). For high-risk subjects, if the surgeons agree, anticoagulation is begun earlier, 12 hours after operation. The patients continue with their elastic stockings until they get up, also with their effective dosage of calciparin. If there is the least suspicion, an ascending angiopneumography is performed [50] and, if the results so indicate, heparin in therapeutically effective doses can be started.

Operative Techniques

The otoneurosurgical routes of access vary greatly, depending on whether the object is to expose the IAM only, the cerebellopontine angle or the posterior lacerate canal. There is a different method for each of these objectives: suspetrous routes to expose the IAM, transpetrous routes to expose the CPA and, by extension, all or part of the posterior aspect of the petrous, and lastly extended transpetrous routes to expose the posterior lacerate canal, the inferior aspect of the petrous and, if need be, also the CPA. The anatomic and pathologic findings result in the great majority of operations requiring a transpetrous approach (90% of our operations), while the suprapetrous and extended transpetrous routes share the rest of the indications about equally. This accounts for the order of discussion of these techniques. As we are concerned here with discussing the otoneurosurgical routes, no reference will be made to the suboccipital or transtentorial subtemporal routes, which are classical neurosurgical approaches. Similarly, we shall not consider the purely otologic routes of access to the middle ear, even if concerned with intrapetrous tumors, as these remain in the field of otology.

Transpetrous Routes

Their principle is to resect by an external mastoid approach one or more of the segments of the petrosal pyramid (Fig. 38), and thus to expose all or part of the anterior aspect of the CPA. Their current conception is the outcome of long maturation, since their first drafts date back to 1904, when Panse [38] suggested approaching tumors of the CPA by directly traversing the petrous. He had even then labeled this procedure as translabyrinthine, resecting the entire mastoid, then the labyrinth (vestibule and cochlea) and removing the facial nerve in the process. Several authors, all German, seem to have tried this technique, but the narrowness of the approach, the risks of hemorrhage and the frequent leakage of CSF and consequent meningitis, all led to its being abandoned. Yet one remains amazed by the temerity of the surgeons of that period. In 1962, with the advent of the operating microscope, House [26] revived Panse's route and soon proved its efficacity. By progressively improving his technique, especially by extending the bony resection to the performance of an actual posterior petrectomy, House finally developed a technique which he called translabyrinthine but which had little in common with that of Panse. It is for this reason that we habitually call this route, extended by ourselves even further, the House extended translabyrinthine route (ETL). It allows the unimpeded approach to and excision of tumors of the CPA whatever their size. Conversely, Panse's route, which is much narrower, is to be discarded for tumors of the angle and should be performed only for section of the vestibulocochlear nerve when indicated in cases of incapacitating Menière's disease. This ETL procedure involves the successive resection of the posterolateral retrolabyrinthine segment and then of the posteromedial vestibular segment. It was thus perfectly conceivable to interrupt the resection at the end of the first stage of the operation in order to expose the outermost part of the CPA while preserving the posterior labyrinth intact: this is the retrolabyrinthine route (RL) proposed in 1972 by Hitselberger and Pulec [24] for trigeminal rhizotomy. It is not indicated for tumors of the angle, except in rare cases where a biopsy is intended. However, we have found some indications for it for tumors arising on the sigmoid sinus, which is why we feel that it should be included with the ETL among the transpetrous routes intended for access to petrosal and peripetrosal tumors. Finally, on the basis of their experience, House and Hitselberger [28] have conceived the forward extension of the petrous resection to expose the entire CPA up to the clivus: this is the transcochlear route (TC).

The Extended Translabyrinthine Route

Preparation

The Skin

Day before operation. In the afternoon, the patient is shampooed with foamy betadine followed by washing. The wet hair is combed carefully upwards and backwards from the ear, after which the skin is carefully shaved to four fingerbreadths above and behind the ear. The future operative field is then degreased with an ether-alcohol mixture and covered with a sterile dressing held in place with a head bandage. The current trend in surgery is not to shave the day before operation to avoid making skin wounds which, though superficial, may rapidly become reservoirs of bacteria. However, we remain faithful to eve of operation shaving, because it is done on a wide-awake alert patient, capable of correcting a nurse who makes a mistake about the side. Despite every precaution and repeated monitoring by the medical staff, one is still haunted by the risk of this error which, however improbable, remains a possibility in a very active unit where several cases for otoneurosurgical operation are hospitalized at the same time.

The upper two-thirds of the pubic hair, and if necessary the abdomen up to the umbilicus, are shampooed and shaved to prepare for the removal of subcutaneous adipose tissue which will be used to fill up the route of access. The abdomen and umbilicus are then cleaned very carefully, and after rinsing with an ether-alcohol mixture an occlusive adhesive dressing is fixed in place. Patient and family are clearly warned of this second incision.

In the evening, before going to sleep, the future cranial operative field is again shampooed and cleaned and a new sterile dressing is applied and held in place with a bandage.

Operating-Room. Further shampooing with foamy betadine, rinsing with sterile water, careful combing of the hair, which is held back from the operative field and "gummed" in place, either with betadine or, better, with tincture of benzoin which "lacquers" the hair effectively. Shaving is completed if necessary, then washing with normal saline and extensive painting with betadine. It is important not to use the ether-alcohol mixture as its accumulation in the EAM and vaporization beneath the drapes may give rise to conflagration and burns when the diathermy knife is used. The suprapubic abdominal field is prepared in the same way.

Operative Arrangements

The patient is placed in dorsal decubitus, with the head in a head-piece or, more simply, flat on the operating-table. In view of the average duration of the operation it is essential to protect all the pressure points with water-bags or silicone gel. A heated mattress for the operating-table may be very useful.

The head is turned towards the opposite side. The mastoid region is placed directly under the sight of the surgeon, who sits at the side of the head of the table, opposite the patient's occiput (Fig. 45). The head should not be fixed, as it may need to be turned during the operation.

It must be possible to tilt the table sideways.

The operating microscope is opposite the surgeon, with its base at the same level but on the other side of the table. The articulated arm bridges over the head. If it is equipped with a television camera or a photographic apparatus, these are placed on the side of the surgeon's best eye.

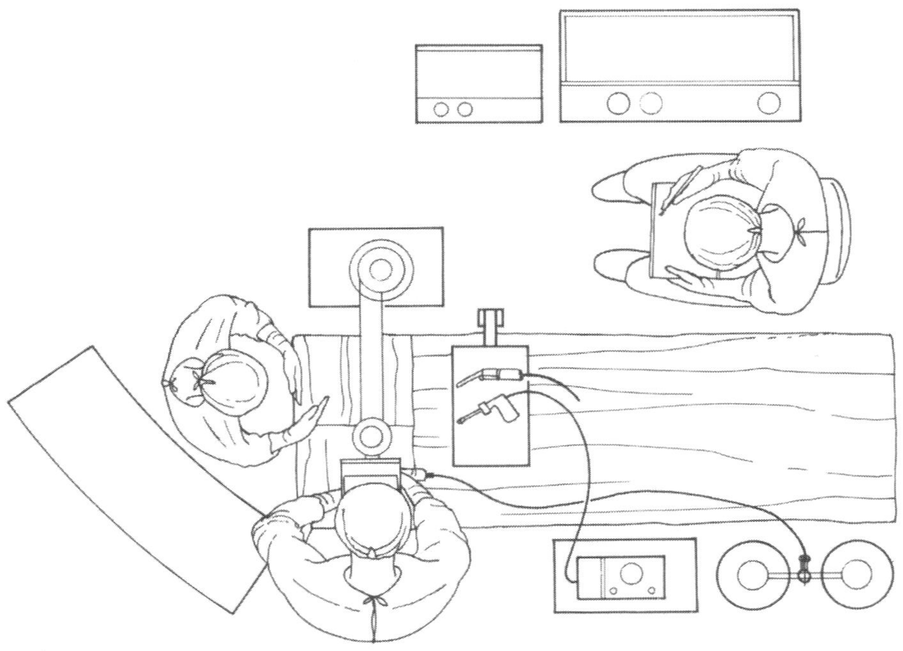

Fig. 45. Operating-room arrangements for a translabyrinthine approach

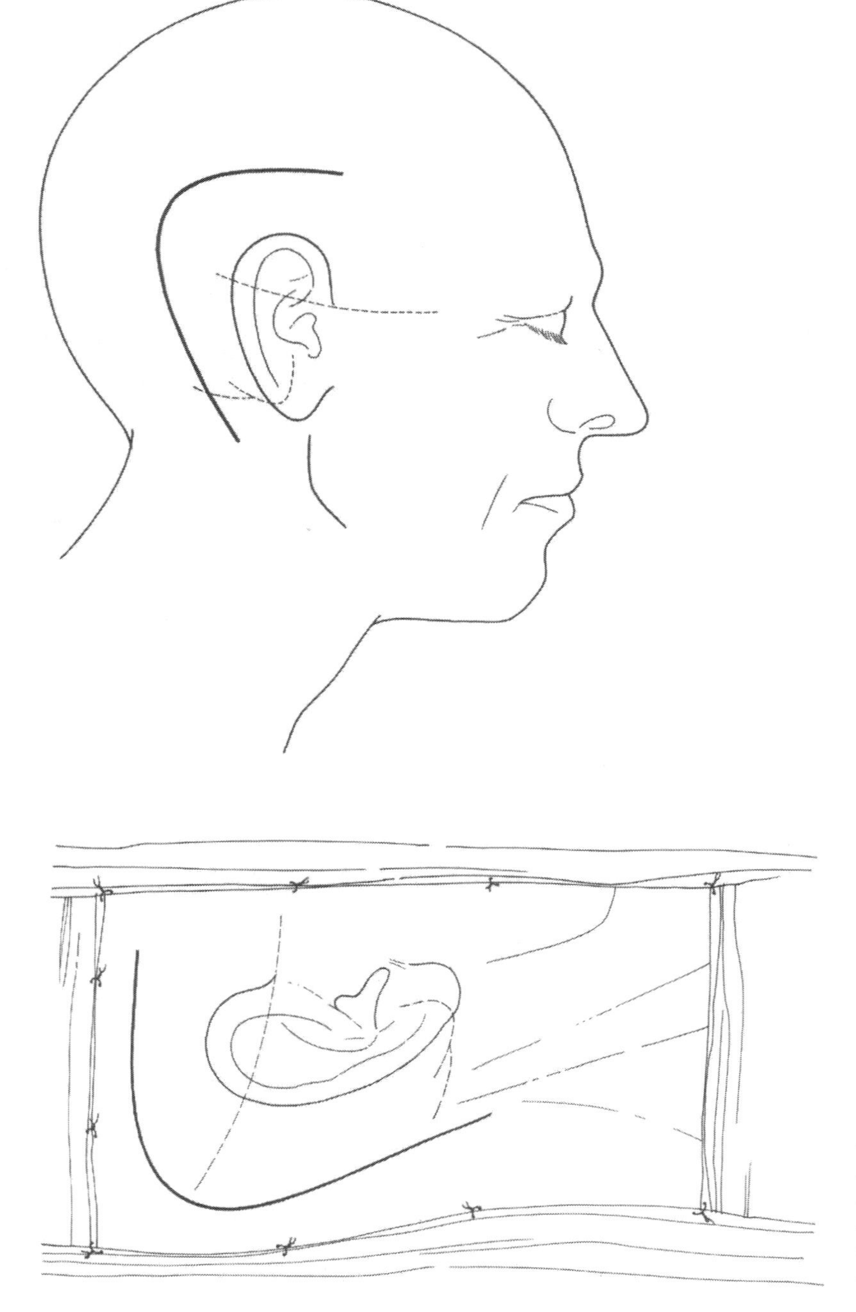

Fig. 46. Extended right translabyrinthine route: line of incision. The discontinuous supra-auricular line indicates the level of the temporal line. The cervical operative field is placed low enough to allow a hypoglossal-facial anastomosis if necessary

The instrument nurse is at the head of the table, seated, slightly turned towards the surgeon so as to be able to serve him but also to comfortably place a hand on the patient's cheek to detect, if necessary, any possible twitches. The instrument table is placed beside her, slightly obliquely to offer one of its angles for the surgeon to lean his elbow.

The anesthetist's place is towards the patient's feet, on the side opposite the surgeon. The anesthetic monitor is turned towards the surgeon.

An instrument tray, supported by a single leg, is placed above the patient's abdomen. On it are placed the motors and the rotary vacuum dissector. This is easily removed when it is time to take the abdominal fat.

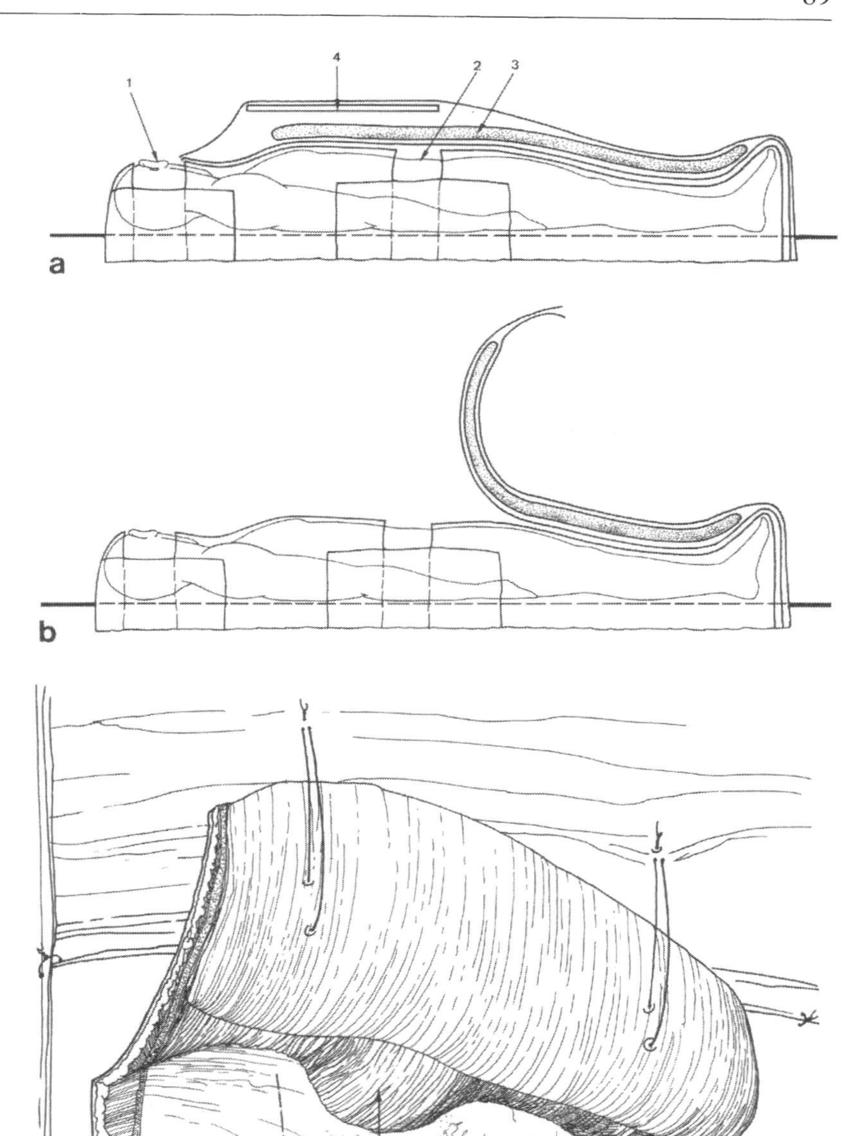

Fig. 47a, b. Extended right translabyrinthine route. **a** Arrangement of operative fields. Head operative field *(1)*, abdominal operative field *(2)*, heated blanket *(3)*, instrument tray *(4)*. **b** Exposure of abdominal field for removal of subcutaneous fat

Fig. 48. Extended right translabyrinthine route. Exposure of operative field. Cartilaginous EAM *(1)*, temporalis muscle *(2)*, removal of fragment of sternocleidomastoid muscle to be used for packing the tympanic cavity at the end of the approach *(3)*, division of attachments of sternocleidomastoid muscle *(4)*, groove of attachment of diagastric muscle *(5)*

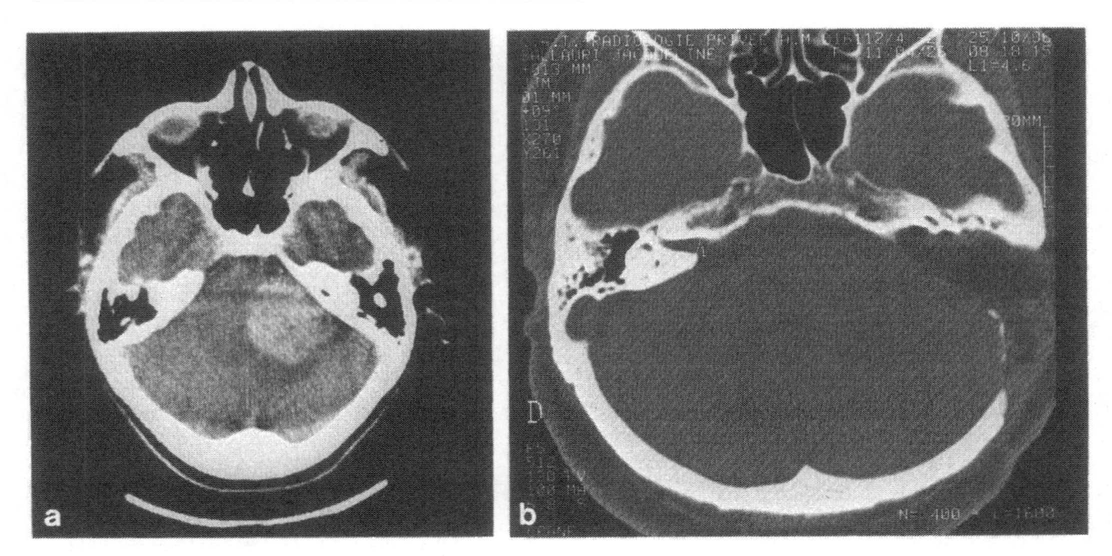

Fig. 49 a, b. Pre- and postoperative CT scans clearly showing the site of the bony opening in an extended translabyrinthine approach

Two cylinders of nitrogen are placed at the foot of the table, on the side of the surgeon. Next to them is a stand to support the irrigation flasks and a small table to place the command module of the rotary vacuum dissector.

Operative Drapes. The temporal field (Fig. 46) is surrounded by 4 sterile drapes centered on the external ear, which is left exposed. The upper and posterior drapes are arranged at the edges of the shaved area. The anterior drape is halfway between the tragus and the outer border of the orbit, leaving the gonion uncovered. The inferior drape is cervical, exposing the region beneath the angle of the mandible so that the posterior limb of the incision can be prolonged downwards if necessary to perform a hypoglossal-facial anastomosis at the end of the operation. A large self-adhesive drape covers the head and head-piece to prevent soaking by the abundant irrigation fluid. The drapes are covered by a drape with an aperture which does not go below the umbilicus (Fig. 47).

The suprapubic abdominal field is prepared similarly with four drapes and a self-adhesive drape, all covered by a large table drape which goes no higher than the shoulders. The instrument tray is then put in place, covered by a rubber sheet and a large table drape (Fig. 47). At the end of the operation it is a simple matter to expose the abdominal field by removing the instrument tray and the heating blanket, but taking care to apply fresh drapes over those already in place.

Operative Technique

Skin Incision and Exposure of Operative Field

After infiltration with xylocaine-adrenalin, the incision is made three fingerbreadths above and behind the external ear. The scalp thus mobilized uncovers a field wide enough to dispense with the use of a superficial retractor, which is always a nuisance (Fig. 46). Behind, this incision must pass at a distance from the posterior border of the mastoid. It begins by traversing all the planes down to the bone, dividing the temporalis muscle above and the sternocleidomastoid behind. Then the entire temporal and mastoid bony surface is exposed by detaching the muscle insertions with a rugine. Behind, the retromastoid region is widely exposed, which usually produces abundant but easily controlled bleeding from the emissary vein. At the lowest part of the incision, near the tip of the mastoid, the posterior part of the groove of the digastric muscle is identified. The auricular musculocutaneous flap, covered with a moist pack, is then reflected forward and held in place by three or four stitches directly through the drapes covering the cheek (Fig. 48). A good fragment of muscle is routinely removed from the sternocleidomastoid; preserved in an antibiotic solution in normal saline, this is used subsequently to occlude the tympanic cavity.

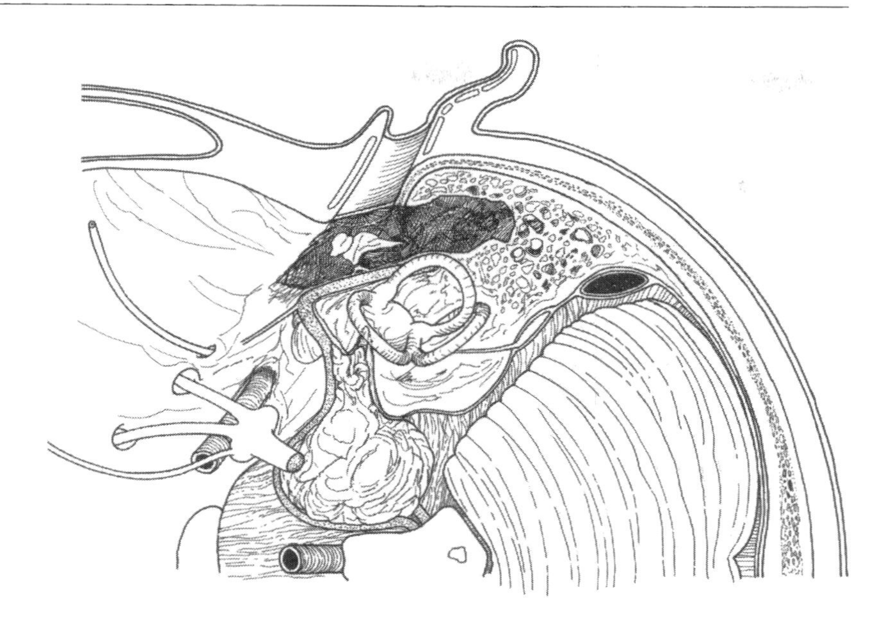

Fig. 50. Diagrammatic horizontal section of a right petrous

Fig. 51. Diagrammatic horizontal section of a right petrous. The extralabyrinthine bony stage is completed

The Bony Otologic Stage

Principles of the ETL

The ETL involves an external posterior petrectomy, allowing access to the tumor by its outer pole, far forward in the posterior cranial fossa (Fig. 49). The three horizontal sections of a right petrous in diagrams 50, 51 and 52 summarize the three successive stages of the ETL:

• *The extralabyrinthine bony stage* (Figs. 50, 51). This is the opening of the middle ear and of the superficial bony plane. It comprises:

- Drilling of the superficial bony covering, widely stripping the dura of the temporal bone and the posterior fossa behind the lateral sinus
- Ablation of all the mastoid cells, exposing the posterior aspect of the EAM, the epitympanic recess, the antrum, the third part of the facial canal, the digastric groove, the labyrinthine core and the sigmoid sinus.
- Ablation of the incus and performance of posterior tympanotomy for good exposure of the tympanic cavity and the second part of the facial canal.

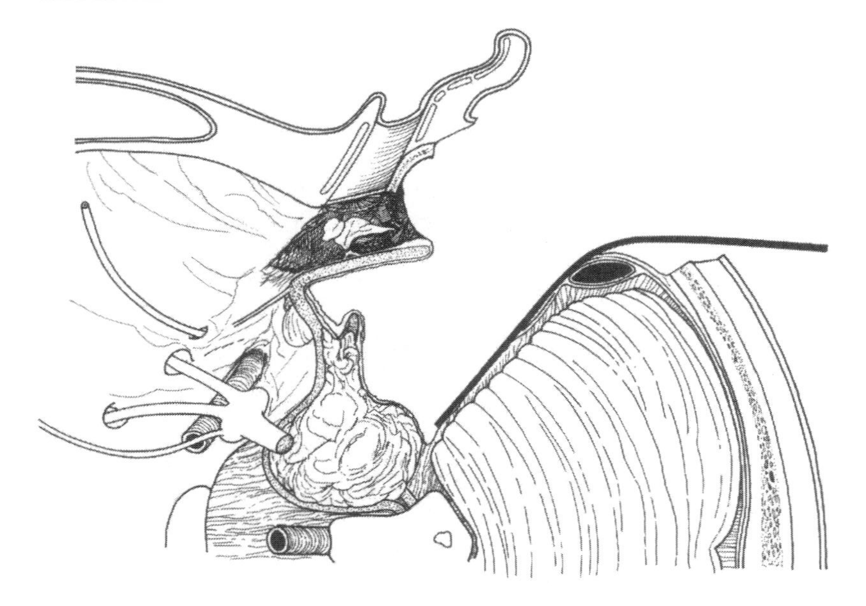

Fig. 52. Diagrammatic horizontal section of a right petrous. The translabyrinthine approach is completed

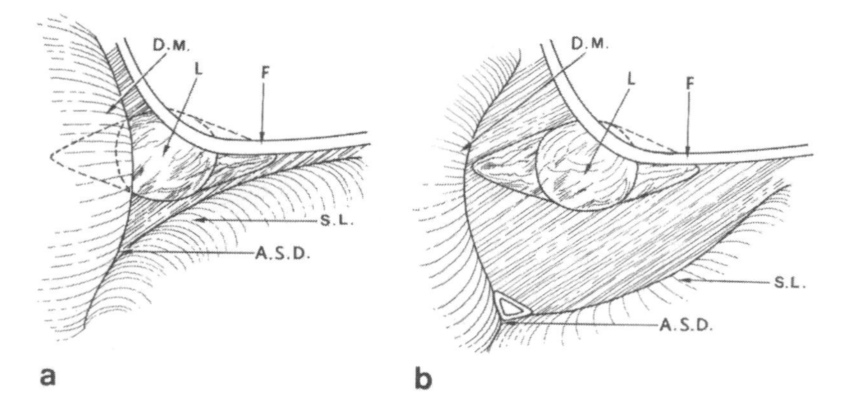

Fig. 53 a, b. Right extended translabyrinthine route. **a** Normally the temporal dura mater and the bend of the lateral sinus conceal the posterior labyrinth and adjacent regions. **b** Elimination of these structures gives wide access to the labyrinth and the cerebellopontine angle. Facial nerve *(F)*, labyrinth *(L)*, temporal dura mater *(D. M.)*, sinusoido-dural angle *(A. S. D.)*, lateral sinus *(S. L.)*

- Impaction of the sigmoid sinus and spreading of the temporal dura mater by the two blades of a Yasargil self-retaining retractor, giving a wide exposure to direct vision of the labyrinthine core and the regions above and below (Fig. 53). The displacement of these superficial structures, besides the fact that it facilitates burring of the labyrinth by avoiding constant manipulation of the microscope, also provides perfect exposure of the tumor in the CPA.

- *The bony labyrinthine stage* (Figs. 51, 52). This is the opening of the internal ear (labyrinthine core) which constitutes the middle bony bloc. At the end of this stage, identification of the fundus of the IAM at the level of the hemispheric and semi-oval fossulae allows exact location of the superior vestibular nerve (SVN) and consequently the first part of the facial nerve which is in immediate contact with it. This prevents going astray during subsequent drilling. Skeletonization of the facial canal is then completed in its second and third portions to give as perfect a view as possible before proceeding to the next stage.

- *The bony pericanalicular stage.* The IAM and its adjacent regions above and below constitute the deep bony bloc. Complete drilling of the superior, inferior and posterior walls of the meatus exposes its contents as well as the neurovascular structures of the CPA (Fig. 54). The dura mater must be freed as widely as possible all around the IAM, below it

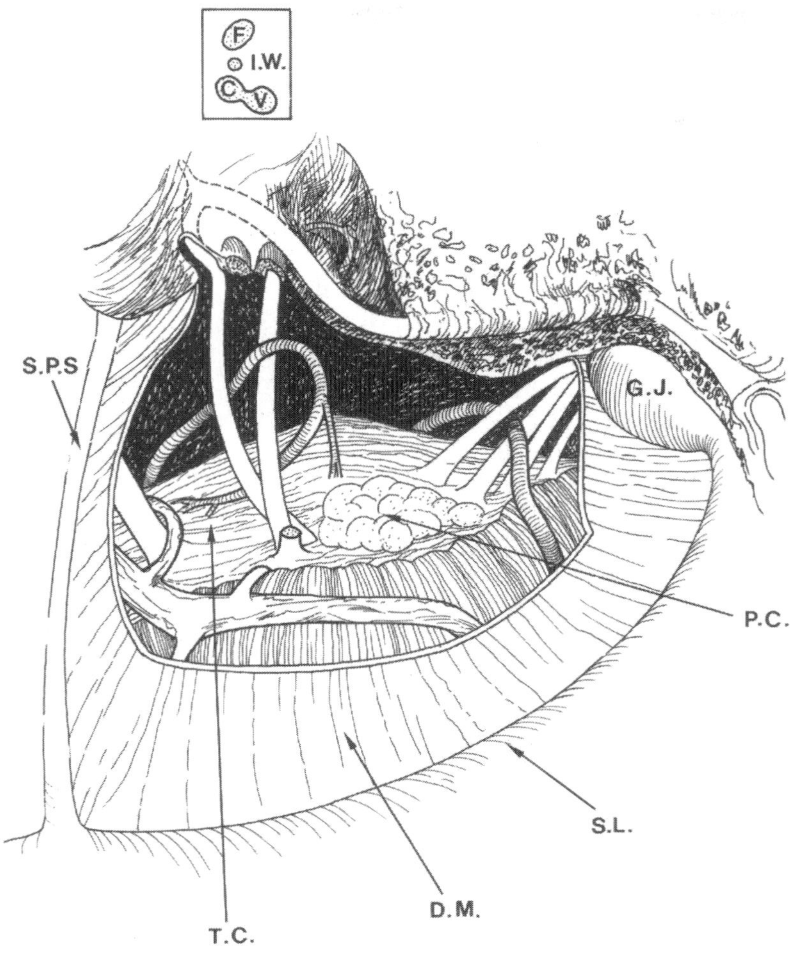

Fig. 54. Right extended translabyrinthine route: arrangement of neurovascular structures in cerebellopontine angle. From left to right there can be recognized successively: the trigeminal nerve with the petrosal vein behind it; the facial nerve and the cochlear nerve with the AICA which makes a loop between the two nerves; the mixed nerves (IX, X, XI) with the PICA which passes between IX and the strands of X. Superior petrosal sinus *(S. P. S.)*, brain-stem *(T. C.)*, jugular bulb *(G. J.)*, lateral sinus *(S. L.)*, choroid plexus *(P. C.)*, dura mater *(D. M.)*,. Box: arrangement of neural components of acousticofacial bundle immediately after their emergence from the brain-stem. Cochlear nerve *(C)*, vestibular nerve *(V)*, nervus intermedius (Wrisberg) *(I. W.)*, facial nerve *(F)*

as far as the jugular bulb and above it to the superior petrosal sinus (SPS). This is the longest and most delicate stage because of the presence of the bulb and the mixed nerves below, the facial nerve immediately under the upper aspect of the IAM and the superior petrosal sinus above. Proper performance of this stage guarantees good access to the IAM and cerebellopontine angle. Preservation of the margins of the IAM by imcomplete drilling is to lose all the advantages of the ETL, and in particular to obtain only an inadequate view of the CPA. It also adds to the risks of damaging the facial nerve at the upper border during manipulation of the tumor.

At the end of the ETL, one should have exposed all the dura mater of the posterior aspect of the petrous (Figs. 66, 67) between the sigmoid sinus and the jugular bulb below and the superior petrosal sinus above.

Performance of the ETL

- *The bony extralabyrinthine stage* (Figs. 55–60).

The greater part of this operative stage is performed without the microscope, using the craniotome and the large "pineapple" burr, so-called because of its shape, under continuous irrigation. The burr is used on the flat, with its "belly" swept to and fro across the bone like a rubber erasing pencil-marks. It should not be used at its tip, which would be too aggressive and dangerous. The advantage of this equipment is twofold: it is quick and it "econ-

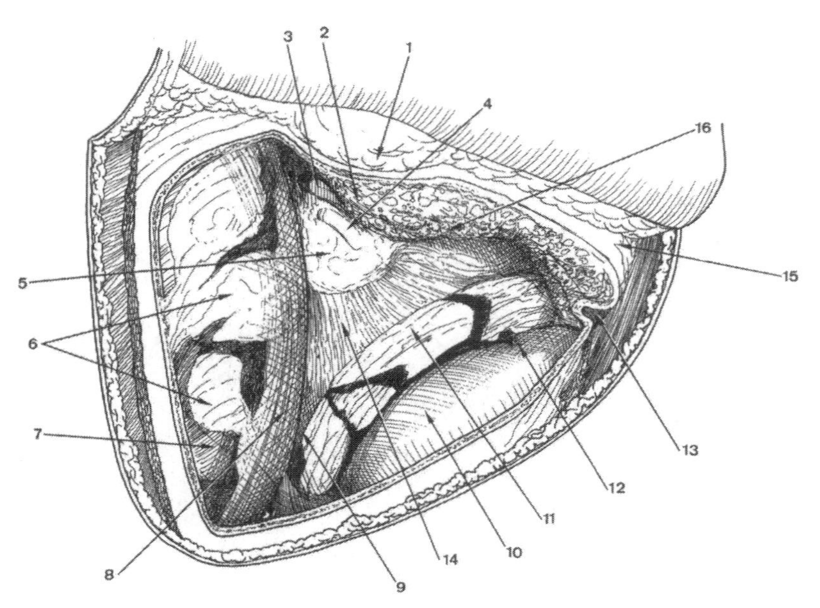

Fig. 55. Extended right translabyrinthine route. Extralabyrinthine bony stage. Appearance of the route of approach after finishing with the craniotome. Cartilaginous EAM *(1)*, posterior face of bony EAM *(2)*, body of incus appearing in posterior attic *(3)*, bulge of lateral SCC *(4)*, bulge of labyrinthine nucleus still covered by some mastoid cells *(5)*, fine bony lamella covering the temporal dura mater *(6)*, temporal dura mater *(7)*, residual shell of upper aspect of petrous *(8)*, sinusoido-dural angle still covered by a fine bony film *(9)*, dura mater of posterior fossa *(10)*, bend of lateral sinus still covered by a thin bony layer *(11)*, mastoid emissary vein *(12)*, groove of attachment of digastric muscle *(13)*, residual shell of posterior face of petrous *(14)*, tip of mastoid *(15)*, approximate situation of 3rd part of facial canal *(16)*

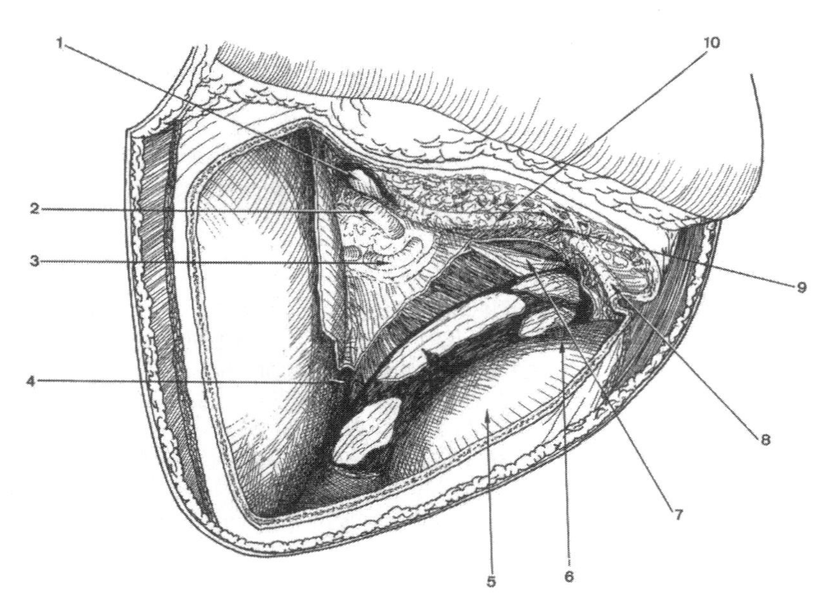

Fig. 56. Extended right translabyrinthine route. Extralabyrinthine bony stage, appearance of route of approach after completion of ablation of mastoid cells. Incus *(1)*, lateral SCC *(2)*, posterior SCC *(3)*, superior petrosal sinus *(4)*, dura mater of posterior fossa *(5)*, hemostasis of emissary vein with a fragment of Surgicel *(6)*, endolymphatic sac *(7)*, bulge of groove of digastric which ends at - *(8)*, stylomastoid foramen *(9)*, third part of facial nerve *(10)*

omises" in the use of the otologic motor in this stage, which carries no particular anatomic risks except to the sigmoid sinus. During this superficial drilling, one should remember to remove the bone-dust which will serve to close the posterior tympanotomy.

In this way one successively exposes the temporal dura, then the dura of the posterior fossa behind the sigmoid sinus as far as the digastric groove, and finally the sinuso-dural angle (SDA), the zone of junction of the sigmoid sinus and the temporal dura mater. Then, still using the large pineapple burr, the mastoid cells are scooped out and the sinus skeletonized until only a thin bony shell is left (Fig. 55).

Throughout the rest of the operation, the operating microscope and the otologic motor are used:

- The posterior surface of the EAM is skeletonized, taking care not to perforate this as the opening created could be a source of infection and especially of secondary cholesteatoma
- The antrum and epitympanic recess are opened to expose the incus and its articulation with the malleus, as well as the lateral semicircular canal (LSCC), always easy to identify
- The 3rd part of the facial canal is skeletonized by drilling the mastoid cells, above and then below the canal, working in the intersinusofacial space which eventually leads to the jugular bulb. The subsequent skeletonization of the digastric groove has a twofold advantage. First, it locates at its anterior end the stylomastoid foramen exactly and consequently the exact site of the third part of the facial canal. Second, it makes possible a good exposure, posteriorly, of the lower part of the bend of the sigmoid sinus which lies under this groove
- The labyrinthine core and the upper aspect of the petrous under the temporal dura mater are perfectly freed.

The "posterior tympanotomy" of the otologists is then performed. In fact, this can be done only partially but it is sufficient to allow a good view of the important structures in the tympanic cavity (Figs. 57 and 58):

- The second part of the facial canal, so that it shall not be damaged when drilling the labyrinthine core
- The stapes and its plate, so as to avoid disturbing it when traversing the labyrinth with the risk of producing an escape of CSF towards the tympanic cavity and auditory tube by this route. However, despite precautions, it often is disturbed and, as will be seen, it is then very important, and simple, to obturate it at the end of the approach.
- The tympanic cavity and tubal orifice, so facilitating the occlusion of these cavities by muscle fragments at the end of the operation.

Still in terms of technique, one first proceeds to ablation of the incus by slipping a fine curved hook into the inco-malleal articulation and pulling the incus backward. Then, with a small-diameter diamond burr one enlarges the aditus ad antrum downwards by drilling the posterior wall of the tympanic cavity between the bend and the beginning of the third part of the facial canal, situated below in the operative position, and the chorda tympani, situated above (Figs. 57 and 58). It is essential to remain in this neural interval. An opening that is too high will enter the EAM externally to the tympanum with the subsequent disadvantages already described. Should this occur, the breach must be carefully closed with a mixture of bone-dust and fibrinogen glue. Once the posterior tympanotomy is done, one can perfectly locate all the structures of the tympanic cavity and the course of the second and third parts of the facial canal (Figs. 58 and 59).

This stage ends with completion of the exposure of the sigmoid sinus and of the temporal dura. The bend of the sinus must be completely bared from the SDA above to its junction with the jugular bulb below. Only a thin bony film should be left on its surface, no thicker than a cigarette-paper. It is better that there should remain no bony film at all, rather than a rigid "bony flap" which might injur the sinus when it is impacted. Behind, the bend of the sinus should be widely freed throughout its length, exposing the dura of the posterior fossa. Of course, the whole of this stage is effected with a large diamond burr. It is absolutely essential not to try to strip the bone fragments with a dural elevator, a maneuver which seems at first easier and quicker but which leads inevitably to opening the sinus. When complete freeing is once effected, it is possible to impact the sinus with the finger. It is then covered with a large swab and kept depressed with the flexible blade of the Yasargil retractor. It is now possible, this time with an elevator and nibbler, to remove the bony film still covering the temporal dura, and then to strip the dura from the upper aspect of the petrous, passing within the arcuate eminence. The temporal dura is then kept held back by means of a flexible blade, the end of which is slipped behind the eminence. Henceforth, there re-

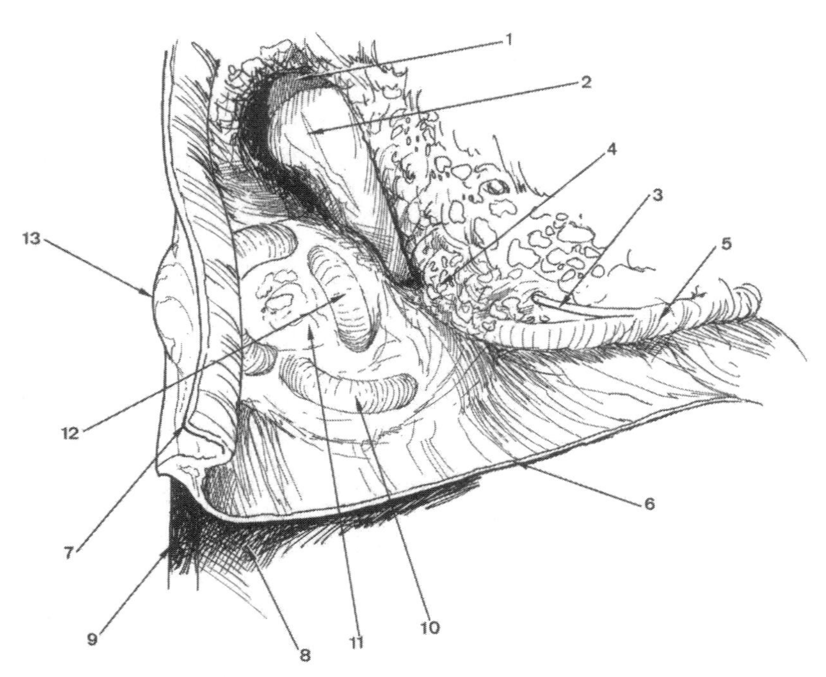

Fig. 57. Extended right translabyrinthine route. Extralabyrinthine bony stage, appearance of attic region before posterior tympanotomy. Head of malleus *(1)*, body of incus *(2)*, chorda tympani *(3)*, site of posterior tympanotomy *(4)*, 3rd part of facial canal *(5)*, residual shell of posterior aspect of petrous *(6)*, residual shell of upper aspect of petrous *(7)*, dura mater *(8)*, superior petrosal sinus *(9)*, posterior SCC *(10)*, labyrinthine nucleus *(11)*, lateral SCC *(12)*, arcuate eminence, bulge of anterior SCC *(13)*

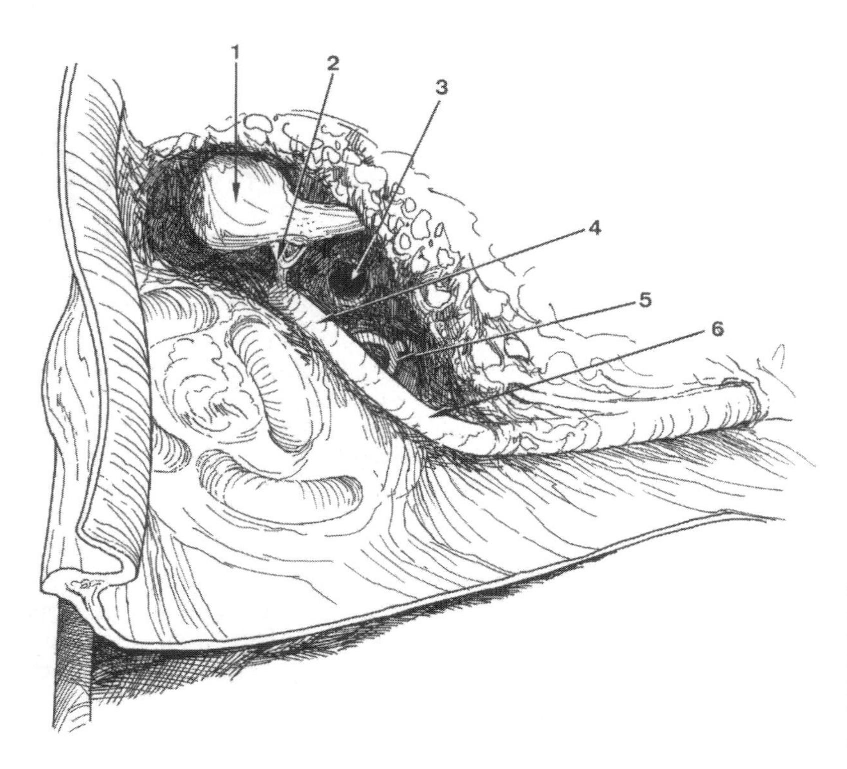

Fig. 58. Extended right translabyrinthine route. Bony extralabyrinthine stage, appearance of attic region after excision of incus and posterior tympanotomy. Head of malleus *(1)*, cochleariform process and tendon of tensor tympani muscle *(2)*, orifice of auditory tube *(3)*, 2nd part of facial canal *(4)*, stapes *(5)*, bend of facial canal *(6)*.

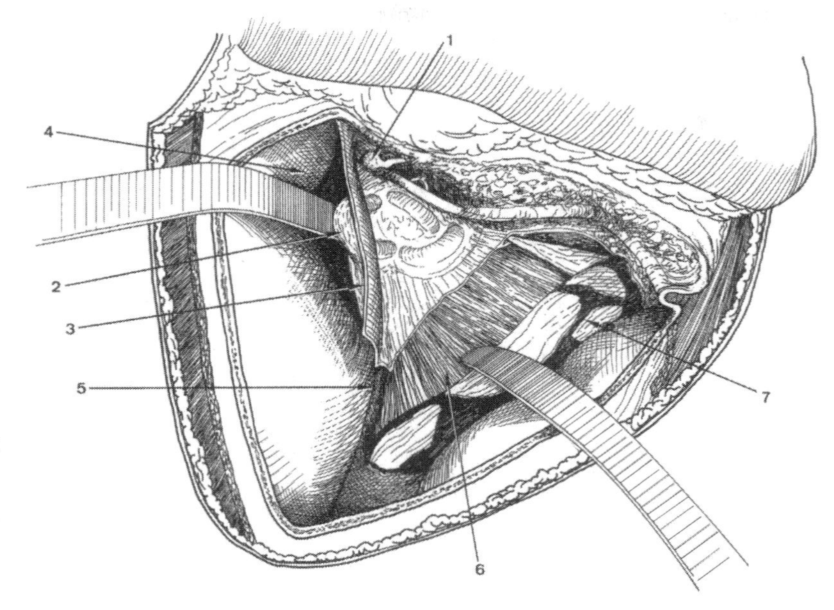

Fig. 59. Right extended translabyrinthine route. Bony extralabyrinthine stage, exposure of labyrinthine nucleus by retraction of the temporal dura mater and the bend of the lateral sinus. Head of malleus *(1)*, arcuate eminence, bulge of anterior SCC *(2)*, upper aspect of petrous *(3)*, temporal dura mater stripped from upper aspect of petrous *(4)*, dura mater which clothed the posterior face of the petrous in front of the sigmoid sinus *(6)*

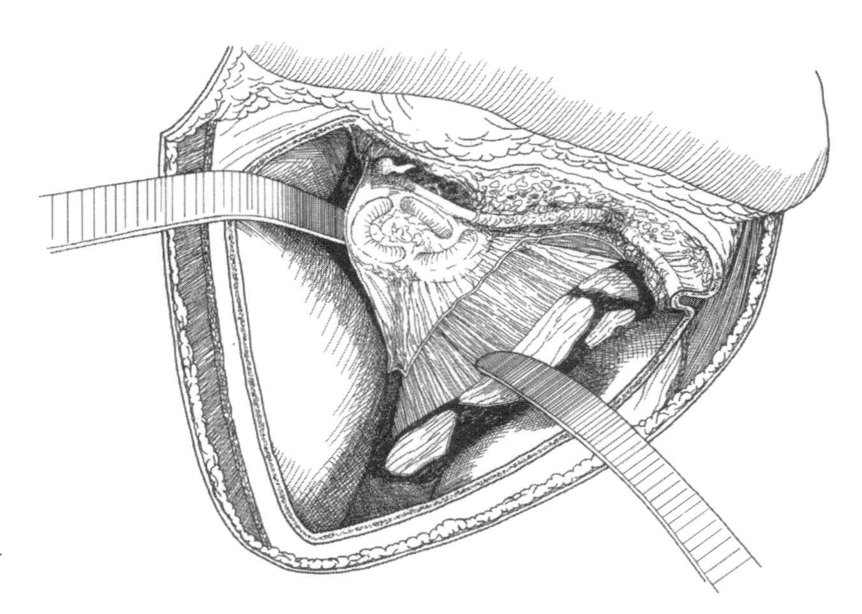

Fig. 60. Extended right translabyrinthine route. Extralabyrinthine stage. The exposure of the labyrinthine nucleus is being completed

main no further superficial obstacles. There is a perfect view of the labyrinthine and subfacial region (Figs. 59 and 60).

• *The Bony Labyrinthine Stage* (Figs. 61 and 62). Generally speaking, from now on the object is to excavate in the labyrinth and adjacent regions, layer by layer, always leaving a bony shell to protect the dura so as to avoid a sudden slip of the burr towards a neurovascular structure. This shell is removed at the end of the drilling.

One begins by opening the semicircular canals, the lateral and anterior canals first, carefully identifying their ampullae, which lie in contact with each other, and then the posterior canal. All these canals are progressively destroyed to open the vestibular cavity, taking good care to preserve:

- In front, the anterior borders of the ampullae of the anterior and lateral SCCs. This landmark avoids going too far forward and injuring the

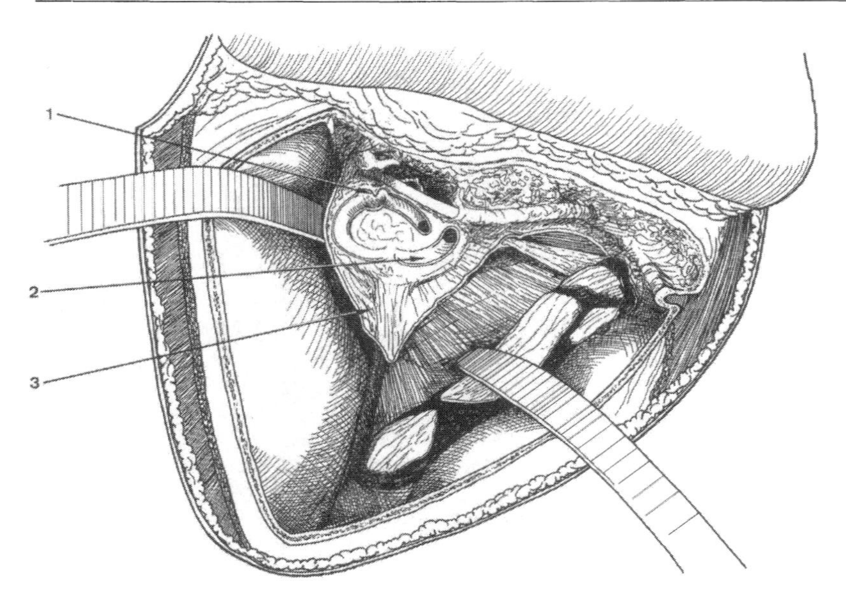

Fig. 61. Extended right translabyrinthine route. Labyrinthine bony stage, opening of semicircular canals. Triangular crest separating ampullae of anterior and lateral SCCs – *(1)*, posterior SCC – *(2)*, fine bony shell protecting the dura mater *(3)*

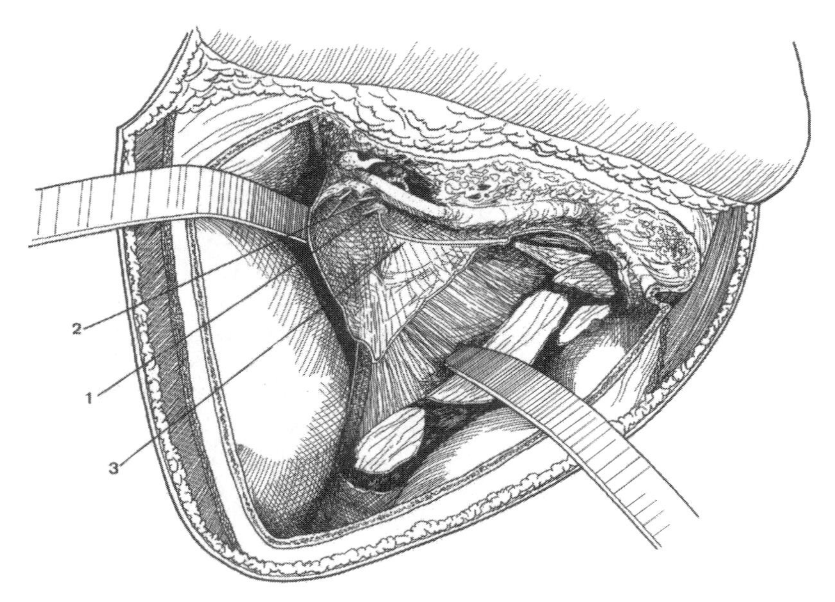

Fig. 62. Extended right translabyrinthine route. Labyrinthine bony stage, opening of vestibular cavity. Hemispheric fossula *(1)*, ampullae of anterior and lateral SCCs *(2)*, endolymphatic canal *(3)*. Section of the endolymphatic sac where it penetrates the bone facilitates stripping of the dura mater of the time of burring in the region of the jugular bulb

second part of the facial nerve, which changes direction at this level

- Laterally, the whole outer border of the anterior limb of the lateral SCC, which skirts and protects the tympanic part of the facial nerve (Fig. 61).

The next step is to open the vestibular cavity (Fig. 62). At the fundus of this can be seen a bluish depression, the hemispheric or saccular fossula, situated under the tympanic part of the facial nerve opposite the vestibular window. Drilling is then done all around the walls of the SCCs so as to obtain a bony cavity the fundus of which consists of the inner wall of the vestibular cavity. In so doing, one has exposed, in the lower part of the field, the endolymphatic canal and sac.

Finally, one skeletonizes again the second and third parts of the facial canal in order to obtain the best possible view of the subfacial region. It is axiomatic always to burr with the end of the burr under vision and not to denude the facial nerve so as not to risk injuring it during tumoral excision with instruments, especially bipolar coagulation.

The Bony Pericanalicular Stage. This starts with identification of the fundus of the IAM. This corresponds very precisely to the vestibular cavity, at the bottom of which the hemispheric or saccular fossula is easily identified by its bluish aspect (Figs. 62 and 63); this corresponds in the IAM to the inferior vestibular nerve (IVN). Immediately external to it can be seen a small vertical crest, the vestibular crest, corresponding to the falciform crest in the IAM. Immediately in contact with it and more lateral is a fossula which is poorly marked, the semi-ovoid fossula which corresponds to the superior vestibular nerve (SVN). This fossula is situated at the level of the triangular crest which separates the ampullae of the lateral and anterior SCCs. Gentle concentric drilling at the level of these fossulae allows identification of all the neural structures of the fundus of the IAM: IVN, falciform crest, SVN (Fig. 64). However, when the neurinoma has invaded the fundus of the IAM, which is very common, one no longer visualizes the vestibular nerves but the tumor, on each side of the falciform crest, which may itself be more or less destroyed. We feel that this identification of the fundus of the IAM is important because it also locates quite precisely the entry of the first part of the facial nerve into the

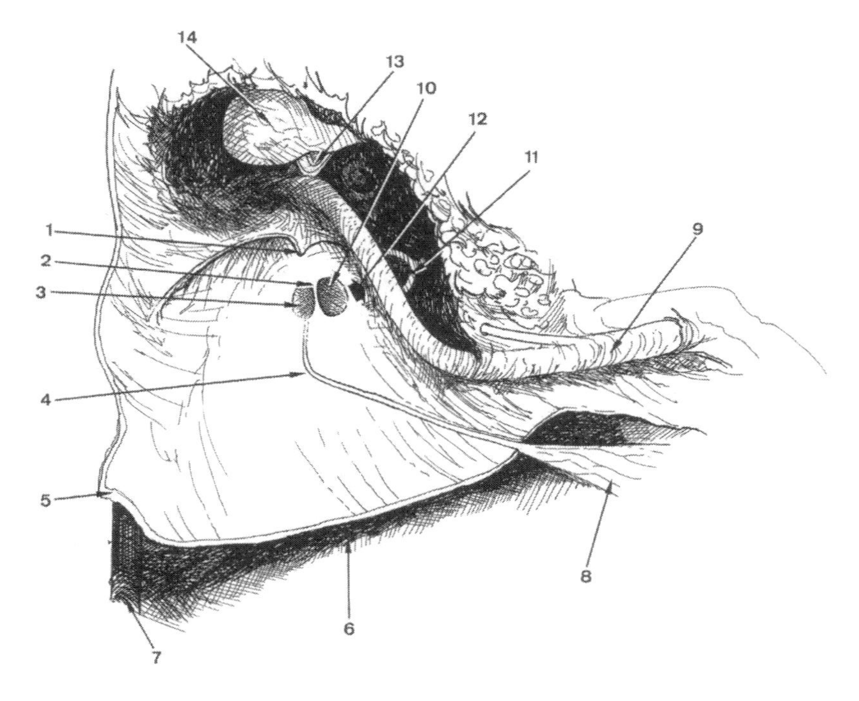

Fig. 63. Extended right translabyrinthine route. Pericanalicular bony stage. Identification of the fundus of the internal acoustic meatus at the level of the vestibular cavity. Triangular crest separating ampullae of anterior and lateral SCCs *(1)*, vestibular crest *(2)*, semi-ovoid fossula *(3)*, endolymphatic canal *(4)*, bony shell *(5)*, dura mater *(6)*, superior petrosal sinus *(7)*, endolymphatic sac *(8)*, 3rd part of facial canal *(9)*, hemispheric fossula *(10)*, stapes *(11)*, arrow indicates site of vestibular window under tympanic part of facial nerve *(12)*, cochleariform process on which is reflected the tendon of the tensor tympani muscle *(13)*, head of malleus *(14)*

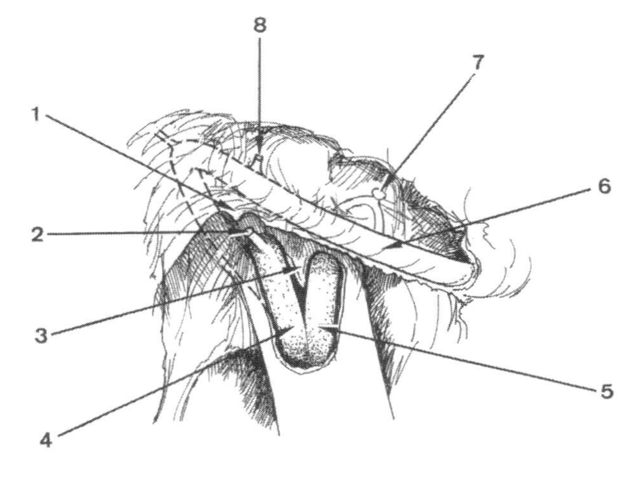

Fig. 64. Extended right translabyrinthine route. Pericanalicular bony stage. Opening of fundus of IAM and location of fossulae of vestibular nerves. Triangular crest seperating ampullae of anterior and lateral SCCs *(1)*, superior ampullary nerve *(2)*, falciform crest *(3)*, superior vestibular nerve *(4)*, inferior vestibular nerve *(5)*, tympanic portion of facial nerve *(6)*, stapes *(7)*, cochleariform process *(8)*

meatus. This occurs in contact with the fossula of the SVN. During the drilling, one should take care not to damage the second part of the facial nerve at its inferior aspect. It is also necessary not to mobilize the plate of the stapes, which lies in the immediate vicinity of the hemispheric fossula. As we have noted, this may cause an escape of CSF towards the tympanic cavity.

Once the fundus of the meatus has been identified, one begins freeing the dura between the IAM and the jugular bulb. This is done with a stout diamond burr and under extensive magnification. This stage is facilitated by preliminary section of the endolymphatic sac where it penetrates the bone and stripping of the bony shell remaining. The dura mater of this dihedron is exposed by following the inferior aspect of the IAM and the summit of the bulb. One has to dip in front of the latter to expose the region of the mixed nerves (Fig. 65). The drilling is stopped in front as soon as the cochlear aqueduct is reached, a vertical groove parallel to the IAM, which is sometimes very wide and allows the

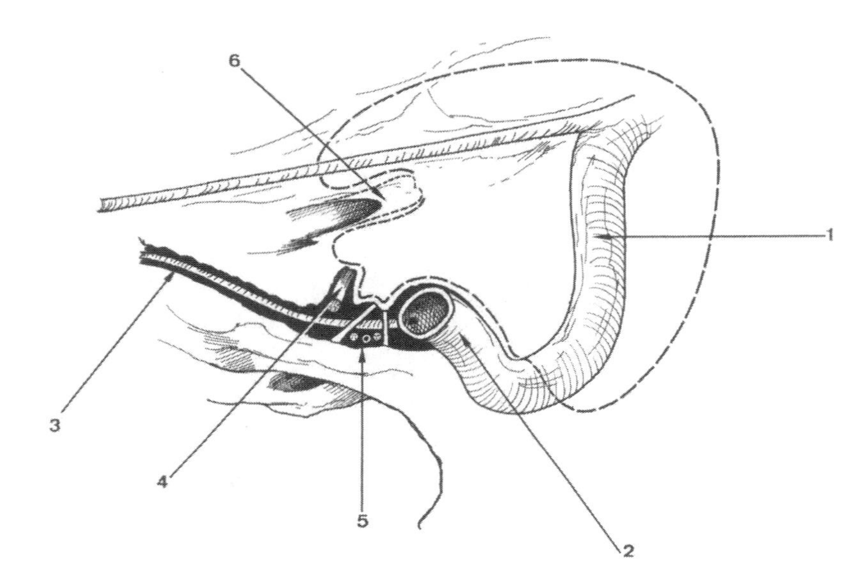

Fig. 65. Extended right translabyrinthine route. Bony pericanalicular stage. Exposure of dura mater between IAM and jugular bulb. The interrupted lines indicate the zones where the dura mater must be exposed by burring. Note that access to the mixed nerves requires burring to be carried forward in front of the jugular bulb. Sigmoid sinus *(1)*, jugular bulb *(2)*, inferior petrosal sinus *(3)*, glossopharyngeal nerve *(4)*, vagus, accessory nerve and ascending pharyngeal artery *(5)*, IAM *(6)*

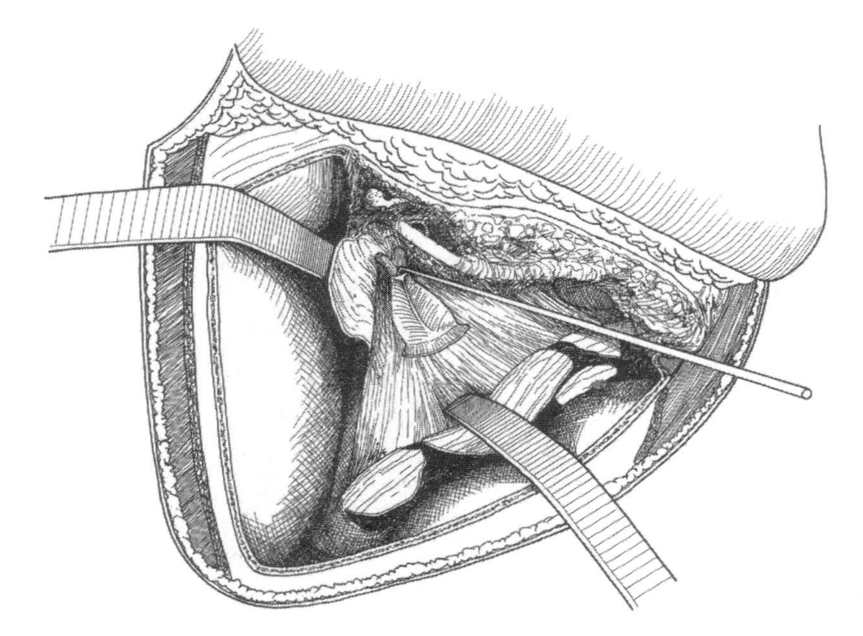

Fig. 66. Extended right translabyrinthine route. Bony pericanalicular stage. End of burring. The protective bony shell at the posterior aspect of the IAM, which has been preserved so far to protect any unwarranted penetration of the burr, is now removed with a fine hook

escape of CSF. Just behind it is the glossopharyngeal nerve. Traveling upward, the dura of the inferior aspect of the IAM is freed as far as the fundus of the meatus, practically under the internal wall of the tympanic cavity (Fig. 66).

When the inferior drilling is completed, the superior stage is begun: exposure of the entire dural surface between the superior aspect of the IAM on one side and the superior petrosal sinus on the other, and of the dura of the posterior fossa above the meatal aperture. It is necessary to work forward to the level of the anterior wall of the IAM and to go beyond the fundus of the IAM above. Usually, the intracanalicular part of the facial nerve is pressed by the tumor against the superior wall of the meatus. Any damage to the dura by the burr at this site damages the facial nerve. It is thus a delicate stage. It should begin near the aperture where the space is widest, allowing for the downward obliquity of the IAM, and continues by elevating the dura from the meatus (Fig. 66). Until now, the bony shell of the posterior aspect of the IAM has been left intact to prevent untimely penetration of the burr. It only remains to remove this last bony fragment with a fine curved elevator to complete the freeing of all of the dura mater of this region (Figs. 67, 68). The facial nerve must then be identified at the fundus of the IAM. Drilling of the superior wall of the meatus allows exposure of the dura mater to which the nerve is applied. The dura is opened vertically at the posterior aspect of the meatus with a stapedectomy hook, from the fundus to the aperture. In the absence of a tumor, identification of the facial nerve poses no problem. It suffices to reflect the dural flap upward and the SVN downward to see the nerve. Displacement of the SVN from the vestibular fossula and freeing of the acousticofacial anastomoses allow complete exposure of the facial nerve (Fig. 69). If the extremities of the vestibular nerves are not tumoral, they can be identified and the procedure is as before. Brackmann [3] ensures that the nerve is further forward by palpating with a blunt hook the crest (Bill's bar) separating the vestibular fossula from the facial fossula, which is lower, towards the meatal aperture, in the operative position of the ETL. When the vestibular nerves are tumoral, the facial nerve can be identified with certainty only after drilling of the first part. This is situated laterally, almost transversely, at a slightly lower level than the vestibular fossula (Fig. 69c). The drilling must be done extremely gently, without pressure, using a magnification of 16 or 25. Once the nerve is identified, the opening of the dura of the IAM is completed in order to free the entry of the nerve. Excision of the tumor should not be performed at this stage; on the contrary, all the contents of the meatus should be respected so as not to modify the planes of cleavage between nerves and tumor. It is enough to identify the facial nerve and then restore all the structures in place. Ablation of the intracanalicular growth will be done during the first stage of the tumoral excision.

Exclusion of the Tympanic Cavity

This is facilitated by the posterior tympanotomy. First, the auditory tube is blocked with a fragment of muscle taken from the sternocleidomastoid. The tympanic cavity is then packed with other muscle fragments mixed with fibrinogen glue. This packing should not be done blindly or forcefully, for fear of tearing the tympanic membrane. The tympanotomy and the epitympanic recess are then occluded with a mixture of bone-dust and fibrinogen glue. Should the stapes have been inadvertently mobilized, a fragment of crushed muscle is stuck to the labyrinthine face of the vestibular window.

The Neurosurgical Tumoral Stage

- *Opening the Dura Mater*

- The microscope is arranged vertically over the dura, with a magnification of $\times 10$.
- The incision is roughly M-shaped; one of the downstrokes is parallel to the superior petrosal sinus, the other to the jugular bulb and the sigmoid sinus. The joining transverse limb of the M skirts the outer border of the aparture of the IAM (Fig. 68). After having elevated the dura with a fine hook, one begins by using the tip of a fine scalpel to make a short incision 1 mm below the superior petrosal sinus, starting at a greater or lesser distance from the sigmoid sinus depending on the size of the tumor. This incision is prolonged with a fine scissors, following the superior petrosal sinus as closely as possible towards the aperture of the meatus. Care must be taken to divide nothing but the dura mater. Through this incision one first sees the cerebellum and then, at a distance depending on its size, the tumor covered with its arachnoid sheath. The borderline is often marked by a bounding artery, the terminal hemispheric branch of the AICA, and a bounding vein, the hemispheric branch of the vein of Dandy. These vessels sometimes exhibit some adhesions to the dura mater which are gently ruptured. As soon as one arrives at the tumor, one must endeavor to gently strip the arachnoid layer before dividing the dura mater. When the meatal

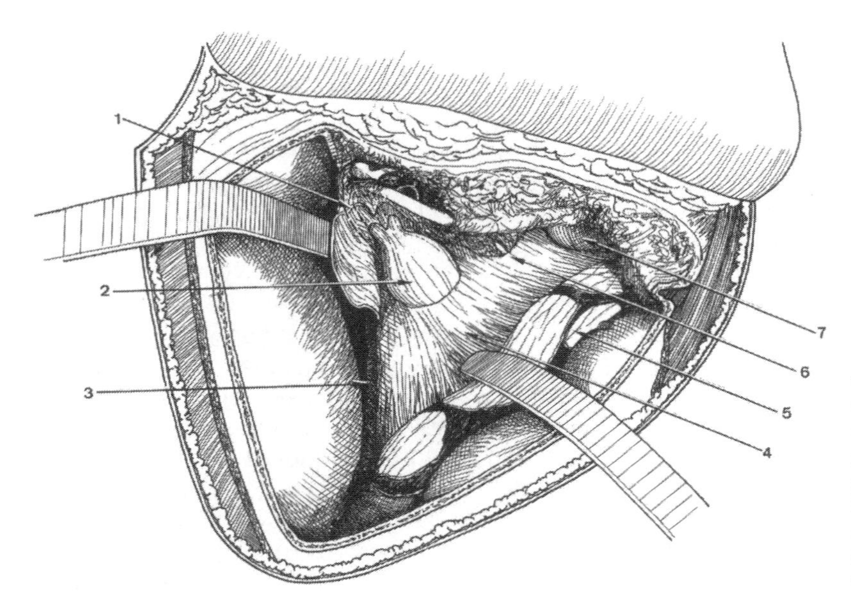

Fig. 67. Extended right translabyrinthine route. Completion of access. Note the complete freeing of the IAM on its three faces, superior, posterior and inferior. Situation of first part of facial canal *(1)*, dura mater of IAM distended by tumor *(2)*, superior petrosal sinus *(3)*, dura mater of posterior aspect of petrous *(4)*, sigmoid sinus *(5)*, cochlear aqueduct *(6)*, jugular bulb *(7)*

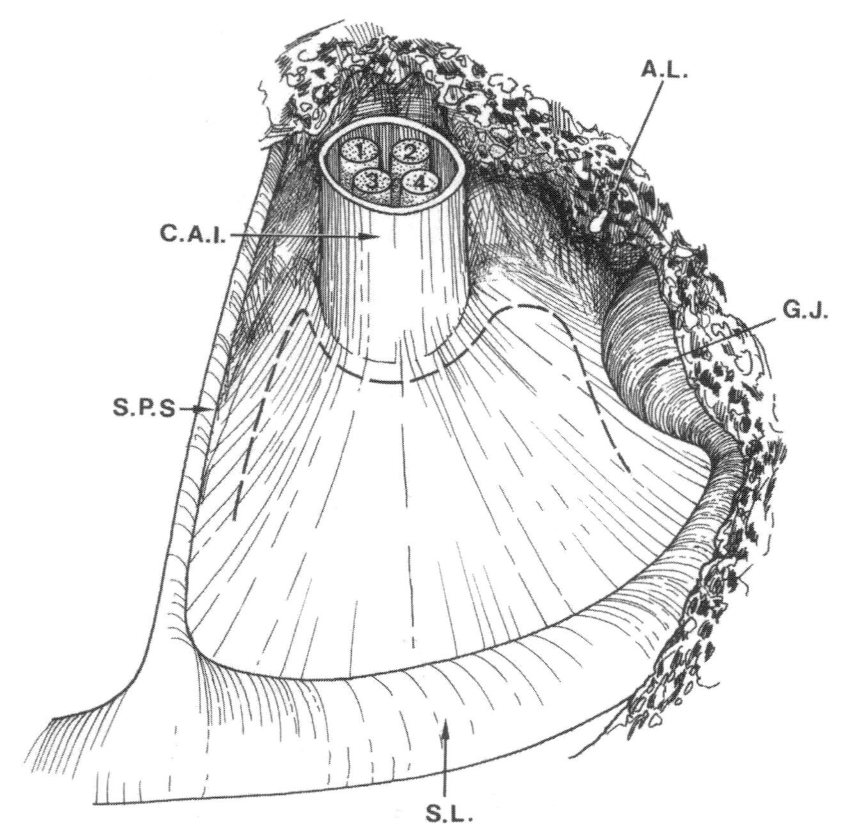

Fig. 68. Extended right translabyrinthine route. Schematic diagram of structures at end of approach. Facial nerve *(1)*, cochlear nerve *(2)*, superior vestibular nerve *(3)*, inferior vestibular nerve 5 *(4)*, dura mater of IAM *(C. A. I.)*, superior petrosal sinus *(S. P. S.)*, sigmoid sinus *(S. L.)*, jugular bulb *(G. J.)*, cochlear aqueduct *(A. L.)*. The *interrupted lines* indicate the line of incision of the dura

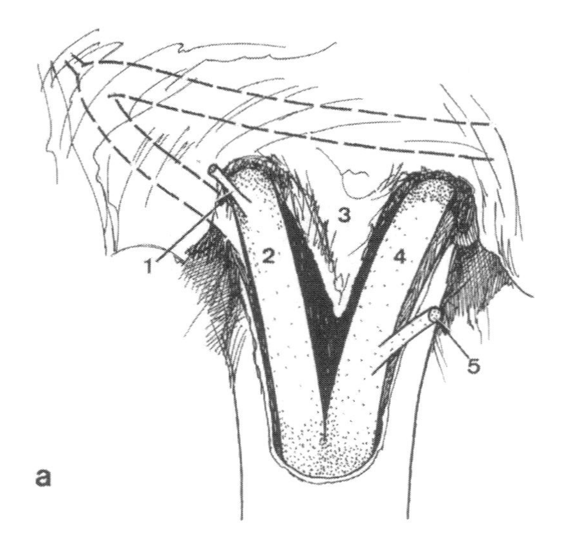

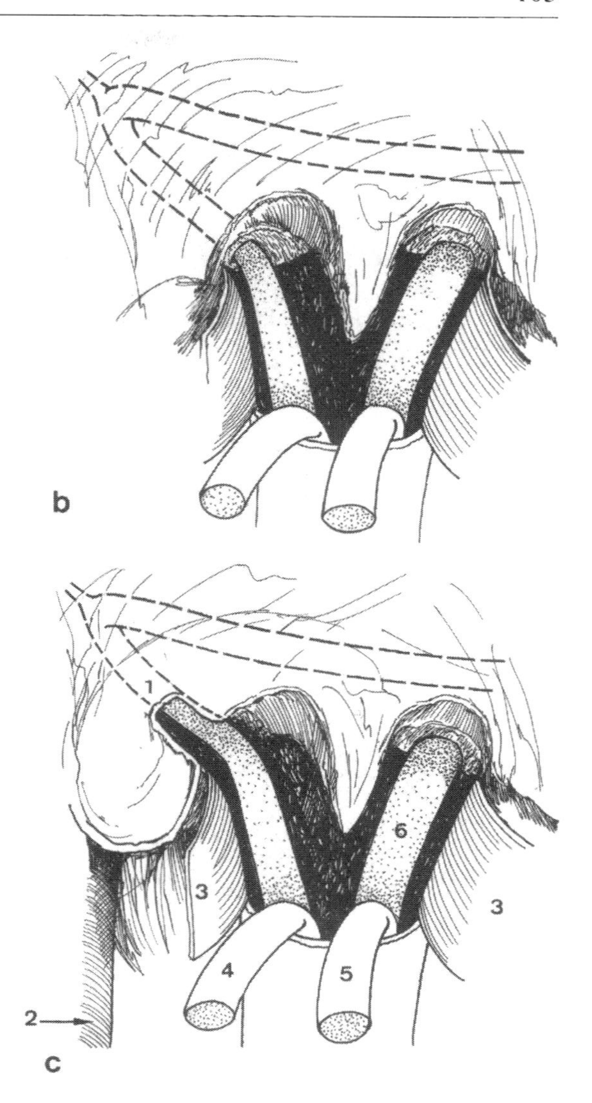

Fig. 69 a–c. Extended right translabyrinthine route. Identification of facial nerve in healthy zone at fundus of IAM. **a** The posterior wall of the IAM has been burred to expose the nerves superior ampullary nerve *(1)*, superior vestibular nerve *(2)*, falciform crest *(3)*, inferior vestibular nerve *(4)*, posterior ampullary nerve (singular) *(5)*. **b** the vestibular nerves have been detached from their superior orifices. Under the superior vestibular nerve, there is a ridge (Bill's bar) behind which the facial nerve narrows as it enters the first part of the facial canal, whose orifice is seen. **c** Exposure of the first part of the facial nerve *(1)*, - superior petrosal sinus *(2)*, dural flap of IAM *(3)*, superior vestibular nerve *(4L)*, inferior vestibular nerve *(5)*, cochlear nerve *(6)*

aperture is reached the adhesions become tight and the incision is halted. Using the same technique, an inferior incision is made, this time just above the sigmoid sinus and the bulb and ending under the inferior border of the aperture. Laterally, especially if the tumor is small, the cerebello-medullary cistern is exposed, the wall of which herniates through the incision. If it is perforated, there is a flood of CSF which relaxes the posterior fossa. If a breach is inadvertently made in a sinus, the simplest course is to occlude it with a fragment of Surgicel. These two incisions are then connected transversely by dividing the dura all around the outer border of the meatal aparture, at a distance of 1–2 mm. This preserves the fibrous diaphragm of the aperture, on the other side of which the tumor is embedded in the CPA. Because of this, the neurinoma is fixed to the petrous and difficult to mobilize. During the maneuvers of scooping out the tumor, this lessens the risks of unfortunate tugging on the acousticofacial bundle. Ultimately, a flap of dura mater has been cut out, hinged posterolaterally on the sigmoid sinus.

• *Exposure of the Outer Pole of the Tumor.* After having loosened the self-retaining retractor holding back the lateral sinus, the end of the blade is bent so that, when it is applied to the dural flap, this is retracted towards the interior of the posterior fossa. By this means, the cerebellar hemisphere is displaced slightly outward and backward, perfectly exposing the posterior and lateral pole of the neurinoma. The dihedron is further opened by using a second retractor to reflect slightly upward the temporal lobe, well-protected by its dura mater (Fig. 70). The neurinoma develops within the arachnoid sheath common to the three nerves of the

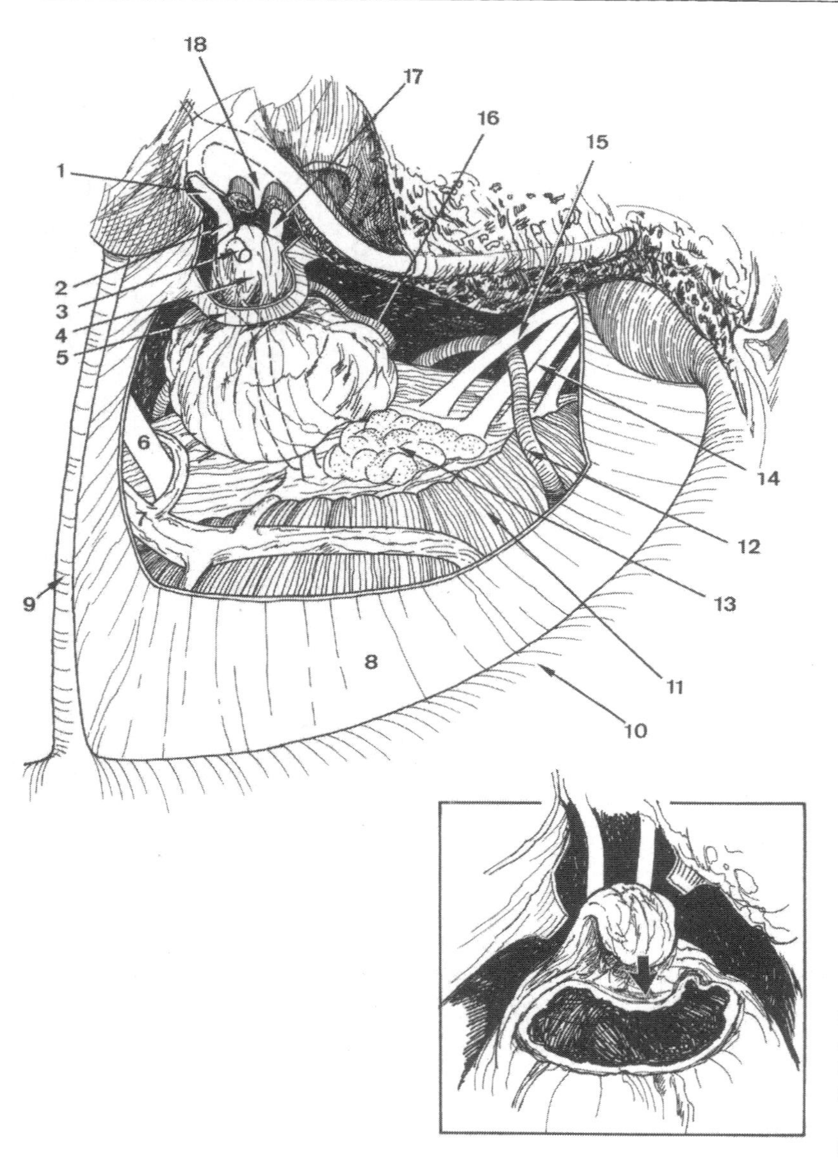

Fig. 70. Extended right translabyrinthine route. Exposure of contents of IAM and cerebellopontine angle with a stage II neurinoma. After opening the dura mater. First part of facial nerve exposed *(1)*, facial nerve in fundus of IAM *(2)*, healthy end of superior vestibular nerve, hooked out of its fossula *(3)*, intracanalicular portion of neurinoma *(4)*, fibrous dural ring preserved at aperture of IAM *(5)*, trigeminal nerve *(6)*, Petrous vein of Dandy *(7)*, dura mater *(8)*, superior petrosal sinus *(9)*, sigmoid sinus *(10)*, cerebellum *(11)*, PICA *(12)*, choroid plexus *(13)*, strands of vagus *(14)*, glossopharyngeal nerve *(15)*, AICA *(16)*, cochlear nerve *(17)*, falciform crest. *(18)*. Box: after evacuation of extracanalicular portion of tumor and section of fibrous ring of aperture of IAM, dissection of intracanalicular growth which is pushed backwards

acousticofacial bundle. This arachnoid sheath completely envelops the tumoral substance and separates it from the other structures in the CPA. However, at the anterior pole of the neurinoma this sheath is interrupted at the constriction formed by the meatal aperture, with which it usually forms tight adhesions. In exposing this pole of the tumor, there is a perfect identification of the arachnoid sheath and its more or less tight neck around the aperture. This neck is carefully dissected with the 90° stapedectomy micro-hook and then, using gentle traction, the whole is extended just as if one were loosening the strings of a purse. This exposes the tumoral pulp, which has no capsule of its own, a maneuver to be performed with care from one border of the IAM to the other. This is an important stage, which gives a good exposure of the arachnoid sac in which the progressive fragmentation of the tumor will be carried out. Success in working properly within this envelope ensures that one will not damage any of the anatomic structures of the angle (cerebellum, brain-stem, cranial nerves, arteries and veins).

- *Evacuation of the Tumor Can Now Begin.* At first, one or more tumoral fragments must be removed with the cutting curette (or a fine scalpel when the tumoral texture is too firm) for histological examination and possibly for study of hormonal receptors. If the tumor is bulky and of suitable texture, continuation for a short time of this fragmentation by curette will speed up the operation. However, to avoid any degree of roughness, one should not take this maneuver too far but resort to the vacuum rotary dissector of Urban House. This instrument combines suction and fragmentation to ensure atraumatic evacuation. This is carried out methodically, first towards the depths by excavating a hole in the tumor, then from its center towards the periphery, progressively directing the suction aperture with the guide-wheel turned between thumb and index, the little projection on this guide indicating the direction of the aperture. The more solid the tumor, the more the rotation must be speeded up. It should also be realized that stopping the rotation immediately causes the suction to stop or decrease. The vacuum dissector is held in the right hand while the left holds a Brackmann suction cannula, whose main fuction is to steady the margins of the tumor or sometimes to retract the cerebellum slightly. The excavation is progressively extended. When the periphery is sufficiently thinned, each suction depresses the residual shell slightly towards the center and thus contributes to its stripping from the cerebellum. The evacuation must not be pushed too far towards the inner pole for fear of damaging the facial nerve. Once the center of the tumor is well excavated, the next task is evacuation of the posterolateral portion. Should there be any intracapsular hemorrhage, hemostasis is achieved by bipolar coagulation or, if this is ineffective, by swab pressure; a swab is placed over the bleeding point and at once covered by a second swab over which gentle pressure with continued suction is applied. Usually, after 1–2 minutes, it is possible to relax the pressure, stop the suction and remove the second swab. The tumor evacuation is pursued elsewhere for a short period before removing the first swab. Sometimes, the texture of the tumor is too firm to allow scooping-out. It is then necessary to fragment the tumoral substance with a fine scalpel, or better by using the diathermy to excavate grooves in which the vacuum dissector can work effectively again. When the periphery is sufficiently retracted, the second stage follows:

• *Peripheral Dissection*. The microscope is still directed vertically. If the tumor has a major lateral extension, the microscope may need to be angled to direct the light-beam into the plane separating the tumor from its cerebellar bed. The angle of vision can also be improved by turning the patient's head slightly or tilting the table toward the surgeon. The dissection is conducted under continuous irrigation. Now, with a Brackmann cannula held in the left hand, the tumoral substance is progressively stripped from the arachnoid sheath using the 90° micro-hook. The freed tumor fragments are removed progressively. If a vessel is exposed, it is dissected a little to verify whether it is extra-arachnoidal and simply adherent to the convexity of the tumor or if it is intra-arachnoidal and penetrating the substance of the neurinoma. In the former case, the artery or vein is dissected until it can be displaced from the tumor; in the second case, the tumoral branch is coagulated and divided. The peripheral covering of the tumor is often folded as a result of retraction, and it is important to dissect each of the folds carefully before resecting the tumoral fragments. It is sometimes simpler to restretch this peripheral cover by means of a swab to give a good view of the plane of dissection between the arachnoid sheath and the tumor. The outer pole is first resected from its cerebellar bed. The dissection must stop as soon as the choroid plexus is seen, if it is identifiable, or when one reaches the origin of the vestibular nerves, which may be recognized easily enough with a little practice. Throughout this stage one is warned of the slightest effect on the facial nerve by an increased electromyographic activity on the anesthesia monitor or by the appearance of facial twitching as detected by the assistant's hand left permanently on the patient's cheek if the equipment is hot available. The resection of the outer pole has allowed upward displacement of the lower pole. The microscope is angled towards the feet to center the vision on this lower pole. If this is still rather bulky, it is further fragmented with the vacuum dissector. Then, still using the same technique, with the Brackmann cannula in one hand and the 90° micro-hook in the other, the peripheral covering of this lower pole is stripped and fragmented. At this site, there are exposed across the arachnoidal layer the more or less spread-out and adherent strands of the mixed nerves. These can usually be isolated without great difficulty, as can the PICA, which usually forms quite a long loop giving it some degree of mobility allowing it to be pulled on if necessary to coagulate and divide a small tumoral branch. Here again, the resection is halted as soon as one approaches the origin of the mixed nerves, situated just below that of the acousticofacial bundle, and as soon as one reaches the proximity of the adhesions at the inferior border of the meatal aperture.

The microscope is then tilted upwards to bear on the upper pole. Using the same technique, the size of this upper pole is reduced if necessary and then the tumoral substance is stripped and resected. By keeping well within the arachnoidal sheath, there is usually no difficulty in freeing the vein of Dandy. It is necessary to stop as soon as one approaches the adhesions at the upper border of the meatal aperture, since the facial nerve here may be so stretched as not to be visible. Here again, the anesthesia monitor or the assistant's hand on the cheek will detect any twitches requiring cessation of all meneuvers in this region.

The problem then is to proceed to:

• *Dissection of the Facial Nerve*. The microscope is lowered and angled horizontally so as to be centered on the intracanalicular fungation. Changing to a magnification of × 16, the facial nerve is once again identified in the healthy zone, just at its entry into the facial canal, right against the falciform crest which separates it from the 8th cranial nerve emerging from the cochlear window.

A magnification of × 10 is used for longitudinal section of the dural sheath of the IAM, if this has not already been split, from the fundus to the aperture, and then the fibrous annulus of this aperture. Still using the 90° micro-hook, the two dural flaps thus freed are readily stripped to expose the intracanalicular fungation of the tumor.

Returning to a magnification of × 19, irrigation is made with the Brackmann cannula and, under constant traction with the 90° micro-hook, the intracanalicular neurinoma can usually be stripped from the facial nerve without much difficulty. The plane of cleavage is under direct vision and it suffices to introduce the micro-hook gently from right to left and back to progressively retract the tumoral mass towards oneself, leaving the facial and acoustic nerves plastered against the anterior wall of the IAM, which, in the operative position, is in fact the ceiling of the route of access. As one is working in the axis of the nerve, its course can be quickly defined. Usually, it remains applied to the anterior wall of the meatus; sometimes it is seen to bend progressively upward or, more rarely, downward. If the tumor has burst the meatus, the intracanalicular extension is often more rigid by virtue of its size. It must then be scooped out with the vacuum dissector if it is to be mobilized. This brings one to the aperture of the meatus and the position of the nerves here has been precieely defined. Under a magnification of × 10, the tight attachments formed between the neurinoma and the meningeal margins of the aperture can be coagulated and sectioned without fear of injuring the nerves. At the inferior margin, quite a large arteriole is often coagulated which is the meningeal branch of the ascending pharyngeal artery. At the upper margin there is also very often an arteriole, usually narrower, arising from the middle meningeal just after its passage through the spinous foramen. After this disinsertion maneuver, the tumor is often much less hemorrhagic.

Returning to magnification × 16, the stripping of the facial nerve ay the convexity of the inner pole is continued for a while sufficiently to specify its position and direction. One is very rapidly arrested by the size of this inner pole and by adhesions of the nerve at the meatal aperture. It is necessary to return to magnification × 10 and to resume evacuation with the vacuum dissector, but now with the knowledge of the position of the facial nerve. If the tumor is rather "moist" or soft, the suction pressure must be reduced by opening the safety-valve so as not to risk transgressing the limits and damaging the often spiderlike fibers of the facial nerve. When the peripheral shell is sufficiently thinned, one resumes magnification at × 16 and dissection of the nerve fibers. One must endeavor to traverse the entire breadth of the shell, gently stripping the tumoral pulp from the facial nerve at the middle and from the arachnoid covering on either side. With a little practice and the alternate use of the 90° micro-hook and the vacuum dissector, the bulk of the residual mass is gradually reduced. Dissection of the facial nerve must then be resumed. This is always tedious at the meatal aperture since the reflection of the arachnoid layer constitutes a zone of tight adhesion to the nerve. Further, there very often exists at this site an exchange of vessels between the tumor and the nerve, each of the fine branches constituting a zone of particular adhesion. This must be scrupulously stripped and the arteriole coagulated and sectioned. It is possible sometimes thus to pursue the progressive dissection of the facial nerve from the aperture to the brain-stem. Once past the zone of adhesion at the neck, dissection is usually easy enough for the facial nerve is outside the tumoral arachnoid covering. It is then relatively easy to reach the emergence of the vestibular fibers, to divide them after coagulation flush with the brain-stem and to remove the last tumoral fragments. A large tumoral venous trunk is often found here draining into the lateral pontomedullary vein, requiring rather tedious dissection and coagulation. In fact, the zone of adhesion at the meatal aperture often seems impassable, the more so since the nerve, which bends sharply over the neck of the intracanalicular growth seems to be lost in the tumor. It is better then not to persist, but to resume the reduction of the internal growth until it is possible to get around it and locate the facial nerve in the angle, usually well under cover behind the arach-

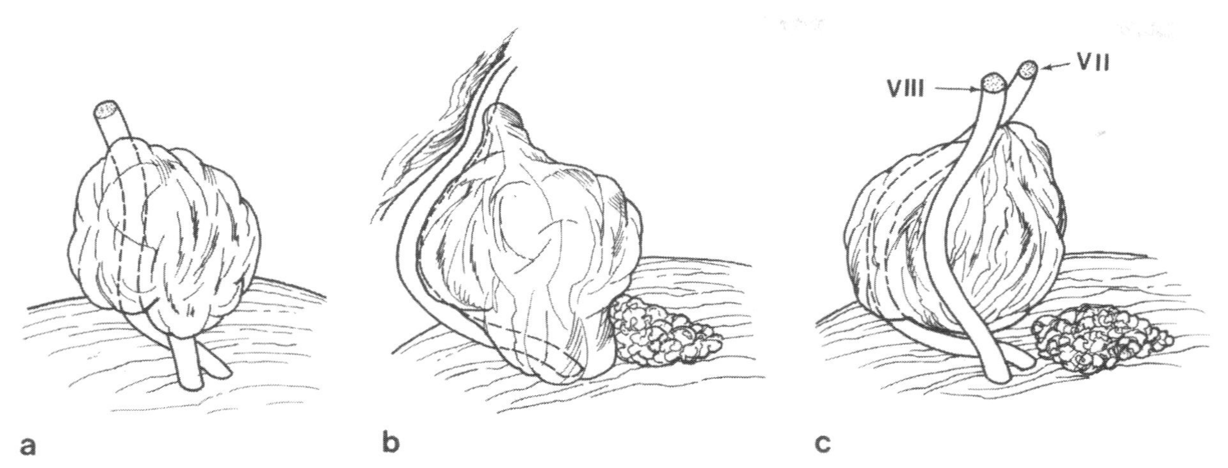

Fig. 71 a–c. Extended right translabyrinthine route. Displacement of facial and cochlear nerves by acoustic neurinoma. **a** Antero-superior displacement of facial nerve. The tumor being small, the facial nerve can be identified where it emerges from the brainstem in front of the cochleovestibular bundle. **b** Anterior displacement and crushing of facial nerve by tumor represented in sagittal section. **c** Representation of dissociated displacement of facial nerve forwards and cochlear nerve backwards. Facial nerve *(VII)*, cochleovestibular bundle *(VIII)*

noid layer. The dissection is then carried towards the emergence of the vestibular fibers, division of which as before allows removal of the greater part of the residual tumor. There now remains only a fragment adherent to the facial nerve near the meatal aperture which is stripped progressively in a centrifugal direction. Sometimes the very tight adhesions cannot be stripped, especially as the nerve is particularly flattened here. The best course then is to coagulate these adhesions after having reduced the bulk of the residual tumor as much as possible. Allowing for the strong magnification used during the dissection, what is involved is in fact only a negligible microfragment once one returns to a magnification of × 10. It sometimes occurs, at the end of the excision, that an arteriole bleeds on the trunk of the facial nerve. Unless coagulation is very easy, it is better not to persist at the risk of damaging the nerve. The best course is to achieve hemostasis with a small fragment of muscle, held in place momentarily with a small swab to which suction is applied after stopping the irrigation. Hemostasis is usually easily so obtained. When the excision is completed, the program then turns to:

• *Inspection of the Excisional Cavity.* This is done under a magnification of × 10.

First, the cavity is washed out with a large quantity of warm saline (37°). Each of the swabs left in place to perfect the hemostasis is then gently removed while constantly moistening it with irrigation by the widely opened Brackmann cannula. If necessary, the hemostasis is completed by diathermy coagulation or tamponnage with a fragment of Surgicel. This last method is only suitable for venous hemorrhage. For arterial bleeding, if tamponnage is all that is available, it is better to use a muscle fragment, whose adhesive qualities are much better. There sometimes persists a minor oozing whose origin cannot be localized. The best course then is to fill the entire operative cavity with saline and then to suck gently while watching the margins for the appearance of the bleeding point, clearly identifiable when it occurs just at the fluid surface.

This stage of hemostasis is of cardinal importance, and must not be halted until the saline remains perfectly clear. Not to respect this imperative is to risk a postoperative hematoma in the excisional cavity whose consequences are most often dramatic. The tumoral stage may be remarkably simplified when the neurinoma is of small size.

In such cases, the CPA is not completely filled and, once the dura has been opened, it is possible to identify at a glance the various structures (acousticofacial bundle, mixed nerves, choroid plexus, anteroinferior cerebellar artery, brain-stem) which, in the cae of a bulkier tumor, can be perceived only after a long stage of evacuation and dissection. The tumor, covered by the arachnoid sheath of the acousticofacial bundle, seems very adherent to the latter, but careful dissection of this sheath gives a good view of the plane of cleavage. The dissection is always conducted very cautiously, and always with the fine 90° hook and the Brackmann cannula. Nevertheless, it is sometimes necessary somewhat to reduce the bulk of the tumor. It is then essential to steady the tumor well with the sucker held in the left hand, while one uses the vacuum rotary dissector in the other hand, since this maneuver may produce serious traction on the nerves and sometimes even a sudden mobilization which is the source of a definitive lesion. With this reduction effected, dissection of the nerve can be undertaken and pursued by the usual technique. The loop of the anteroinferior cerebellar artery is usually stuck to the tumor, but its freeing poses no great problem. The appearance and texture of the acousticofacial bundle are much better, but experience has shown that the dissection must be conducted with just as much gentleness and caution as if it were very laminated, especially since the small bulk of the tumor allows possibly more aggressive mobilization.

Closure of the ETL

Removal of Abdominal Fat. An incision, generally suprapubic, 8 to 10 cm long, is made down to the rectus aponeurosis and a quantity of adipose subcutaneous tissue the size of a hand is removed. Hemostatis of the resulting cavity must be scrupulous and a suction drain is inserted routinely to avoid the formation of a large hematoma, the source of local complications and fever. The sutures are removed at the same time as those of the ETL, at the 12th day.

Packing of the Mastoid Cavity. This packing is done after fragmentation of the block of fat into 3 or 4 portions. It must be complete, but not tight, and go beyond the bony cavity. There is no advantage in trying to construct a hammock in the depth with the dural residues or some form of synthetic material. This is one of the causes of escape of CSF. It is better for the deep fragment to penetrate partly at its end into the CPA like the cork of a champagne bottle.

Closure of the Auricular Flap. The auricular flap is folded back and the first step is to construct a deep and firm watertight muscular plane with a slowly-absorbed stitch, taking the muscular plane of the auricular flap with that of the temporalis muscle above and the sternocleidomastoid behind and below. The skin plane is closed with a continuous suture which allows good apposition and hemostasis of the margins. No drainage need be employed.

Dressing
A biogauze wick is inserted in the EAM and the concha. The operative field is then covered with sterile compresses, lined with a thick layer of cottonwool (American dressing) which acts both as an absorbent and a protector of the skin, especially the external ear, from the necessary pressure exerted by the bandage. The stitches are removed at the 12th postoperative day.

Variants

Modifications of the customary technique are required for anatomic variations, especially for venous and/or meningeal prolapses, or for special aspects of particular tumors.

Venous and Meningeal Prolapses
In 60% of cases, these structures are situated at a distance from the EAM and IAM, permitting a naturally wide route of access. At the surface, the temporal dura mater and the lateral sinus are very removed from the EAM (Fig. 72). In depth, the superior petrosal sinus and the jugular bulb are far from the IAM (Fig. 73).

In 40% of cases the prolapses, isolated or in combination, restrict access.

- Superficially, access is restricted in 25% of cases by a prolapse of the sigmoid sinus, whether or not associated with a prolapse of the temporal meninges (Fig. 72 B and C). The sigmoid sinus is regarded as prolapsed when it approaches to within 15 mm of the EAM. In 4 to 7% of cases, its first bend may even be directly in contact with the posterior wall of the EAM, leaving no natural space for performance of the ETL. That the superficial access is narrowed can be foreseen from Schiller's radiologic view, and better still in the sagittal tomograms of the middle ear and IAM which should always be made preoperatively. The existence of a restricted superficial space sensibly modifies the operative tactics. After performing the superficial bony stage with the craniotome the following procedures are performed in succession:

- Freeing of the temporal dura mater, its extensive stripping allowing the insertion of a retractor behind the arcuate eminence so as to place the operator, as it were, in a "suprapetrous position" and working from above downwards
- Resection of the upper aspect of the petrous to abrade a passage between the EAM and the bend of the sinus
- Thinning of the posterior wall of the EAM and exposure of the greater part of the sigmoid sinus
- Impaction of the sinus is now possible, thus opening up access towards the labyrinthine bloc

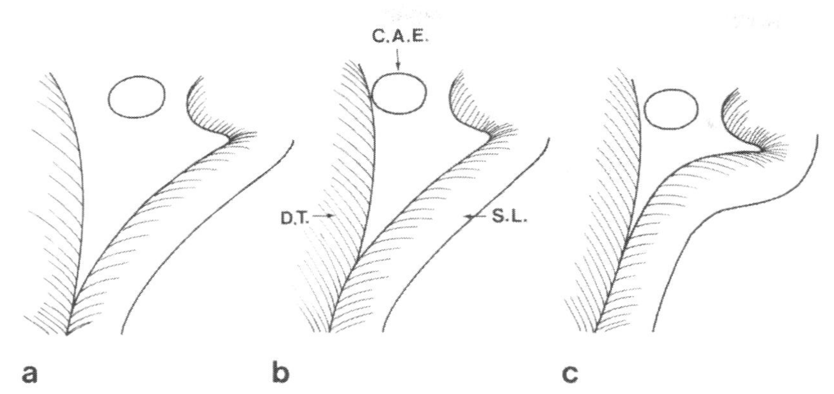

Fig. 72a–c. Extended right translabyrinthine route. Situation of temporal dura mater and bend of lateral sinus. **a** Normal situation leaving a wide superficial space between the sinus, dura and IAM **b** Prolapse of temporal dura mater which comes into contact with IAM and restricts the superficial space. Temporal dura mater *(D. T.)*, lateral sinus *(S. L.)*, internal acoustic meatus *(C. A. E.)*. **c** Prolapse of temporal dura mater and of lateral sinus, the bend of which approaches the IAM and further narrows the superficial space

in a virtually normal way. However, the lower part of the sigmoid sinus near the tip of the mastoid and under the digastric groove is not cleared or held back because of uncertainty as to the exact position of the mastoid part of the facial nerve and the frequent association of a major prolapse of the jugular bulb which may be situated in immediate contact with the nerve
- The performance of posterior tympanotomy then allows visualization of the second part of the facial canal and, in consequence, the safe skeletonization of the third part. From here on, the situation is now normal, with either normal deep access or a more or less pronounced prolapse of the jugular bulb which must be reduced.

- In depth, access is restricted in 30% of cases, either by a prolapse of the superior petrosal sinus (10% of cases) or of the jugular bulb (29% of cases). These are only rarely combined, for there is generally a balance: when one space is wide the other is narrow. By prolapse of the superior petrosal sinus, admittedly a rather inappropriate term, we refer to the closer position of the venous sinus, which sometimes comes into immediate contact with the upper aspect of the IAM and prevents any access to its roof. However, it is essential to free the entire dural dihedron at this site, not only to give adequate access towards the CPA but especially to free and preserve the facial nerve in the meatus. It is in these cases that the end of the flexible blade fixed to the Yasargil retractor plays an important part in holding back the dura of the upper aspect of the petrous and protecting the superior petrosal sinus from the burr. Should the sinus be opened, hemostasis is easily obtained by packing it with a strip of Surgicel applied superficially with a small swab. Prolapse of the jugular bulb (PJB), encountered by us in 29% of cases, poses trickier problems. Normally, there exists enough space between the inferior aspect of the IAM and the summit of the bulb to pass in front of the latter towards the cochlear aqueduct, free all the dura of this dihedron and gain good access to the CPA and mixed nerves (Figs. 65 and 74). Normally, this space varies between 1 and 14 mm, with an average of 6.5 mm (Fig. 73). Wide opening at this site is absolutely essential. To leave a PJB in place, even if minor, results in a very re-

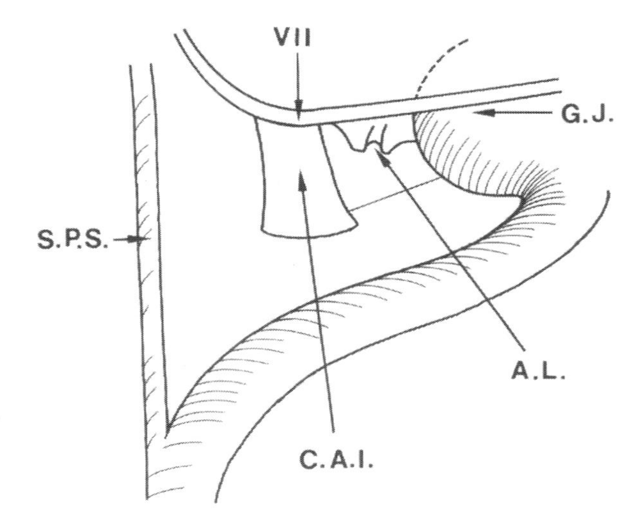

Fig. 73. Extended right translabyrinthine route. Normal position of superior petrosal sinus *(S. P. S.)* and jugular bulb *(G. J.)*. Facial nerve *(VII)*, internal acoustic meatus *(C. A. I.)*, cochlear aqueduct *(A. L.)*

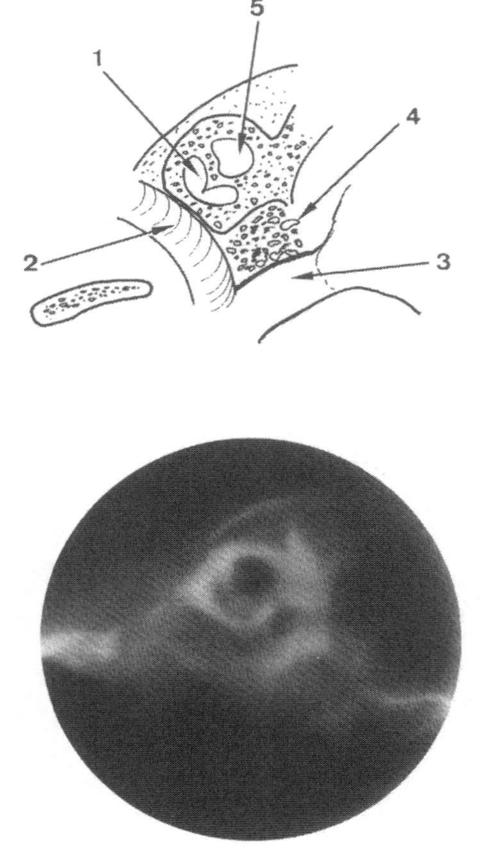

Fig. 74. Sagittal tomogram and diagram showing the normal position of the jugular bulb, at a distance from the IAM. Cochlea *(1)*, internal carotid *(2)*, jugular bulb *(3)*, cells separating jugular bulb from inferior aspect of IAM *(4)*, IAM *(5)*

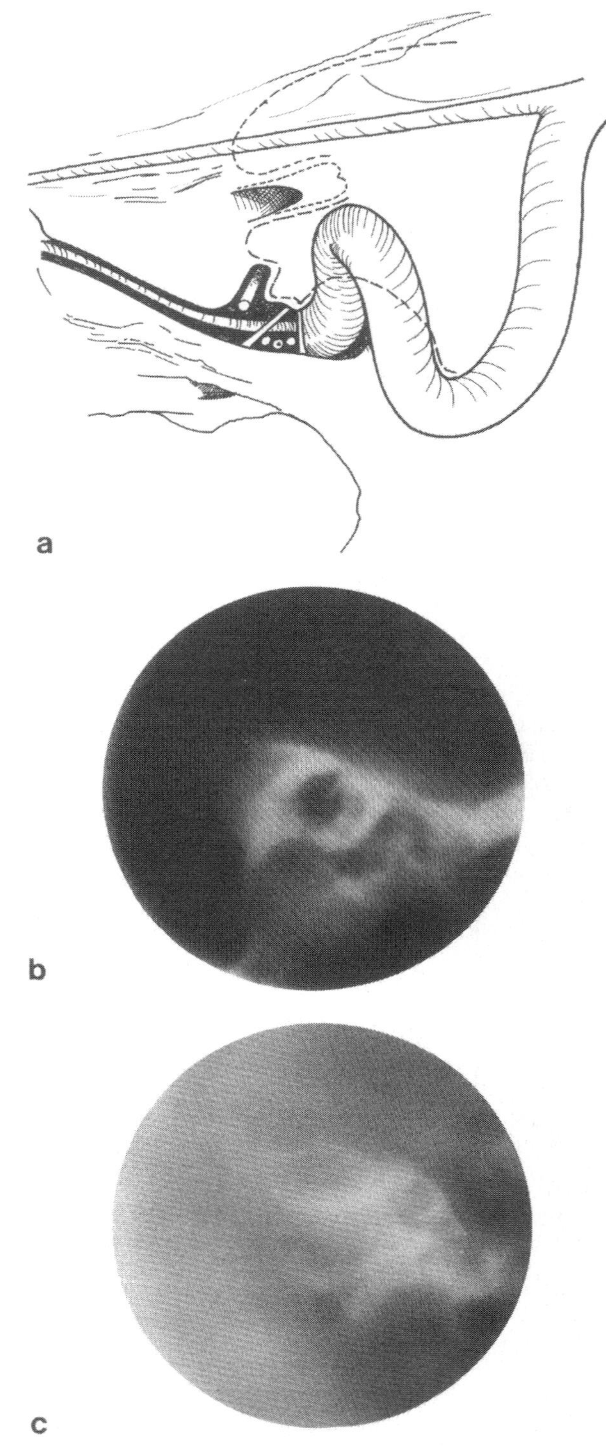

Fig. 75a–c. Prolapse of right jugular bulb. Here there is a simple contact of the dome of the bulb with the inferior aspect of the IAM. **a** It is evident that the jugular bulb must be depressed downwards to reach the region of the mixed nerves. **b** Lateral tomogram showing contact between the meatus and the bulb. **c** Note in this horizontal tomogram that the prolapse of the bulb is surrounded by a particular bony shell

stricted ETL approach without a proper view of the greater part of the CPA and only allows the safe excision of intracanalicular tumors (Fig. 75 A). It is therefore imperative to completely reduce these PJBs if one does not want to operate "through a keyhole". Of the 29 prolapses encountered in 100 successive ETLs, we can report that:

- in 16 cases there was simple contact of the summit of the bulb with the floor of the IAM (Fig. 75 and 76). Such cases pose few problems. It is enough to expose and free the bulb with a large diamond burr and then gradually push it back by the mediation of a fine bony film left in contact with it (what Brackmann calls "Bill's island"). As at the sigmoid sinus, this bony film should be imperceptible. It is depressed as the drilling proceeds and behaves like a barrier that itself maintains reduction of the PJB. After this there is normal access to the inferior aspect of the IAM, the cochlear aqueduct and the mixed nerves.

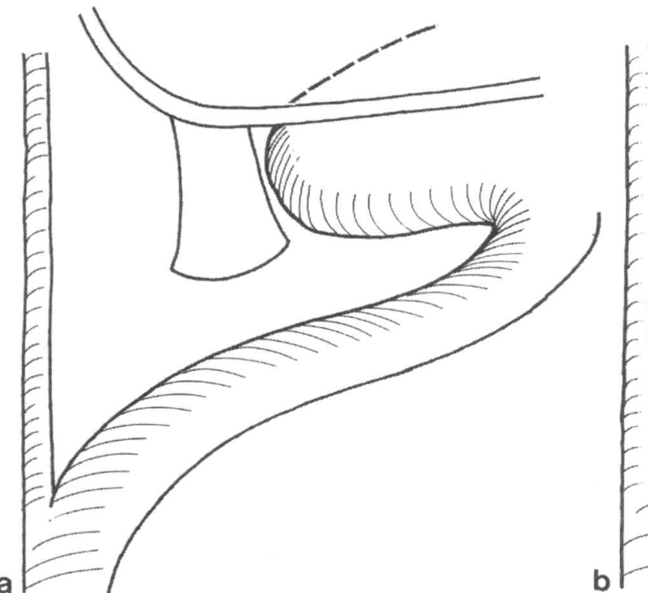

Fig. 76a-c. Prolapse of right jugular bulb. **a** Simple contact between inferior aspect of IAM and summit of bulb. **b** And **c** Major prolapses ascending behind the IAM to mask it partly or completely

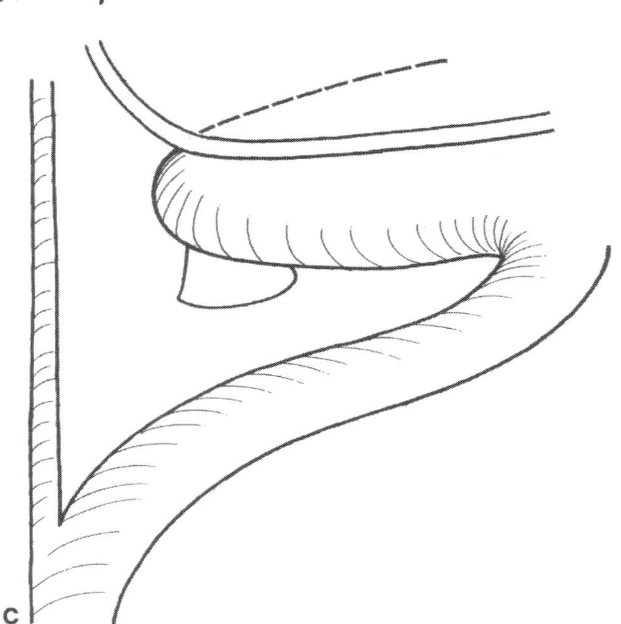

- In 13 cases, the PJB was large, ascending behind the IAM, covering it partly in 8 cases and completely in 5 cases; in one of the latter the summit of the bulb was even in contact with the superior petrosal sinus (Fig. 76). The narrower the superficial space between the sigmoid sinus and the EAM, the greater the probability of a PJB: 20% if this space is wide, 50% if it is narrow and 66% when the space is rendered only potential by direct application of the bend of the sinus to the posterior wall of the EAM. It is quite obvious that access to the IAM is impossible by the suboccipital route in such cases, whereas it is possible by adapting the ETL. Moreover, these prolapsed bulbs come to lie under the second and third parts of the facial nerve (Fig. 77). It is essential that they be diagnosed preoperatively, which is always possible in lateral tomograms of the IAM (Fig. 78) and sometimes, when very large, in the scan (Fig. 79). If preoperative diagnosis is not made, there is every chance of penetrating the venous lumen after abrasion of the mastoid cells under the facial nerve; packing the PJB with Surgicel as required for the abundant hemorrhage makes subsequent displacement very difficult or even impossible. Therefore, the existence of a large prolapse modifies the operative tactics, especially as there will be every chance of encountering a space that is narrow superficially. In such cases, one must be prepared for a longer and more delicate phase of access, and here a high-performance motor proves very useful. The successive procedures are:

- To obtain the widest possible superficial access using the technique described above
- To perform a good posterior tympanotomy, clearing the second and third parts of the facial nerve as closely as possible, these being in contact with the outer aspect of the PJB, as well as the digastric

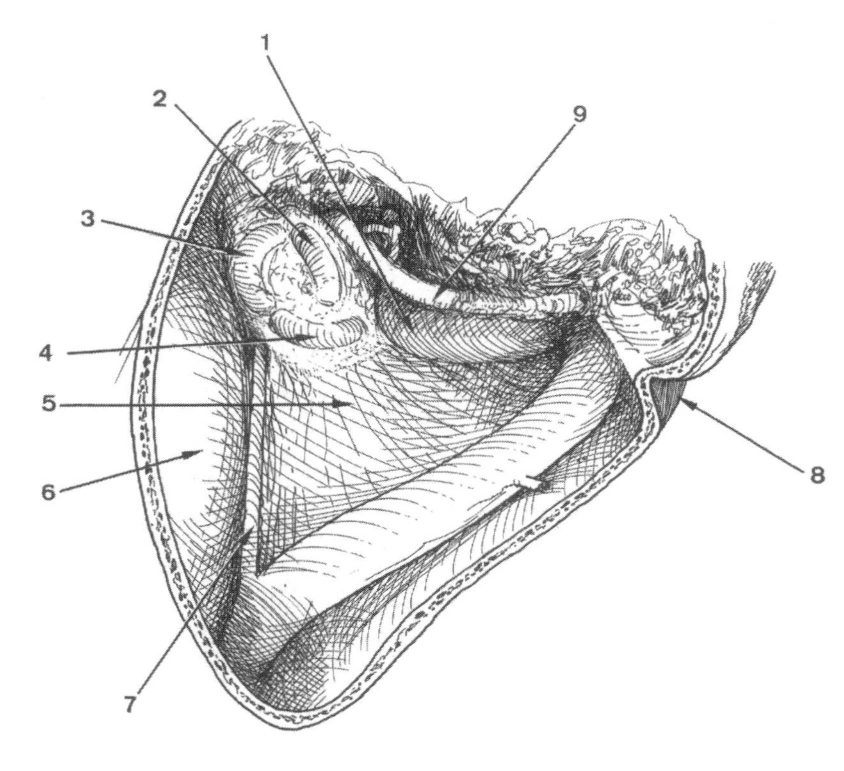

Fig. 77. Extended right translabyrinthine route. Position of jugular bulb in relation to facial nerve where a major prolapse partly covers the nerve. The bulb comes to lie immediately under the internal aspect of the tympanic cabity and under the 2nd and 3rd parts of the facial nerve. Jugular bulb *(1)*, lateral SCC *(2)*, anterior SCC *(3)*, posterior SCC *(4)*, dura mater at posterior face of petrous *(5)*, temporal dura mater *(6L)*, superior petrosal sinus *(7)*, digastric groove *(8)*, facial nerve *(9)*

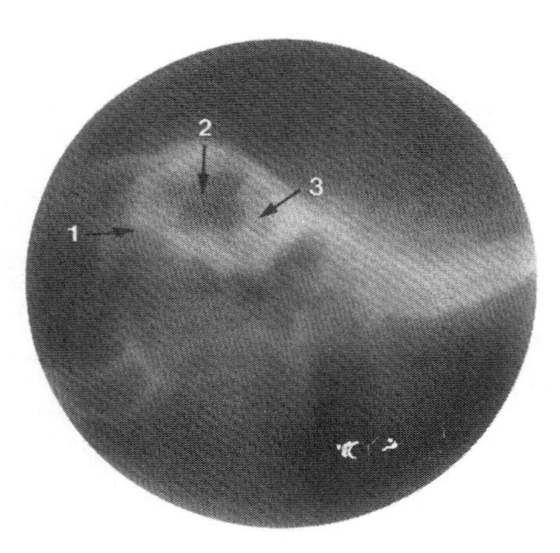

Fig. 78. Tomogram of petrous (lateral). Major prolapse of jugular bulb partly covering posterior aspect of IAM. Summit of jugular bulb *(1)*, IAM *(2)*, cochlea *(3)*

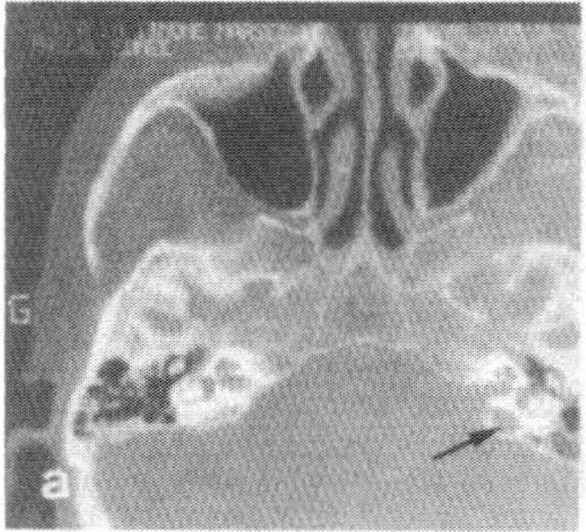

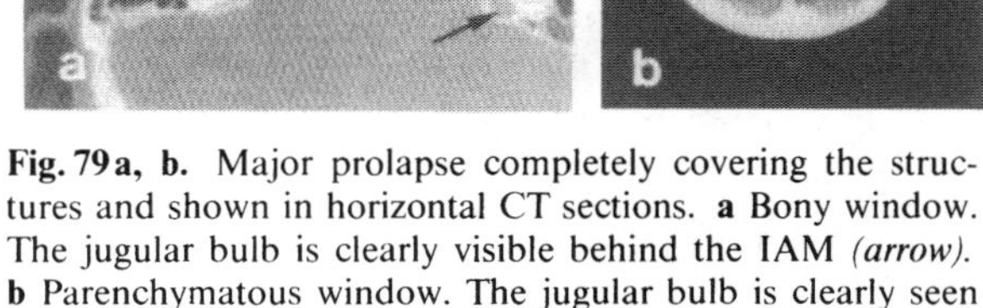

Fig. 79 a, b. Major prolapse completely covering the structures and shown in horizontal CT sections. **a** Bony window. The jugular bulb is clearly visible behind the IAM *(arrow)*. **b** Parenchymatous window. The jugular bulb is clearly seen surrounded by tumor

groove, and to completely clear the sigmoid sinus up to its junction with the lower part of the PJB

- To drill the labyrinthine nucleus and identify the fundus of the IAM, then to start the pericanalicular stage at its upper part by widely clearing the upper aspect of the IAM
- To work subsequently from above downwards, freeing as widely as possible the entire bony shell surrounding the PJB, particularly between it and the facial nerve superficially and the dura of the posterior fossa in depth. The presence of the dura mater protects the posterior fossa from hemorrhage due to rupture of the bulb, a common accident. Here again, the ETL has an immense advantage. At the summit of the prolapse, a very fine bony lozenge is left in place (Figs. 80 and 81)

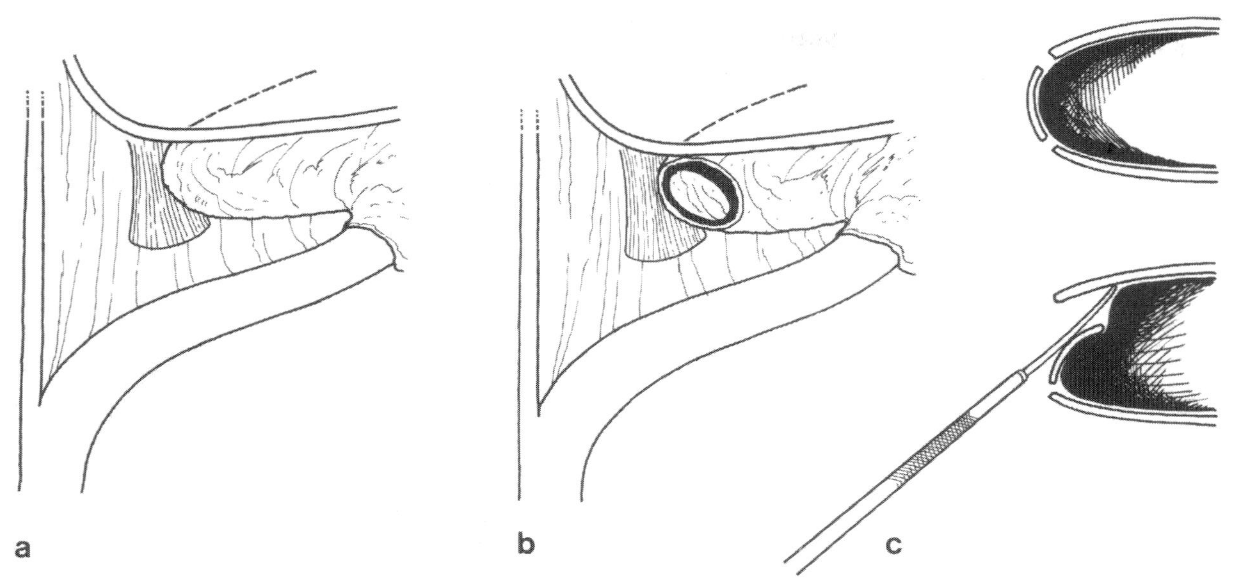

Fig. 80a–c. Extended right translabyrinthine route. Reduction of a major prolapse of jugular bulb. **a** Appearance of prolapse partly covering the IAM **b** Liberation of small bony flap at summit of prolapse. **c** Separation of lateral sinus from its bony wall with blunt dissector

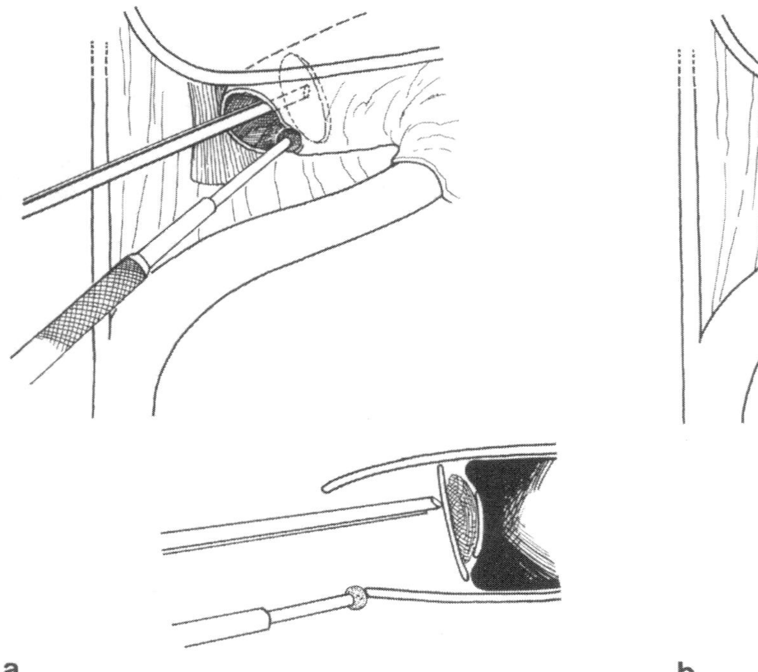

Fig. 81a, b. Extended right translabyrinthine route. Reduction of a major prolapse of jugular bulb. **a** By means of a fine interposed bony lozenge, the freeing of the lateral sinus and the burring of the surrounding bony cylinder around it are progressively completed. A small protective Surgicel pack is placed between the bony lozenge and the bony islet. **b** After its complete reduction, the prolapse is blocked by a larger bony lozenge

– Finally, to practise the freeing and depression of the PJB. This maneuver is easy when the PJB is narrow and spread over the inner aspect of the tympanic cavity. Using Brackmann's suction, one presses against the protective bony lozenge and progressively strips the vein wall from the bony box that still covers it. The stripping is ended when one reaches the horizontal part of the sinus below, always clearly visible in front. It then suffices to maintain the reduction by means of some fragments of Surgicel covered by a fine bony lozenge, removed during the freeing of the temporal dura mater and slipped under the anterior border so as to give wide access. Usually, the cohesion of this arrangement is ensured by fibrinogen glue placed in the Surgicel and around the bony lozenge. It remains only to burr the rest of the bony box covering the prolapse to find oneself under normal conditions, with wide access under the IAM. When the prolapse is large, it is impossible to see under the tympanic cavity for safe stripping of the vein wall. It is then necessary to prepare in advance some small fragments of Surgicel and a bony lozenge adapted to the opening of the bony box. Holding back the vein initially poses no problem at the anterior part, nor at the internal (towards the dura) and posterior parts, but the vein wall is habitually torn as soon as one strips under the facial nerve towards the internal aspect of the tympanic cavity. The sinus should not then be packed with a long wick of Surgicel, which would make it impossible to press it backwards. The small fragments of Surgicel already prepared are applied one by one to the bleeding region with the aid of a swab of moderate size. Once hemostasis is obtained, it is supplemented with fibrinogen glue and then the bony lozenge is applied overall. There is no advantage in further stripping behind and at the inner aspect of the tympanic cavity. Pressing on the bony lozenge, the stripping of the remaining bony box is completed forward, where there is a clear view of the inner wall of this box, until one reaches the lower part, marked by a ledge beneath which the anterior part of the lozenge is slipped. The fibrinogen glue ensures that reduction of the PJB is maintained, and one is now in the same situation as before. More exceptionally, the PJB is very large, extending behind the labyrinthine nucleus as far as contact with the superior petrosal sinus. It is then necessary to begin by freeing its entire posterior aspect, then its outer aspect all around the facial nerve. Next, still with the help of a fine bony lozenge placed near its summit, supplemented by Surgicel and fibrinogen glue, the prolapse is partly held back so as to clear the labyrinthine nucleus. It is then possible to proceed as before.

Combination of the ETL with a Suboccipital Route. Certain tumors of the CPA exhibit adhesions around the posterior lacerate foramen or at the petrous apex, medial to the aperture of the IAM. These are rarely neurinomas of the mixed nerves, more usually meningiomas. Others, usually bulky acoustic neurinomas or meningiomas, are wedged in the most lateral part of the posterior fossa. If such tumors are to be completely excised, it is necessary to simultaneously perform a classical suboccipital approach. For this, it suffices to enlarge the bony opening with a gouge to expose the occipital dura mater behind the sigmoid sinus and then to open the posterior fossa. In these cases, major reduction of the tumoral volume, already performed by the ETL route, simplifies the suboccipital approach for traction or pressure of the cerebellum is no longer needed to expose the tumor. At the end of the operation, the dura is sutured continuously to leave it water-tight (Fig. 82).

The Retrolabyrinthine Route

This consists of resecting the posterolateral segment of the petrous pyramid to gain access to the outer part of the CPA by working in front of the sigmoid sinus, while respecting the labyrinthine massif which is exposed but not opened. This is, in fact, the first, extralabyrinthine, stage of the ETL route. The preparation, theater set-up and operative stages remain identical but the burring ceases at the outer wall of the labyrinthine bloc. This exposes the dura mater clothing the posterior aspect of the petrous in the triangular space bounded externally by the vertical portion of the sigmoid sinus, above by the superior superior petrosal sinus and internally by the labyrinthine massif. The internal limit of the dural exposure is situated at the outer border of the endolymphatic sac. To respect this, the dura is opened by cutting a flap hinged medially (Fig. 83), to be carefully closed by a fine continuous suture at the end of the operation. The apparent analogy with the start of the ETL route means that the points of difference must be clearly defined. First, it should be stressed that during this approach one must take particular care not to open the cavities of the posterior labyrinth, particularly the summit of the loop of the posterior SCC, as the prime object is to safeguard hearing, which, in cases calling for this access route, is usually intact. Despite this fear, drilling should be taken as far as possible so as to give per-

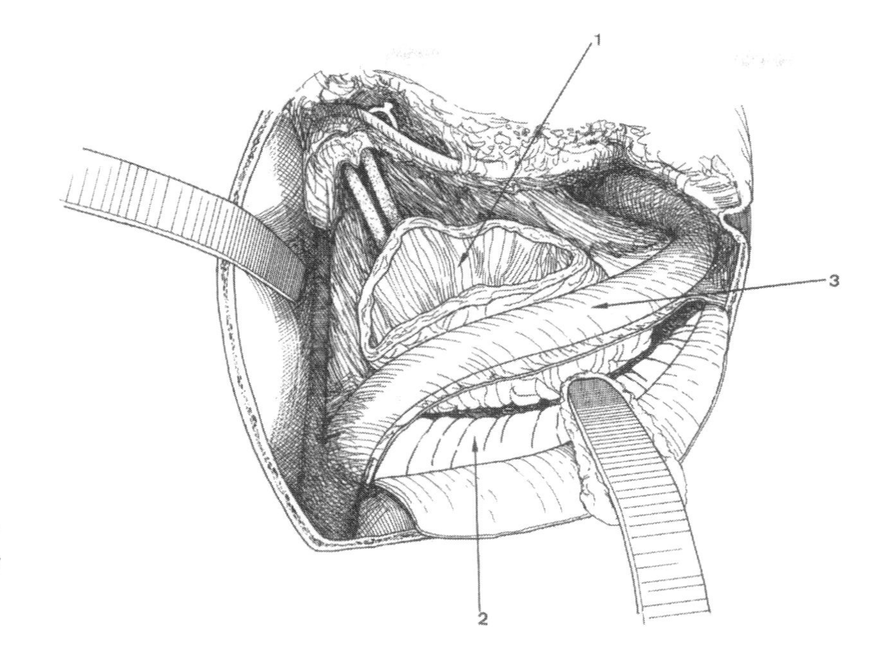

Fig. 82. Extended right translabyrinthine route associated with opening of the dura mater behind the sinus. Partly evacuated meningioma *(1)*, cerebellum *(2)*, sigmoid sinus *(3)*

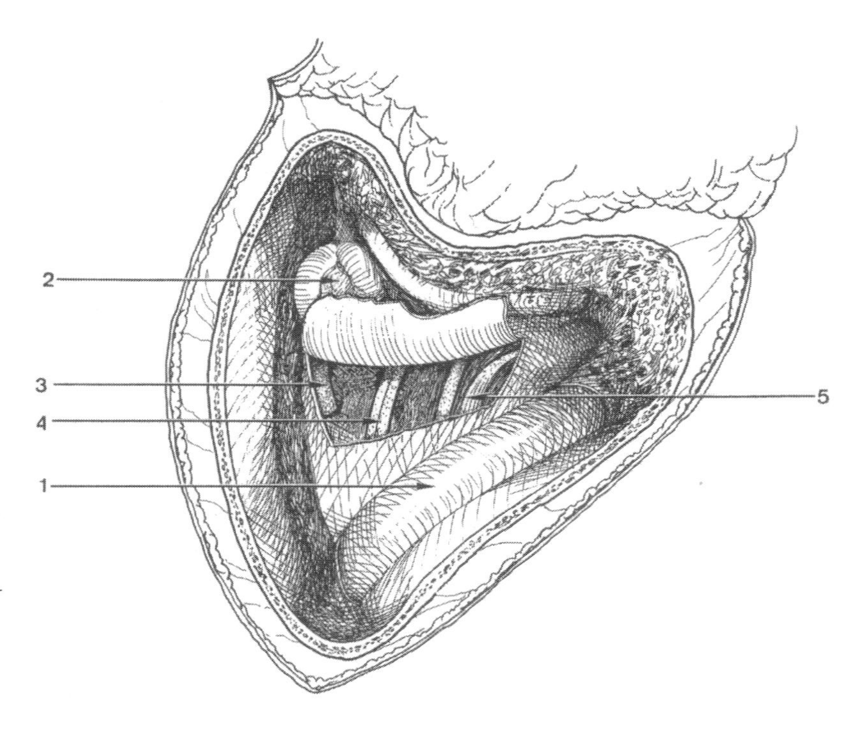

Fig. 83. Right retrolabyrinthine route. The mastoidectomy is done as in the bony extralabyrinthine stage of the extended translabyrinthine approach, then the dura mater of the posterior fossa is opened between the sigmoid sinus *(1)*, and the labyrinthine block *(2)*. Through the opening are seen the trigeminal nerve *(3)*, the acousticofacial bundle *(4)*, and the mixed nerves *(5)*

fect exposure of the dura and provide the best view of the outer part of the CPA. Finally, since hearing is preserved, it is not possible to pack the auditory tube and tympanic cavity. Particular care must be given to avoiding any leakage. The aditus ad antrum is closed with a fragment of aponeurosis stuck around the margins of the orifice, and then the entire bony trench is layered with bone-dust mixed with fibrinogen glue. The dura is carefully sutured and finally the cavity of the mastoid eradication is filled with fragments of fatty tissue. This approach exposes the outer part of the CPA, permitting biopsy of a tumor at this site. However, such a procedure is not commonly required. This approach has the special advantage of giving an excellent exposure of the vertical portion of the sigmoid sinus and adjacent dura. It thus permits a proper excision of lesions at this site (meningioma, cavernous angioma ... see Figs. 127 and 130 in the subsequent section).

The Transcochlear Routes

In addition to the two posterior segments, these allow for resection of the anteromedial segment so as to completely expose the dura mater of the posterior aspect of the petrous, thus providing an approach to well-developed tumors, especially if embedded in front of and internal to the aperture of the IAM. This near-total petrectomy completely clears the facial nerve, the three segments of which, once extracted from the facial canal, can be mobilized upwards, the more easily since straightening of its two angles gives plenty of length: this is the rerouting of the facial nerve. In order to preserve impermeability, House and Hitselberger [28] preserve the skeleton of the EAM and tympanic membrane. Since 1972 [39] we have practised this approach while completely resecting the EAM, so necessitating obstruction of the latter. This technique gives a much wider superficial exposure and is the one we shall describe. The start of the operation closely resembles the ETL route, but is followed by the properly so-called transcochlear stages.

Initial Stages

The preparation of the patient, theater set-up, preparation of the operative field and instrumentation are identical with those for the ETL.

Skin Incision. This is performed as for ETL. It is simply extended at its two extremities, above the ear towards the temporal region and over the mastoid towards the cervical region (Fig. 84).

Exposure of the Operative Field. For House [28] this is identical to the ETL. It is preferable, after having stripped the musculo-aponeurotic plane in one piece as for the ETL approach, to divide the EAM at its osteocartilaginous junction. This perfectly exposes the zygoma, the anterior temporal region, the tympanal part of the temporal bone and the mandibular condyle (Fig. 85). The meatus of the EAM is obliterated at the outset as indicated in the figure. The destruction of the bony EAM and the occlusion of the cartilaginous meatus leave no disadvantages, save for the absence of the external meatus which is a curiosity rather than a true cosmetic defect. On the other hand, the performance of the route of access and the tumoral excision are greatly facilitated by it.

The Bony Translabyrinthine Stage. The ETL is performed as for an acoustic neurinoma, with exposure of the dura of the temporal bone and posterior

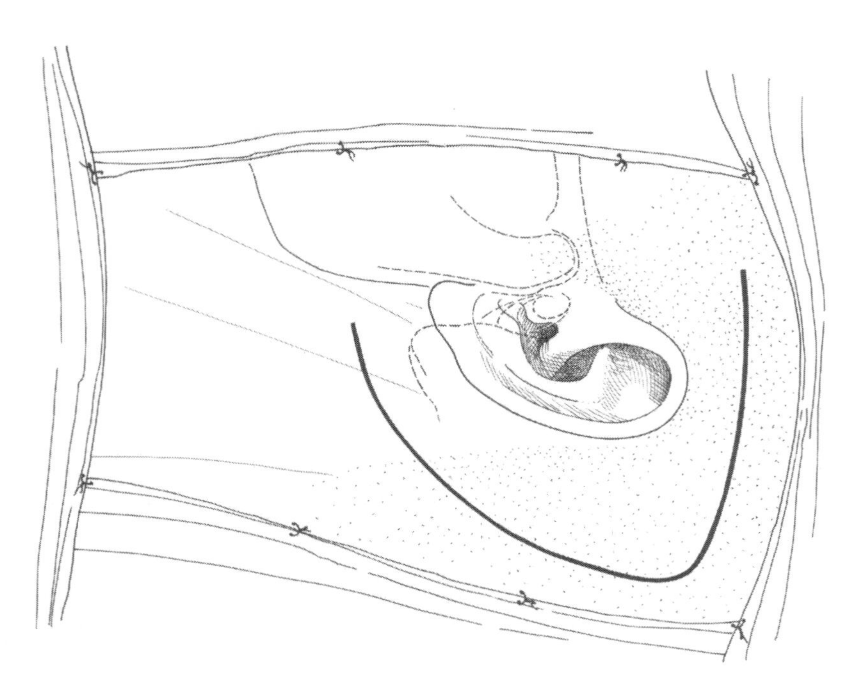

Fig. 84. Left transcochlear route. Line of skin incision. Note that it extends in front of the EAM above. The cervical region is exposed in case a hypoglossal-facial anatomosis is necessary

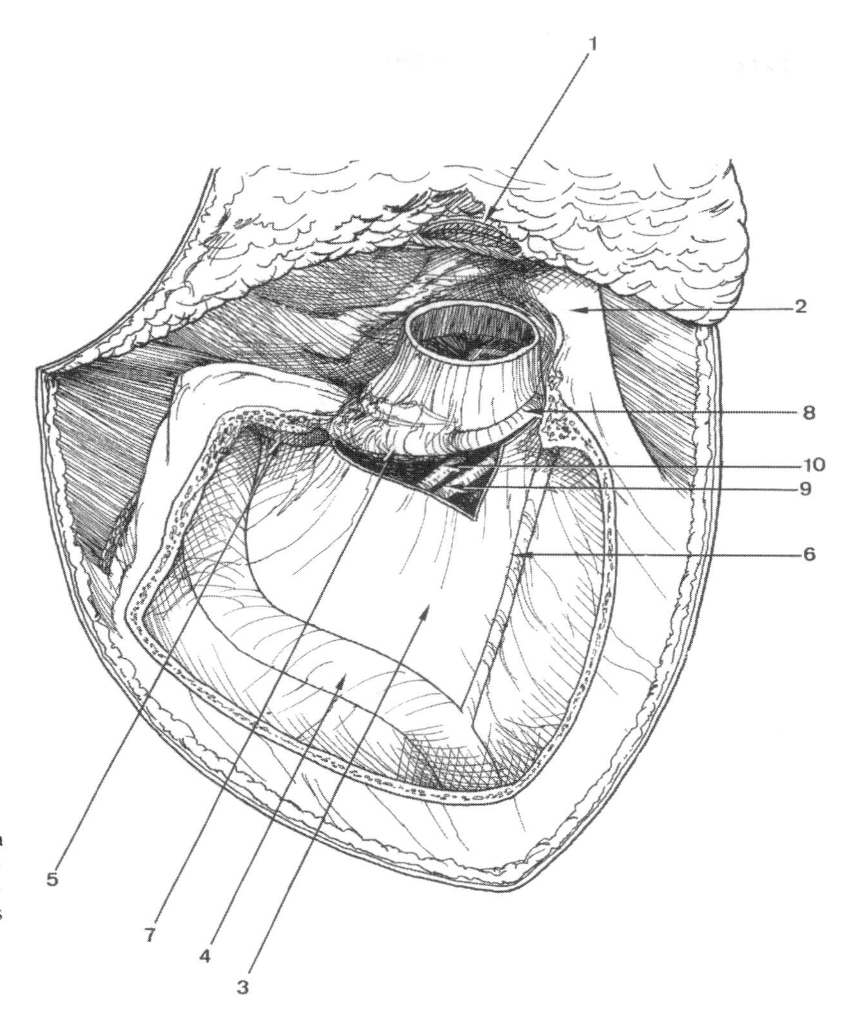

Fig. 85. Left transcochlear route. The membranous EAM has been resected and its outer orifice *(1)*, sutured. The posterior part of the zygoma *(2)*, is exposed. The extended translabyrinthine route has been made, showing the dura mater of the posterior aspect of the petrous *(3)*, the sigmoid sinus *(4)*, the jugular bulb *(5)*, the superior petrosal sinus *(6)*, the facial nerve, 3rd part *(7)*, 2nd part *(8)* intracanalicular part *(9)* and cochlear nerve *(10)*

fossa as well as the structures of the IAM. However, the bony excision towards the tip of the mastoid and the region of the jugular bulb is more complete (Fig. 85).

The Transcochlear Stages

Ablation of the External Acoustic Meatus and Skeletonization of the Facial Nerve. The entire tympanal portion of the temporal bone is excised, opening access to the posterior lacerate foramen and the internal carotid. At the same time, the EAM is excised and the facial nerve skeletonized from the stylomastoid foramen to the IAM (Fig. 86). It is also necessary to drill the root of the zygoma for good visualization of the upper part of the tympanic cavity. After section of the greater superficial petrosal nerve, the facial nerve is dislodged from the facial canal and displaced backwards (Fig. 87). This type of rerouting completely clears the operative field and facilitates exposure of the jugular bulb and mixed nerves.

The Bony Cochlear Stage. The cochlea is opened by drilling from its windows and carrying the bony ablation towards the tip of the petrous and the region of the bulb. In front and somewhat overhanging it, one exposes the internal carotid which has a bluish appearance seen through the bone. The arterial coat is very thick and no risk is incurred by drilling with the diamond burr up to contact with it. Drilling is then done medial to the carotid in the direction of the inferior petrosal sinus, which, anatomically, is lower down and more internal. This opens adequate access towards the clivus and especially exposes the external oculomotor nerve at the lower part of the route of access (Fig. 89). This space is likely to house the petrosal extensions of large neurinomas of the mixed nerves.

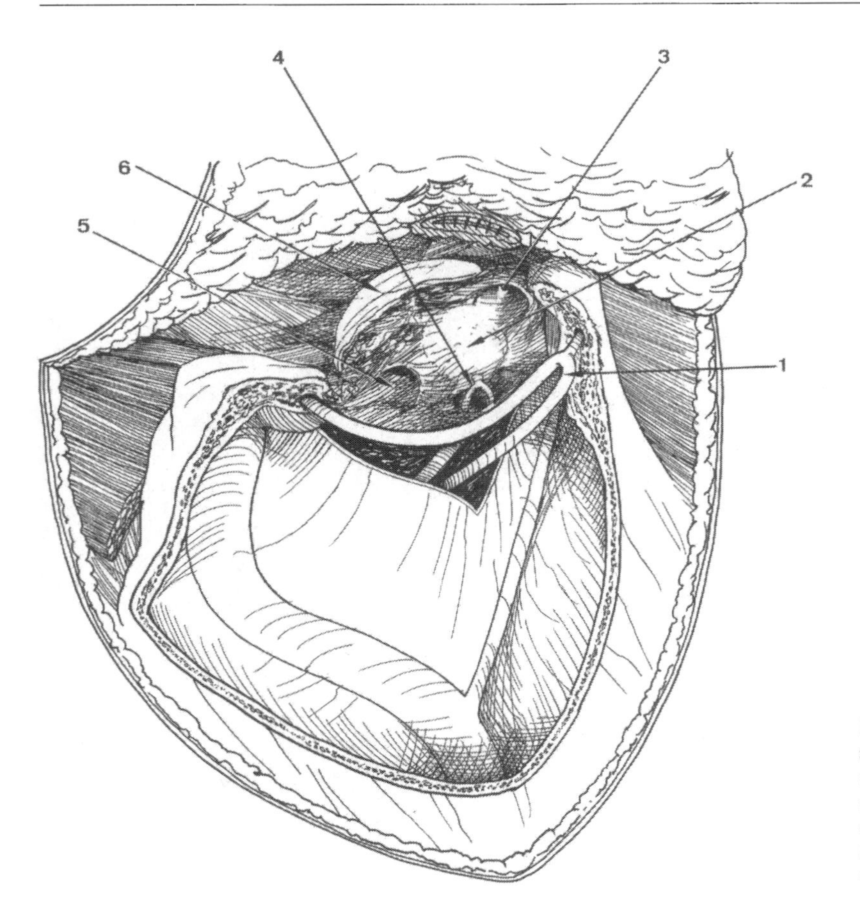

Fig. 86. Left transcochlear route. Resection of bony external acoustic meatus and freeing of entire facial nerve. Internal wall of tympanic cavity exposed *(1)*, promontory *(2)*, orifice of auditory tube *(3)*, stapes *(4)*, cochlear window *(5)*, remains of tympanum *(6)*

The Tumoral (Neurosurgical) Stage. At the end of the bony stage, the entire dural dihedron clothing the upper and posterior aspects is exposed (Fig. 87). In dealing with an intrapetrous, extradural tumor (primary cholesteatoma of the petrous, tumor of the glomus jugulare, intrapetrous meningioma, etc.), the stage of excision takes place within this dural dihedron. If, on the other hand, it is a matter of a petrosal tumor extending into the angle (jugular glomus tumor, neurinoma of the mixed nerves), or a tumor of the anterior part of the CPA, the dura must be opened. For tumors of the petrosal apex, the dural incision is made from behind forward, starting from the IAM, and without risk since there are no subjacent neural structures. If the tumor has developed around or in the IAM, the incision must begin outside this and be extended medially towards the apex. It is then necessary to identify the facial nerve so as not to damage it. The exposure of the brain-stem and structures of the CPA is more anterior and direct than by simple ETL (Fig. 89). When the tumoral excision so requires, the region of the mixed nerves can be perfectly exposed between the internal carotid in front and the jugular bulb behind (Fig. 88). The mixed nerves pass over the posterior bony margin of the inferior orifice of the posterior lacerate canal. Drilling in this region must be particularly careful to avoid damaging them against the bone. If the tumor is small, the neurovascular structures are remarkably well exposed by this approach, particularly the basilar trunk and its branches (Fig. 89), the abducent nerve emerging from the brain-stem and disappearing in front under the dura of the clivus, and the mixed nerves below. When the tumor is bulky, all these structures are masked by it and the same principles of excision employed for the ETL route are applicable here: identification and dissection of the facial nerve in the first place, then of the mixed nerves, preservation of the arachnoid layer of the various cisterns, progressive fragmentation and excision of the tumor with the Urban House vacuum dissector while monitoring any changes in the pulse, arterial pressure, respiratory rhythm and facial EMG. Hemostasis must of course be perfect at the end of operation. The external cutaneous orifice of the EAM has already been closed (Fig. 106). After blocking what remains of the auditory tube in

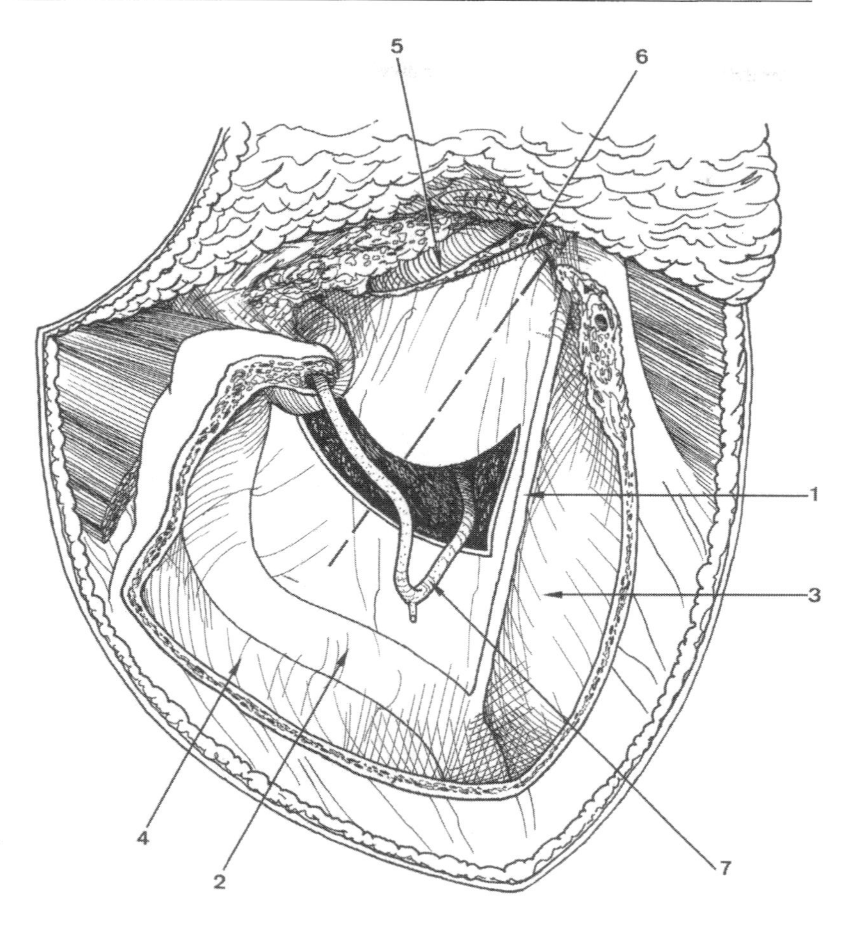

Fig. 87. Left transcochlear route. The bony stage is complete. Opening of dura mater of posterior aspect of petrous. Superior petrosal sinus *(1)*, sigmoid sinus *(2)*, dura mater of upper aspect of petrous *(3)*, occipital dura behind sigmoid sinus *(4)*, intrapetrous internal carotid *(5)*, inferior petrosal sinus *(6)*, facial nerve *(7)*

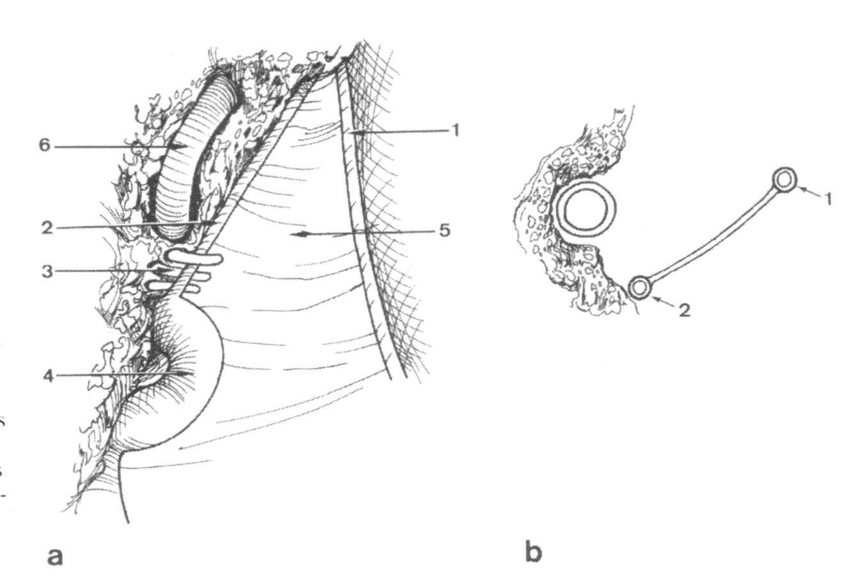

Fig. 88 a, b. Left transcochlear route. Burring towards apex of petrous to free the intrapetrous internal carotid (6).
a Diagram of operative view. Superior petrosal sinus *(1)*, inferior petrosal sinus *(2)*, mixed nerves *(3)*, jugular bulb *(4)*, dura mater of posterior face of petrous *(5)*. **b** Frontal section of operative cavity. Superior petrosal sinus *(1)*, inferior petrosal sinus *(2)*

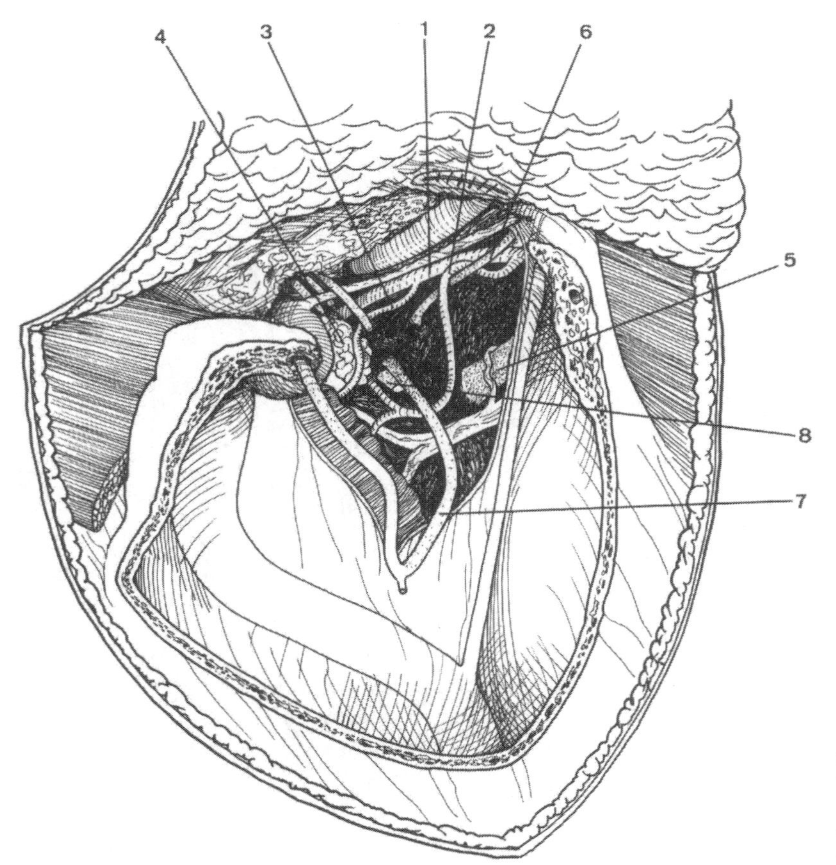

Fig. 89. Left transcochlear route. Exposure of cerebellopontine angle. Basilar trunk *(1)*, AICA *(2)*, PICA *(3)*, choroid plexus *(4)*, trigeminal nerve *(5)*, external oculomotor nerve *(6)*, facial nerve *(7)*, cochlear nerve *(8)*

front with a muscle fragment, the operative cavity is filled with abdominal fat and completion is as for ETL.

Variant

This is the transotic route. In 1980, Fish and Jenkins [30] proposed a route of access, modified in 1983 with Gantz [19], designed to approach acoustic neurinomas by passing across the cochlea. This route, which they called "transotic", is very nearly comparable to the transcochlear route, with the difference that the second and third parts of the facial nerve are skeletonized without rerouting in the hope of avoiding the facial palsies which are not infrequent after operative handling of the nerve. The drawback of this technique is that the skeletonized nerve constitutes a permanent obstacle in the midst of the approach throughout the operation. Further, the magnitude of the approach seems unjustified for acoustic neurinomas, which can all be removed by the ETL route.

The Suprapetrous Routes

The aim of these is to expose the superior aspect of the petrous by an extradural route, as was done from the beginning of the century by neurosurgeons intending to divide the root of the trigeminal ('the Spiller-Frazier operation'), and then, with this exposure made, to drill above the IAM so as to approach this at its roof. It is to House [25] again that we owe the first account of this approach to the contents of the IAM via the middle cranial fossa. Necessarily conducted under the operating microscope, this technique must follow precise rules. It is essential to respect the facial nerve and the cavities of the inner ear embedded under the upper aspect of the petrous; and the drilling of the latter must be rigorously conducted with reference to the only really precise landmark, the loop of the anterior SCC, which the otologist must learn to discover by transparency while cautiously drilling the internal slope of the arcuate eminence under continuous irrigation. This perception of the "blue line" is an essential stage in this operation, and our own view is that any other method is hazardous, particularly if it marks out the drilling by various axes and measurements which can never be considered as having real stereotactic precision during operation. Otologists currently perform this operation for section of the vestibular nerves, but strictly intracanalicular procedure must be considerably extended and must also include opening of the meatal aperture when the aim of the operation is to extirpate a tumor arising in the meatus, even if this remains more or less contained within the limits of the meatus and certainly if the tumor extends into the CPA. This is why, as well as the classic suprapetrous route adapted to vestibular neurectomy, it is necessary to describe the extended supratrous route recently recommended by Wigand [56] for rather bulky neurinomas.

The preparation and arrangements are very comparable to those used for the transpetrous routes: the same skin preparation, dorsal decubitus, complete but unforced rotation of the head so as to present the temporo-auricular region approached to the surgeon's direct vision, the biauricular axis orthogonal to the plane of the table. Should this cervical rotation be limited, it must not be forced but compensated by slight lateral tilt of the table. Three slight but important differences are to be stressed:

- The surgeon stands at the end of the operating-table, not at the side but at the head of the patient, with the microscope on his left and the assistant on his right
- The cervical rotation must be accompanied by slight flexion of the head on the trunk, either by means of a head-piece or by placing one or two drapes under the back of the neck (Fig. 90). This makes it possible to orientate the upper aspect of the petrous approached as nearly as possible opposite to the surgeon. Of course, the pre- and supra-auricular temporal region is shaved over four finger-breadths.
- Some longer burrs of 10 cm must be added to the usual instruments, and some surgeons will prefer a self-retaining retractor adapted to this approach, of which there are several models [15, 29, 54]. We ourselves remain faithful to Yasargil's self-retaining retractor.

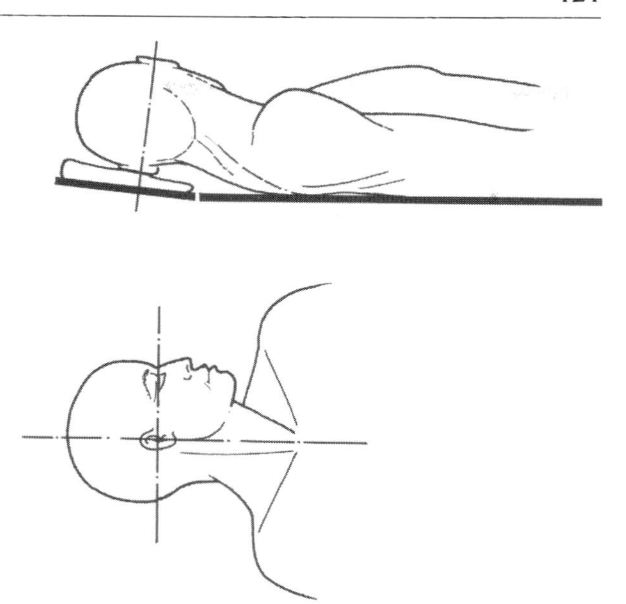

Fig. 90. Right suprapetrous route. Position of head, slightly flexed on trunk and in complete rotation

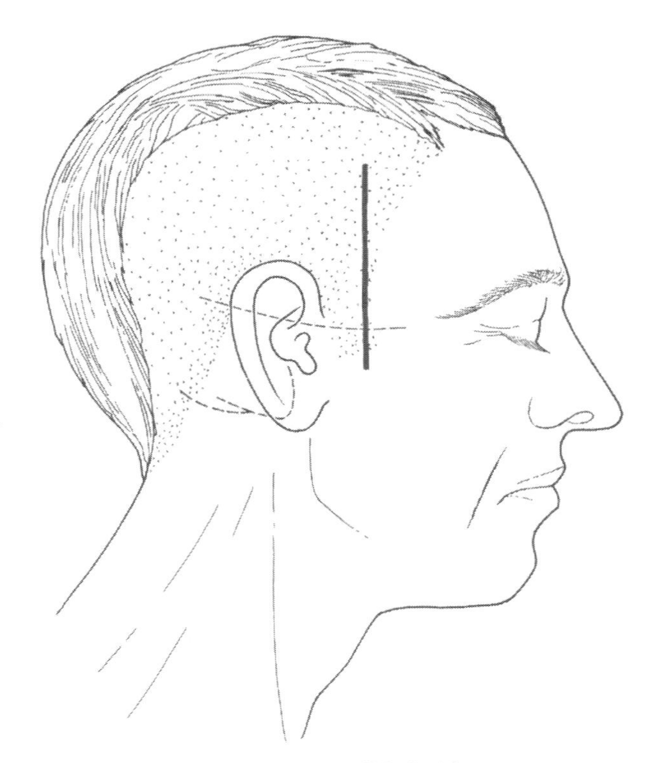

Fig. 91. Right suprapetrous route. Skin incision

Exposure

Skin Incision

This is vertical and 8 cm in length, and made 1 cm in front of the tragus. In its lower part it descends below the zygoma to reach the base of the ear-lobe (Fig. 91). A self-retaining retractor put in place widely exposes the plane of the temporalis muscle and its aponeurosis. A large graft is taken from this temporal aponeurosis as well as fragments of the pad of fat at the anterior part of the operative field.

A T-shaped incision is then made in the temporalis muscle. Its transverse limb divides the temporal aponeurosis at the level of the zygoma, which must be widely cleared, without going in front beyond the middle of the line from tragus to eyebrow which marks the course of the frontal branch of the facial nerve, while behind it reaches to just above the EAM. The vertical limb of the T is situated on the projection of the skin incision. The temporalis muscle is detached. The temporal squama is then widely exposed by a single self-retaining retractor.

Craniotomy

With a 2 mm cutting burr, a rectangular flap 4 cm by 4 cm is made, centered at the level of the skin incision and situated as low as possible. Its lower border must be at the level of the middle cranial fossa and its anterior two-thirds must be in front of the EAM. Any cells that have been opened at the root of the zygoma are packed with Horsley's wax (Fig. 92). It is often necessary to use the nibblers to enlarge the craniotomy below and in front to gain good access to the region of the middle meningeal artery and the petrosal nerve.

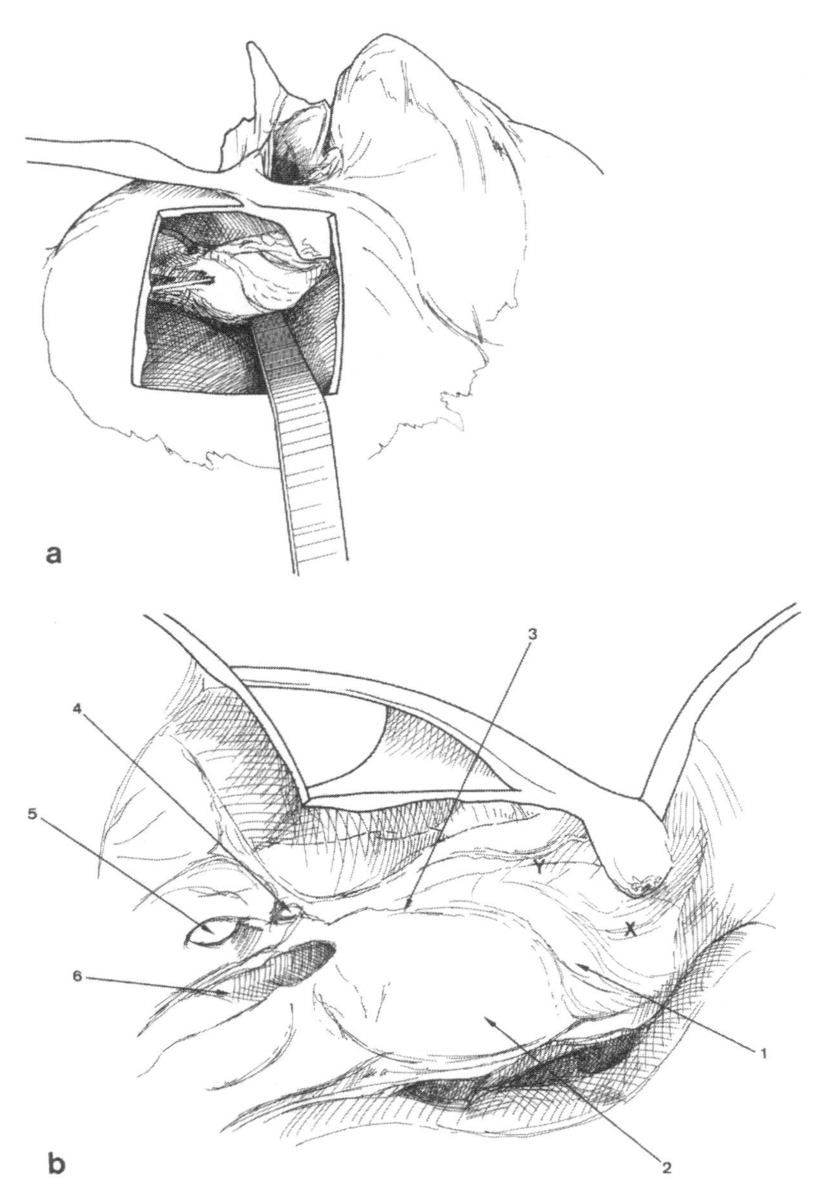

Fig. 92 a, b. Right suprapetrous route. **a** The bony flap and exposure of superior aspect of petrous. The tip of the retractor blade catches the petrous crest. **b** Close-up view of upper aspect of petrous. Arcuate eminence *(1)*, meatal field *(2)*, hiatus of canal of petrosal nerve *(3)*, spinous foramen *(4)*, oval foramen *(5)*, carotid canal and anterior lacerate foramen (foramen lacerum) *(6)*. Zone *XL* supralabyrinthine cells. Zone *Y* gives access to attic

Stripping of the Dura Mater

This must be scrupulous, with particular attention to hemostasis. The least bleeding risks to impair the subsequent stage of bony drilling. Osmotic diuresis is not necessary. Slightly negative respiratory pressures are sufficient. One begins by making a 1 cm incision in the exposed dura to empty the CSF and relax the temporal lobe. The stripping must always be extensive. It starts behind by seeking the posterior border of the petrous and of the groove for the superior petrosal sinus. This posterior border is followed forward along its axis, angled by 45° towards the patient's nose. This axis is fundamental, the key to the suprapetrous route and the essential landmark from which the other structures are identified. First, one reaches the convexity of the arcuate eminence, more or less marked but always visible; and, once past this, the meatal field is identified at a lower level. This is a smooth triangular bony surface, the base of which corresponds to the petrous crest and posterior border of the arcuate eminence; the anterior border is barred by the fold of dura mater stretching out from the petrosal nerve. It must be remembered that the geniculate ganglion may be very superficial, sometimes even laid bare across a dehiscence in the roof of its compartment. For a classic suprapetrous approach in a vestibular neurectomy the stripping would stop here, but for an extended approach the stripping must be taken further, in front of the meatal field. To do so requires obliteration of the dural fold which still bars progress in front of the meatal field by dividing the middle meningeal artery after its coagulation and plugging of the spinous foramen. The dura can then be reflected, carefully freeing the petrosal nerve which is adherent to it, as far as the foramen ovale. The superior petrosal sinus is cautiously disengaged from the petrosal rim. Dissection must be gentle throughout this zone, any necessary hemostasis being done by bipolar coagulation and applications of Surgicel. One must use tamponnage with neurosurgical swabs, irrigate with the double-flow cannula and be patient. The following stage of drilling is not to be undertaken until the bleeding is completely dried-up. When the stripping is completed, the self-retaining retractor must be inserted to elevate the temporal lobe. We place a retractor blade fixed by the Yasargil retractor and its tip is hooked on the posterior border of the petrous.

The Classic Suprapetrous Route

The Burring Stage

Identification of the IAM

The method varies with different authors (see Fig. 39, Chapter I):

- The first is that of House [25]. It consists in identifying the greater petrosal nerve and then, by drilling, to ascend progressively to the geniculate ganglion, then to the first part of the facial nerve and then to the IAM. This is the simplest technique, but it is also the most risky for the facial nerve, which is very vulnerable at this site, and also for the cochlea which is 1 mm in front.
- The second method is that of Fisch [14], which locates the IAM on an axis making an angle of 60° with the plane of the anterior SCC. The canal is discovered by looking for the blue line of the endosteum of the canal by drilling under continuous irrigation in the dense compact zone of the arcuate eminence.
- Other methods endeavor to approach the IAM directly in its middle third, the "silent zone", far from any dangerous structures.
  - Portmann [40] drills at 8 mm in front of the arcuate eminence
  - Sterkers [53] drills at 28 mm from the inner border of the temporal squama on the biauricular axis (landmark of Narcy), keeping in mind that the summit of the loop of the anterior SCC is 24 mm from the temporal squama (measurement of Clerc and Batisse [7]).
  - Garcia-Ibanez [20] locates the IAM at the bisector of the angle formed by the direction of the petrous and that of the ASCC, the latter being simply identified by the situation of the arcuate eminence without drilling. The drilling of the meatus is done in the middle part of this bisecting line.
- All these landmarks are certainly valid and worth knowing, but they are only approximately reliable. Experience has taught us that exposure of the anterior SCC and not of its blue line is by far the most reliable method. It is necessary to have a mental picture of the different structures shown diagrammatically in Fig. 93:
- Behind and below, the arcuate eminence and the meatal field
- In front and above, the petrosal nerve, which can be slightly cleared just at the start of the geniculate ganglion
- The biauricular axis which marks the direction of the IAM.

Two pneumatized zones need to be identified, sometimes after a light drilling with the diamond

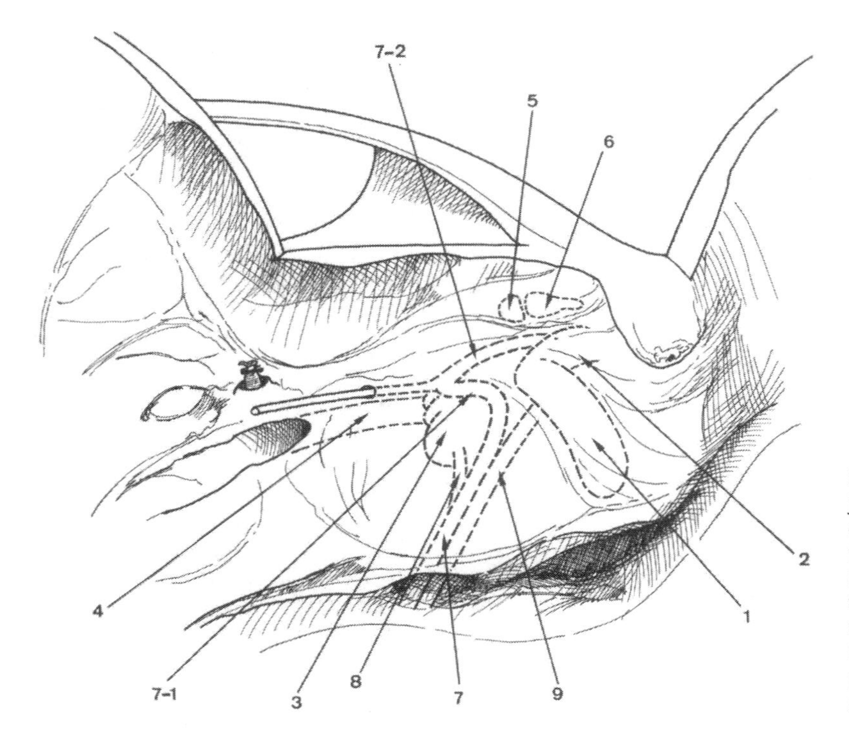

Fig. 93. Right suprapetrous route. Projection of intrapetrous structures that can be mentally localized by reference to visible structures. Anterior SCC *(1)*, lateral SCC *(2)*, cochlea *(3)*, intrapetrous internal carotid *(4)*, head of malleus *(5)*, incus *(6)*, intracanalicular facial nerve *(7)*, first part *(7-1)*, second part *(7-2)*, cochlear nerve *(8)*, vestibular nerve *(9)*

burr: zone X, corresponding to the supralabyrinthine cells, and zone Y, corresponding to the anterior part of the epitympanic recess.

Exposure of the IAM

- *Freeing of the anterior SCC and its ampulla* forms the first stage. Drilling in zone X, from behind forwards, exposes the posterior flank of the anterior SCC and then its apex. Its loop is followed towards zone Y, drilling of which allows exposure of the head of the malleus, the incus and the origin of the lateral SCC (Fig. 94). This supralabyrinthine drilling allows quite easy recognition of the essential landmarks. Fisch [15] also seems to recommend this same technique. In spite of everything, identification of the blue line remains indispensable for opening the IAM to its fundus quite safely and without damaging the facial nerve. In seeking it, the burring under continuous irrigation must be gentle and light, as if one were "rubbing out" in the axis of the anterior SCC and on its anterior aspect. There should be no drilling at its apex, where the risk of opening and therefore of deafness is greatest. The appearance of the blue line must be watched for "under water". It is enough simply to spot it and it is important not to persist. To see it again, if need be, one need only replace the fluid.
- *Drilling in the safety zone,* situated in the 60° angle of Fisch drawn from the anterior SCC, can then be undertaken. A 3 mm diamond burr should be used, and the attack made in the middle part of the meatal field and close to its posterior border. One is then sure of being in the silent zone at the level of the superior vestibular nerve, the facial nerve remaining protected in front. The bluish surface of the dura appears more or less rapidly, depending on the thickness of the bone, along the line bisecting the angle between the petrosal nerve and the anterior SCC (Fig. 95). One should not be deceived by opening an air-cell overhanging the IAM. The existence of this latter should have been recognized on the tomographs before operation, hence the importance of these in the preoperative assessment.
- *Opening of the IAM.* This is to be done over the entire extent of its superior aspect. To identify the structures contained in the IAM with certainty, it is essential for the meatus to be opened up to its fundus, i. e., up to Bill's bar which separates the vestibular from the facial nerve. To this end, the drilling ascends gently along the posterior border of the meatus up to the vestibular fossula situated under the anterior SCC (Fig. 95). The work is then taken forward in the direction of the geniculate ganglion, cautiously drilling the region of the ampulla of the anterior SCC, which may mask Bill's bar. The latter is deeply embedded between this ampulla behind and the labyrinthine part of the facial nerve and the

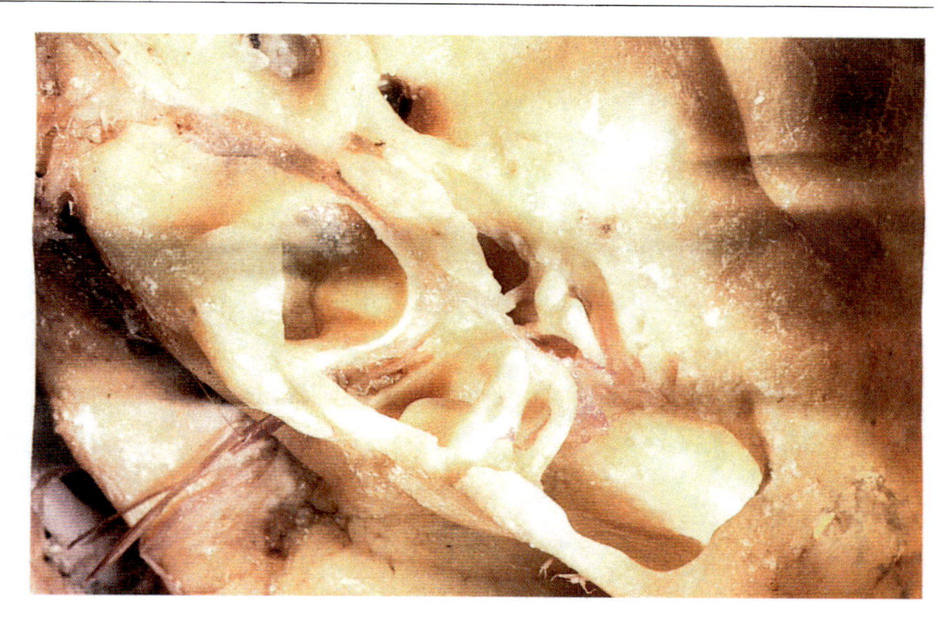

Fig. 94. Dissection of petrous. Burring of upper aspect of petrous and mastoid evacuation exposing intrapetrous cavities. See legend to Fig. 14

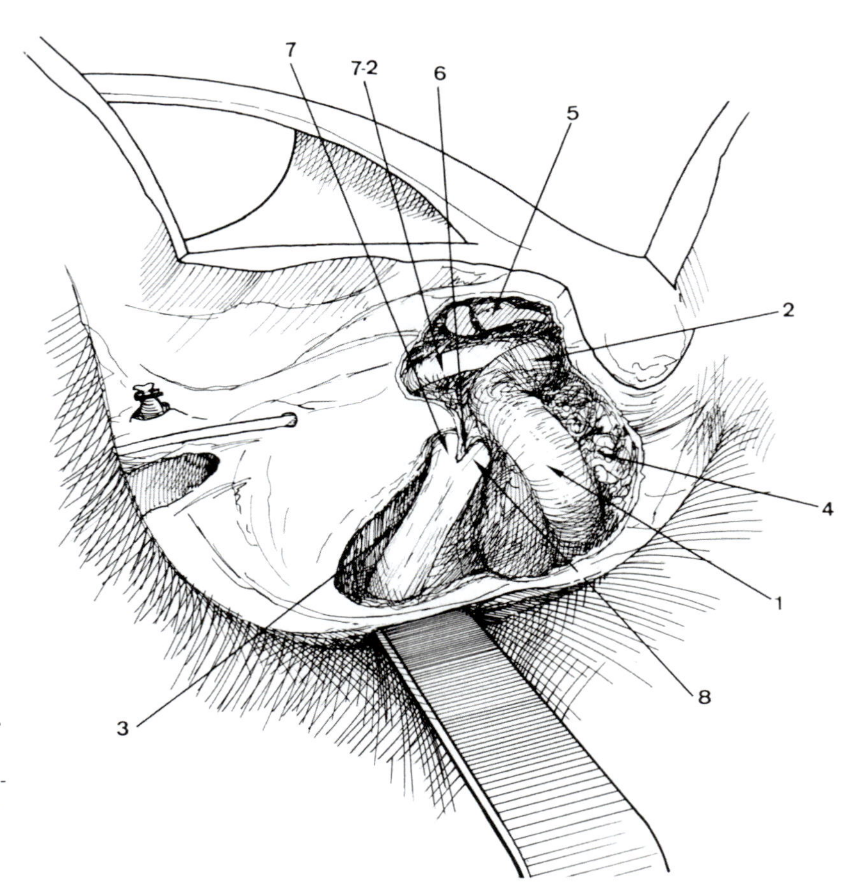

Fig. 95. Right suprapetrous route. Opening of roof of IAM. Anterior SCC *(1)*, lateral SCC *(2)*, anterior wall of IAM *(3)*, supralabyrinthine cells *(4)*, ossicles *(5)*, Bill's bar *(6)*, facial nerve entering first part of facial canal *(7)*, second part *(7-2)*, superior vestibular nerve *(8)*

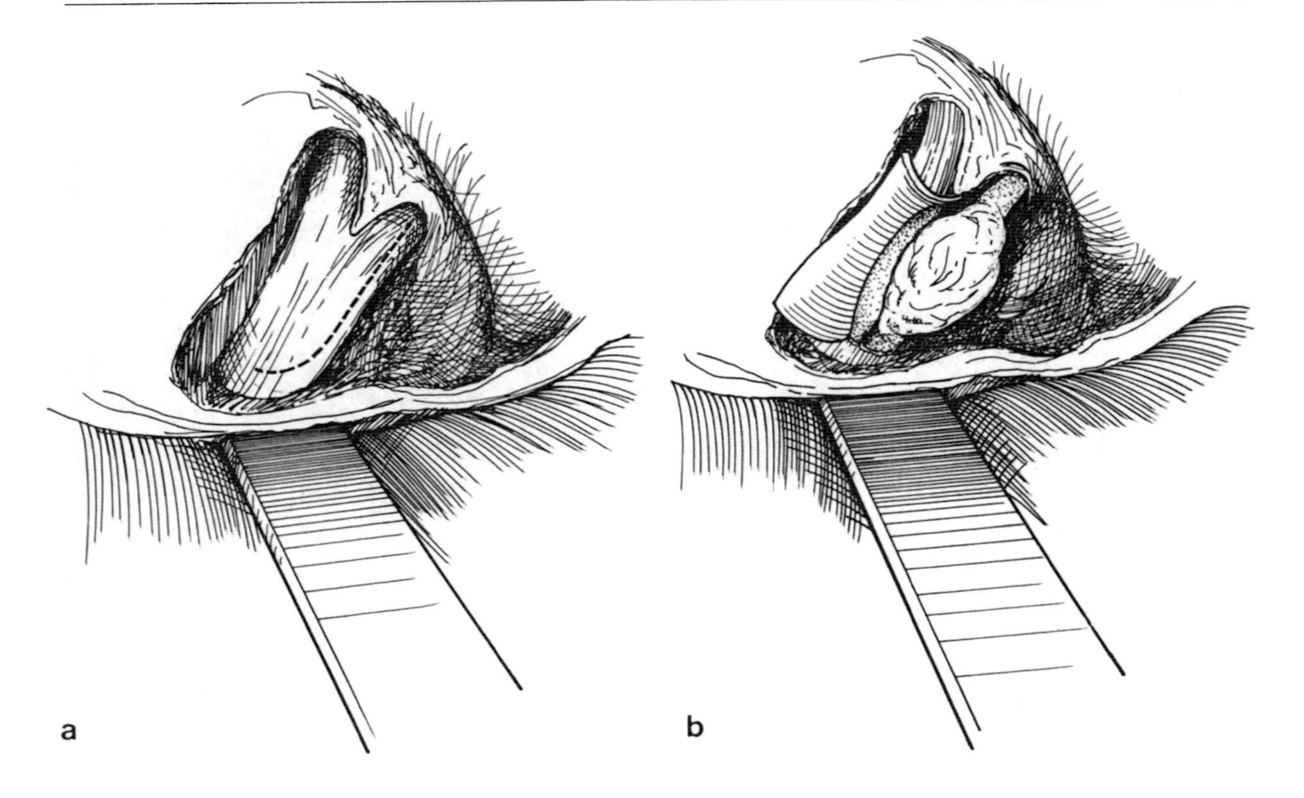

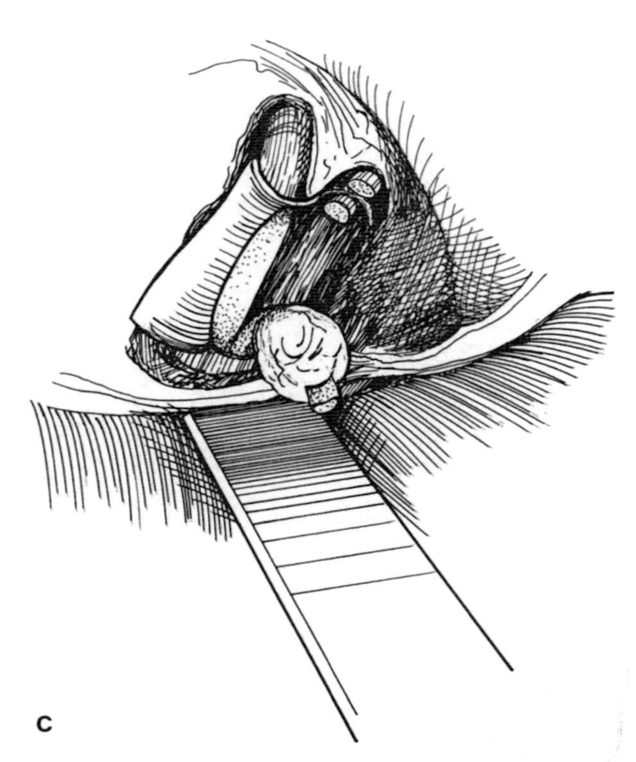

Fig. 96 a–c. Right suprapetrous route. Excision of an intracanalicular acoustic neurinoma. **a** Opening the dura mater of the IAM **b** Exposure of intracanalicular tumor **c** Section of vestibular nerves and excision of tumor

cochlea in front (Fig. 95). Drilling is then resumed along the anterior border of the IAM in the direction of its aperture, taking care not to open the dura as the facial nerve is theoretically immediately subjacent. One must ensure removal of the entire superior aspect of the meatus, and contact with the anterior and posterior walls is sought while reflecting the dural sheath with a blunt spatula. The bony opening so constructed has the final shape of a triangle whose apex corresponds to the fossula of the facial nerve. The anterior and posterior margins, thin near the fundus, thickened as the aperture is approached. The base is marked by the bony border that separates the superior petrosal sinus from the aperture, which has not been opened.

Opening the Dura Mater

This must be done along the posterior margin of the meatus, so as to avoid as far as possible touching the facial nerve, which is in an anterosuperior position in the meatus, also to avoid the blood-supply to the cochlea which follows the cochlear nerve immediately subjacent to the facial nerve. A fine hook is used to elevate and then to tear the dural layer, progressing from the superior vestibular fossula to the aperture. The dural flap so freed is turned forward, so exposing the contents of the IAM (Fig. 96 a, b).

The Intracanalicular Stage

This usually involves division of the vestibular nerves. It is exceptional to use this approach for a tumor as this usually fills the entire meatus up to its aperture, so much so that the region has to be more widely opened. However, it may happen that the neurinoma measures only a few mm, and that it can then be extirpated by this limited approach (Fig. 96). Under stronger magnification ($\times 16$), the facial and cochlear nerves are progressively stripped using Fisch's stapedectomy hook. The best course is to begin by detaching the vestibular nerves from their respective fossulae and then to extract the tumor progressively from the meatus until exposure of the vestibular nerve, to which it is appended. This nerve is then divided above the tumor. However, such an opportunity has only been granted us on one occasion. Of course, one must respect the arterioles passing to the cochlea and even, as far as possible, avoid mobilizing them.

Closure

A fragment of temporal muscle or fat is placed in the trench that led to the IAM and is stuck down with biologic glue. The opening at the roof of the epitympanic recess and the supralabyrinthine cells is occluded by bone-dust mixed with glue. A large graft of temporal aponeurosis is placed on the entire upper surface of the petrous and also glued down. The retractor is removed and the negative pressure of the sucker stopped. After checking for hemostasis, the dural bottonhole made at the start of the operation is closed with one or two stitches. The dura is carefully suspended, preferably through holes pierced in the temporal squama at the periphery of the flap. This is replaced and fixed by two or three stitches at the lateral bony margins. The rest of the craniectomy is packed with bone-dust. The muscle plane is carefully reconstituted. The skin is closed with separate sutures. If hemostasis is quite satisfactory, there is no point in drainage. If this is not the case, a suction drain can be placed between the skin and the muscle plane. This avoids the risk of palpebral ecchymosis. An American dressing is held in place by a wide bandage.

The Extended Suprapetrous Route

The guiding principle is not to limit the superior approach to the IAM to mere drilling of its roof, but to push to the maximum the bony resection at its anterior and posterior faces. It is thus possible to construct on the posterior border of the pyramid a large trench of triangular prismatic shape (Fig. 97), limited externally by the plane of the internal aspect of the posterior labyrinthine bloc, formed roughly by the fundus of the meatus and the plane of the anterior SCC; in front by the base of the first turn of the cochlear spiral, prolonged inward by the posterior wall of the carotid canal in its horizontal portion; behind by the dura of the posterior aspect of the petrous; and below by the plane tangential to the floor of the meatus. In contrast with the classic suprapetrosal approach, it is not possible to gain very much externally as one very soon comes up against the posterior labyrinthine bloc. On the other hand, inwards one can push the resection very far along the petrous ridge, as far as the cave of Meckel and even slightly below it. This exposes all the dura mater which lines the posterior face of the petrous, from the floor of the IAM to the superior petrosal sinus and from the inner border of the anterior SCC to the level of Meckel's cave. Splitting this dura mater after ligature of the superior petrosal sinus gives wide exposure of the CPA.

The superficial stages are identical with those of the classic approach. The only slight difference is that it is essential in this case to divide the middle meningeal artery at its exit from the spinous foramen, in order to be able to strip the dura properly up to the foramen ovale.

The identification of the anterior SCC, followed by drilling at the roof of the IAM, are identical in every way, but once the dura of the meatus is exposed, the surgeon must make his best efforts to resect the bone so as to completely expose the upper, anterior and posterior faces of this canalicular dura mater and to expose the dura of the posterior aspect of the petrous, lateral, above and medial to the aperture of the meatus. To do this, it must be clearly realized that the plane of the anterior SCC, perpendicular to the upper aspect of the petrous, is found in the operative position in a plane oblique downwards and towards the patient's feet. To gain plenty of room externally and reach the fundus of the IAM, the surgeon must drill under the loop of the anterior SCC, which now somewhat overhangs the excavation above. At this site, at the end of drilling, it forms a somewhat rounded blunt projection. Just internal and in front of this there is the projection of the first turn of the cochlear spiral, itself also blunt but in the opposite direction. In front of and medial to the cochlea, one can drill without risk until one perceives the carotid canal in its horizontal portion, the wall of which can be thinned

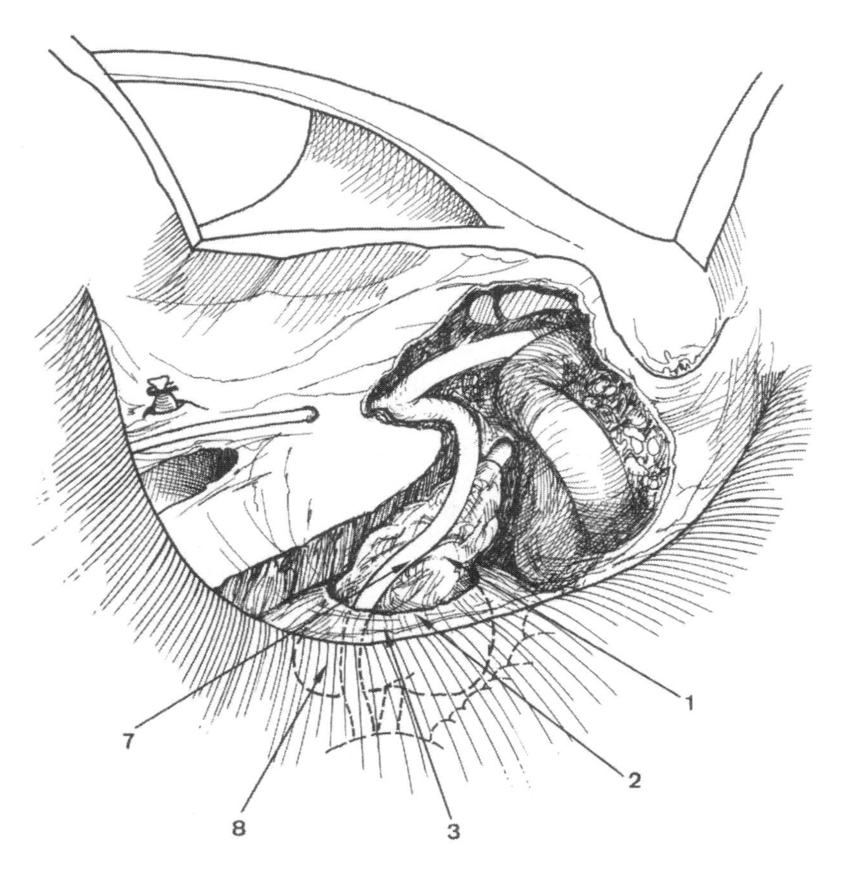

Fig. 97. Extended right suprapetrous route. Exposure of stage II acoustic neurinoma. Intracanalicular growth *(1)*, aperture of IAM *(2)*, dura of posterior face of petrous around IAM *(3)*, facial nerve *(7)*, tumoral bud in cerebellopontine angle seen by transparency *(8)*

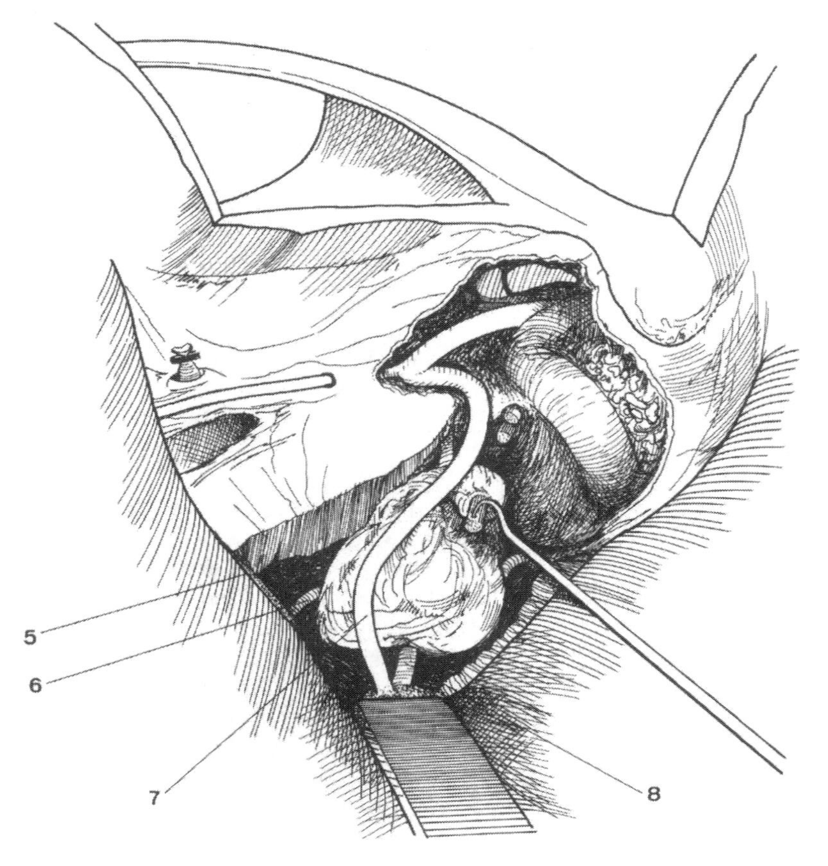

Fig. 98. Right extended suprapetrous route. The dura mater has been incised. Exposure of the growth developed in the cerebellopontine angle. The hook is elevating the superior vestibular nerve which has been divided. Edge of trigeminal nerve *(5)*, AICA *(6)*, facial nerve *(7)*, cochlear nerve *(8)*

down safely enough provided one uses a diamond burr of thick enough caliber (usually 3 mm). This burr is similarly used to completely denude the dura of the posterior aspect of the petrous. In this way, one succeeds in opening a breach about 25 mm broad, giving at last quite a wide, very anterior, view of the CPA. Wigand [56] claims to be able to approach and extract stage III or even IV neurinomas by this route with the advantage of perfectly respecting the labyrinth and exposing the entire IAM up to its fundus, while exposing the facial nerve in a healthy zone below the tumor as by the transpetrous route. We have not had sufficient experience of this route to assess its effective importance, but we well recall very well that Billet, at the December 1977 meeting of the *Société de Neurochirurgie de Langue Francaise*, had already recommended an approach comparable in every point and with an operative success which, to our knowledge, he unfortunately failed to publish. Extraction of the tumor is performed very much as by the translabyrinthine route, with dissection of the arachnoid layer, tumoral evacuation and extraction of its periphery. However, the more reduced size of the tumor does not usually permit the use of the rotary vacuum dissector. One can only fragment it progressively with the diathermy or use an ordinary sucker if its texture is not too solid. The position of the facial nerve in the IAM is mainly responsible for the operative difficulties. When the nerve is applied to the anterior wall of the meatus, dissection is relatively easy. On the other hand, when it is elevated to the ceiling of the meatus it is constantly interposed in the midst of the operative field, with a marked increase of the risk of damage (Fig. 98).

Closure is performed in the same way as for a classic suprapetrous approach; nevertheless, one must take special care to seal the cells of the petrous apex, if possible with bone-dust mixed with biologic glue.

The Extended Transpetrous Routes

The principle here is to combine with a more or less extended petrous resection a wide exposure of the upper latero-cervical regions, immediately subtemporal. Their value lies in allowing perfect control of the neurovascular axes traversing these infratemporal regions and completely exposing lesions arising at the inferior aspect of the pyramid and the neurovascular axes and lesions which remain quite inaccessible to the strictly transpetrous routes.

Some tumors remain localized to the infratemporal region. To approach these, it suffices to remove all or part of the bony arch (mastoid, tympanal and ramus of the mandible) which closes the region externally. This is the principle of the infratemporal routes proposed by Fisch [16]. The problem lies with the facial nerve which runs through the region and which must be exposed and rerouted.

Other, not exceptional, tumors, usually benefitting from the natural passage provided by the posterior lacerate canal, develop astride the posterior fossa and the infratemporal regions. Their one-stage excision calls for simultaneous exposure of both parts of the double sac. The key to the problem consists in wide resection of the pyramid, which is interposed between the two regions. By combining an infratemporal and a transcochlear route, a good exposure is obtained of both the CPA and the infratemporal region. This is what we call the extended transcochlear route, a route we used for the first time, without naming it, in 1972 [39] and which we have since employed on several occasions

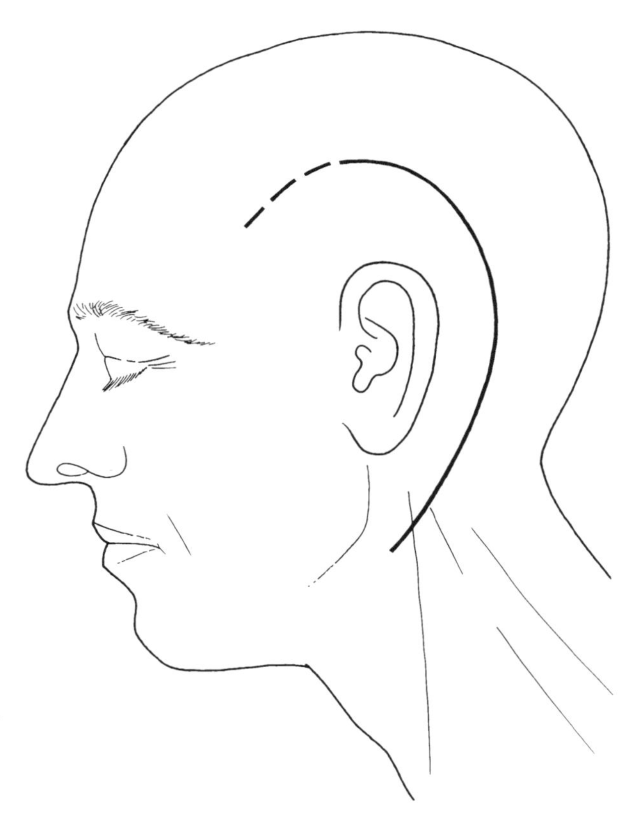

Fig. 99. Left infratemporal route. Skin incision. *Continous line:* incision for Fisch's type A procedure, *interrupted line:* incision for Fisch's type B or C procedure

[5]. Quite recently, Sekhar and Estonillo [48] have proposed a transtemporal route which is altogether comparable.

The preparation and arrangement of the patients are practically identical with those used for the strict transpetrous routes, except for the detail that in these cases we prefer to shave the patients completely, since the extent of the incision calls for shaving of the hemicranium and this implies the loss of the esthetic value of minimal shaving as recommended by the Reims school [47].

The Infratemporal Routes

The aim of these is to allow control of lesions developing around the carotid and jugular axes in their intrapetrous and infratemporal portions. From the date of his first publication [16], Fisch already foresaw three types of exposure, determined by whether he wished only to reach the inferior aspect of the petrous and the infratemporal region (type A) or the clivus (type B) or even as far as the parasellar region (type C). Only type A procedure is generally used.

Infratemporal Route, Type A

This route is well adapted to the excision of tumors arising in the infratemporal region, usually tumors of the glomus jugulare. This route respects the integrity of the cochlea but removes the tympanum and chain of ossicles. It sacrifices the air transmission of sound but theoretically preserves its sensory perception. We describe the approach as proposed by Fisch [17].

Exposure Stage

The curved retroauricular incision (Fig. 99) mobilizes a large flap which is turned forward. The meatus of the EAM is at once closed (Fig. 106) and the meatus sectioned at the osteocartilaginous junction.

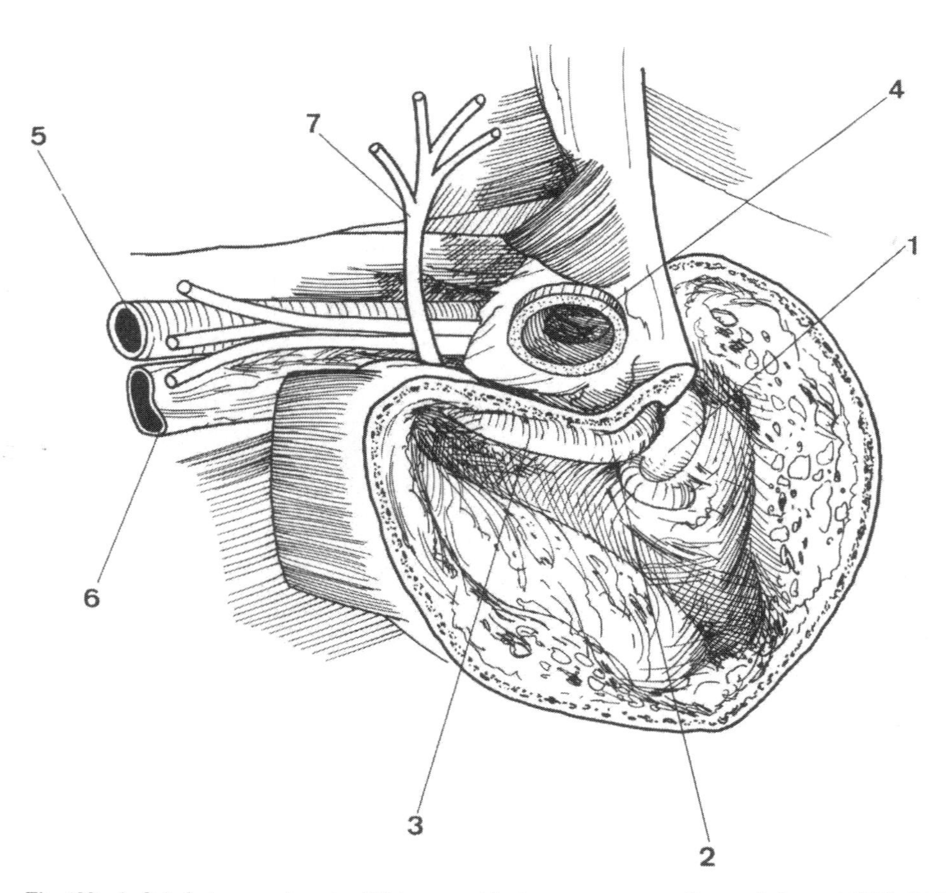

Fig. 100. Left infratemporal route. Wide mastoidectomy respecting labyrinthine massif. The external acoustic meatus had previously been divided and the cervical neurovascular axes dissected. Lateral SCC *(1)*, bend of facial canal *(2)*, 3rd part *(3)*, EAM *(4)*, internal carotid artery *(5)*, internal jugular vein *(6)*, facial nerve *(7)*

A superficial parotidectomy exposes the trunk and branches of the facial nerve which has been exposed at its exit from the stylomastoid foramen. The vessels and nerves are identified in the neck and the external carotid ligatured above the lingual, as also the branches going to the tumor, particularly the ascending pharyngeal. The skin of the EAM, the tympanum and the handle of the malleus are excised *en bloc* after incudostapedial disarticulation to avoid mobilizing the stapes. Mastoidectomy (Fig. 100) and resection of the EAM (Fig. 101) are followed by rerouting of the facial nerve (2nd and 3rd parts), which is displaced towards the upper part of the approach in a bony groove drilled in the epitympanum, between the geniculate ganglion and the root of the zygoma. When there is localized invasion of the nerve by the tumor, excision and grafting is performed at the end of the operation. Ablation of the styloid process completes free access to the jugular bulb. The ramus of the mandible is dislocated forward, either after disarticulation or after resection of the condyle, depending on the tumoral extension. This dislocation is maintained by a retractor. The sigmoid sinus is ligated below the emissary vein and the bony covering of the posterior lacerate canal is drilled. At this stage the upper and posterior poles of the tumor will have been freed. The intrapetrous part of the internal carotid is exposed by carrying the drilling forward. When the tumor has extended far forward, it is preferable to resect the condyle and to drill the glenoid fossa of the

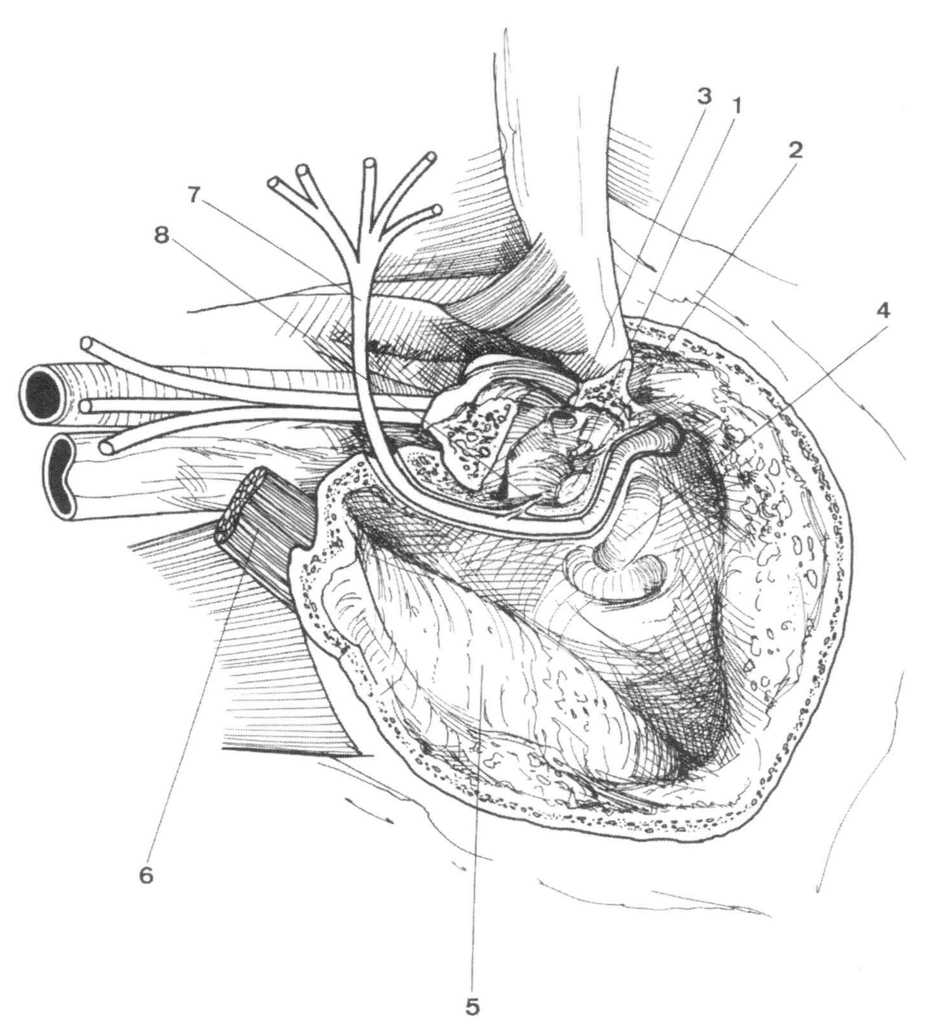

Fig. 101. Left infratemporal route: resection of external acoustic meatus, then freeing of facial nerve from third and second parts of facial canal. Cochleariform process and tendon of tensor tympani muscle *(1)*, vestibular window and base of stapes *(2)*, orifice of auditory tube *(3)*, lateral SCC *(4)*, groove of sigmoid sinus *(5)*, digastric muscle *(6)*, facial nerve *(7)*, chorda tympani *8*

temporal bone. It now only remains to free the anterior pole of the tumor (Fig. 102).

Tumoral Stage

Tumoral excision begins in front by freeing it from the internal carotid, with diathermy coagulation of the pedicles arising from it. This dissection is continued backwards in contact with the dura mater, as far as the jugular bulb. When the latter is invaded, the termination of the sinus and the bulb with the tumoral content are resected after ligature of the internal jugular and its dissection from below upwards (Fig. 103). Hemorrhage from the inferior petrosal sinus is dealt with by packing it with a fine layer of Surgicel. Resection of the mixed nerves, especially the vagus, is often needed because of their invasion by the tumor. The tumoral fragments still adherent to the dura are carefully stripped off or coagulated on the spot. In principle, if the tumor is bulkier and exhibits extension into the CPA, Fisch prefers to divide the constricted part of the bilocular sac and allow for a second, neurosurgical, approach by the suboccipital route. Should the intracranial growth be large, the first stage would be neurosurgical and excision of the infratemporal fragment would be undertaken secondarily.

Closure

Closure is effected by occluding the auditory tube by a fragment of muscle and fascia, followed by packing the operative cavity with abdominal fat and rotation of a flap of the temporalis muscle, which is sutured to the sternocleidomastoid and digastric. The skin is closed so as to be water-tight; we do not use suction drainage, so that hemostasis must be perfect.

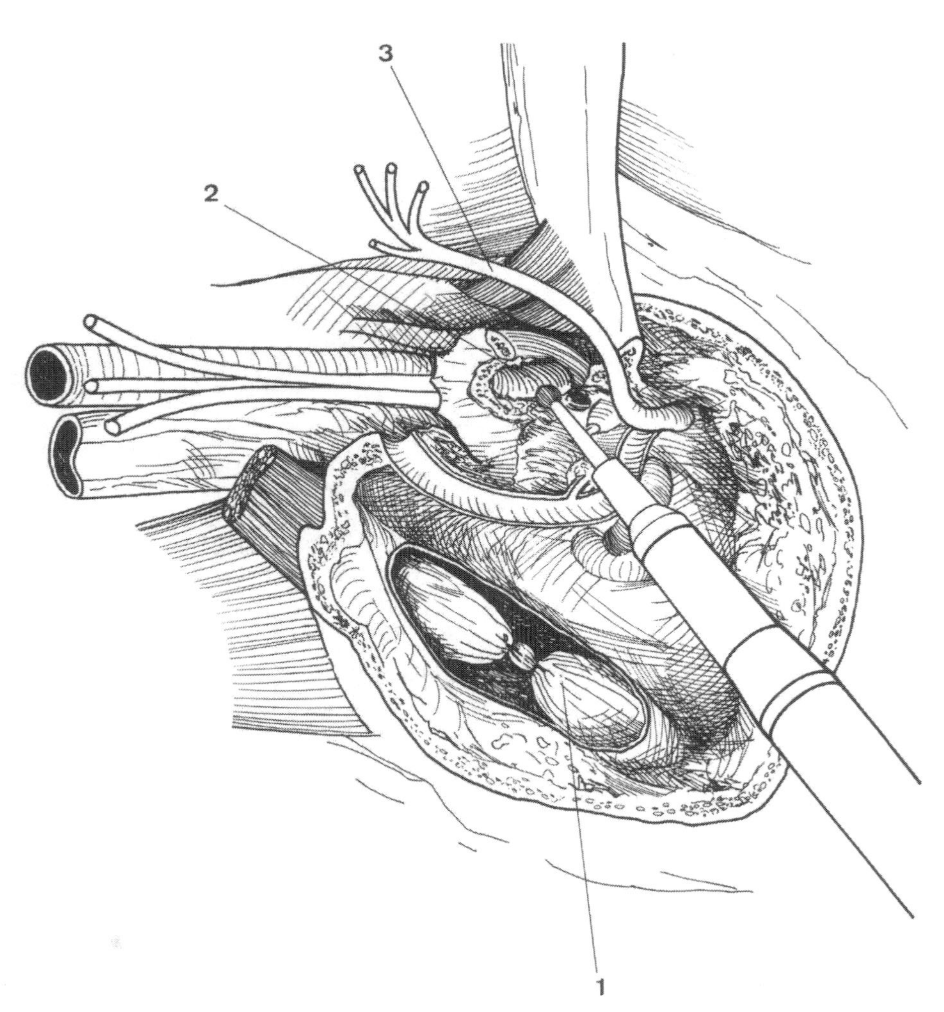

Fig. 102. Left infratemporal route: after rerouting facial nerve, burring towards the petrous apex then towards the jugular bulb. Sigmoid sinus exposed, then ligatured *(1)*, intrapetrous internal carotid *(2)*, facial nerve rerouted upwards *(3)*

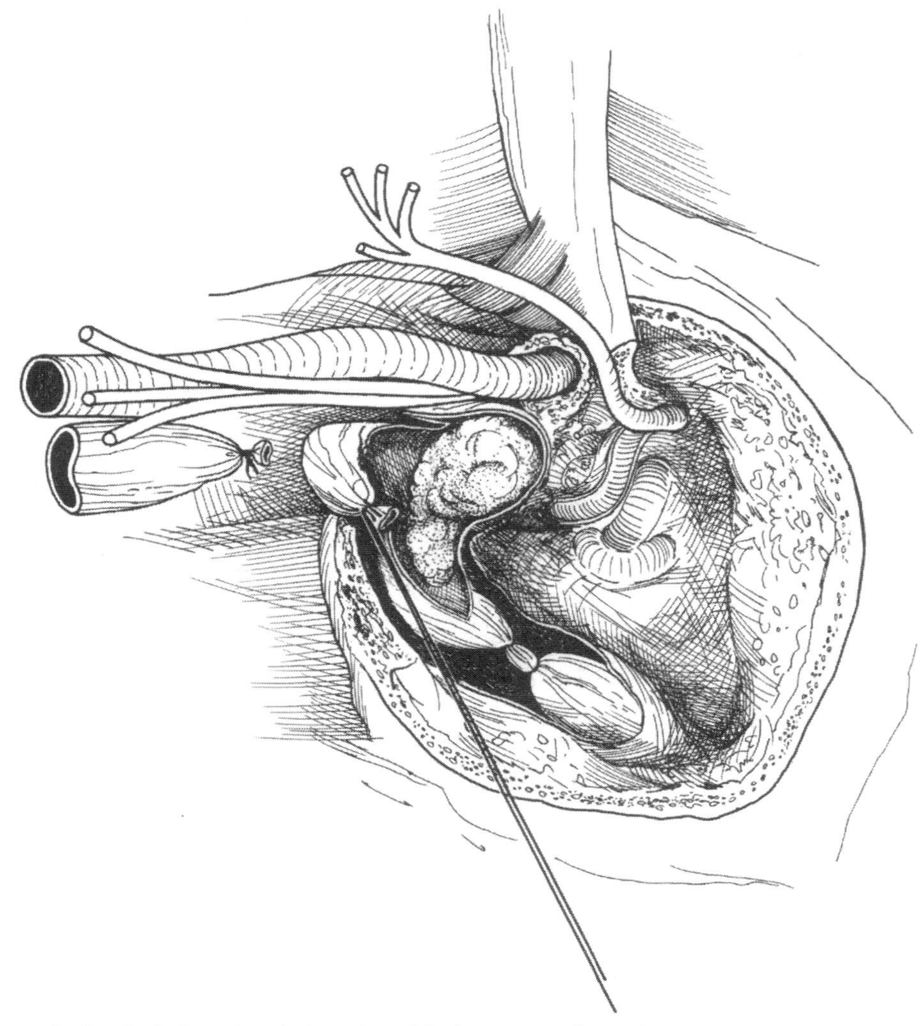

Fig. 103. Left infratemporal route: the jugular bulb and vertical portion of the intrapetrous internal carotid artery are exposed. The internal jugular vein is divided and the tumor exposed in the jugular bulb

The "Economic" Infratemporal Route

Starting from the principle that many infratemporal tumors, particularly the type C or D jugular glomus tumors of Fisch, invade the tympanic cavity not at all or only slightly and leave the chain of ossicles intact, Farrior [13] suggests limiting the external approach to respect this chain and preserve a better quality of hearing. According to him, since it is essentially the third part of the facial nerve that hampers the infratemporal approach rerouting of this third part alone should suffice to allow good exposure of the tumor. The technique as proposed by this author is as follows:

- Retroauricular incision following the hair-line and extending to the upper part of the anterior border of the sternocleidomastoid
- The flap so fashioned is turned forward after dividing the EAM at its osteocartilaginous junction. The skin of the meatus is incised horizontally at 2 o'clock and 10 o'clock, from the tympanum to the external orifice. These two incisions are joined by an incision circumscribing the tympanic membrane all around its inferior circumference. The skin of the inferior two-thirds of the EAM is thus resected and preserved to be replaced at the end of the operation.
- The inferior two-thirds of the tympanal part of the temporal bone and the tip of the mastoid pro-

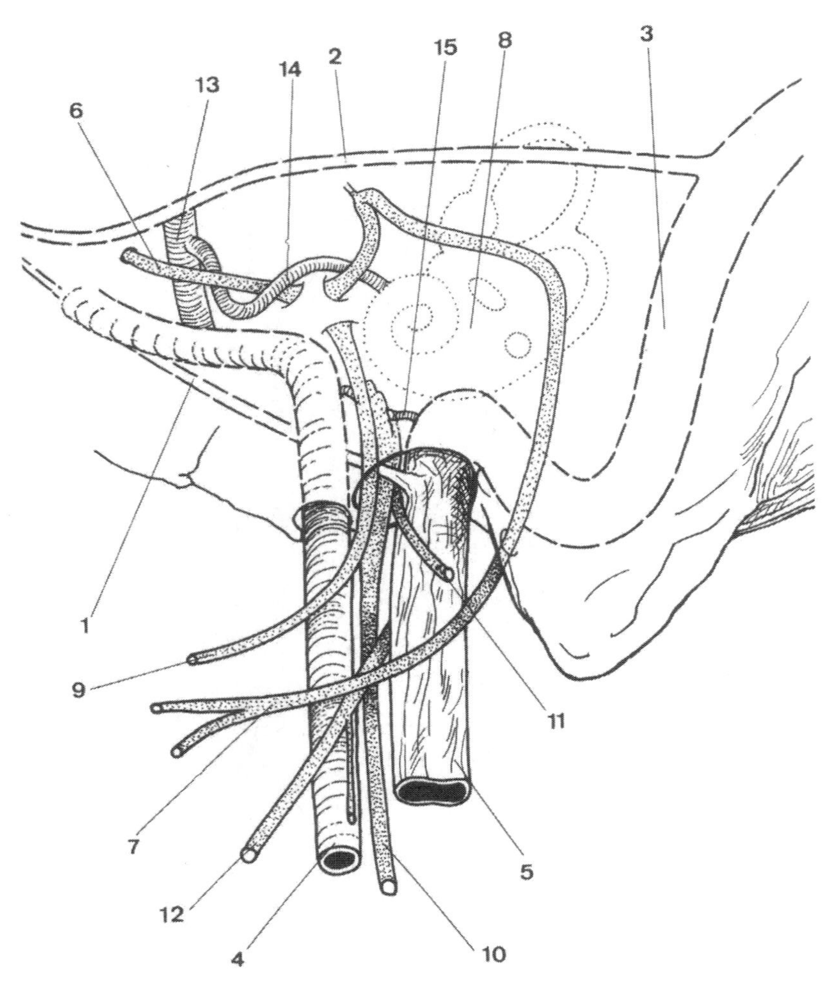

Fig. 104. Left extended transcochlear route. Anatomic diagram of different structures to be exposed. Inferior petrosal sinus *(1)*, superior petrosal sinus *(2)*, sigmoid sinus *(3)*, internal carotid artery *(4)*, internal jugular vein *(5)*, external oculomotor (trochlear) nerve *(6)*, facial nerve *(7)*, labyrinth *(8)*, glossopharyngeal nerve *(9)*, vagus nerve *(10)*, accessory nerve *(11)*, great hypoglossal nerve *(12)*, basilar trunk *(13)*, AICA *(14)*, PICA *(15)*

cess are resected by drilling under the operating microscope. The suprameatal spine marks the upper limit of the mastoid resection, which, in depth, remains below the level of the fossa incudis. The whole of the third part of the intrapetrous facial nerve is cleared and its main divisional branches are then dissected in the parotid. The facial nerve can then be rerouted upwards. The sternocleidomastoid is turned downwards. The digastric muscle is sectioned at the level of its intermediate tendon and then the styloid process is resected, well exposing the infratemporal region. The tympanic membrane, freed over the lower two-thirds of its periphery, is then gently reflected upwards over the handle of the malleus. The hypotympanum and the tumor it contains are then well exposed. The tumor can be gently stripped from the walls of the tympanic cavity and extracted. The chain of ossicles is carefully preserved. The infralabyrinthine cells are then burred to give a good exposure of the jugular bulb and, more deeply, of the carotid canal and auditory tube. The sigmoid sinus is denuded at the level of the mastoid resection and then occluded, and the jugular vein is divided between two ligatures and the upper segment is turned up

Fig. 105 a, b. Left extended transcochlear route. Skin incision. **a** Recommended incision, especially after previous embolization; the *arrow* indicates a possible extension for better exposure of the middle cranial fossa. **b** Conley's incision

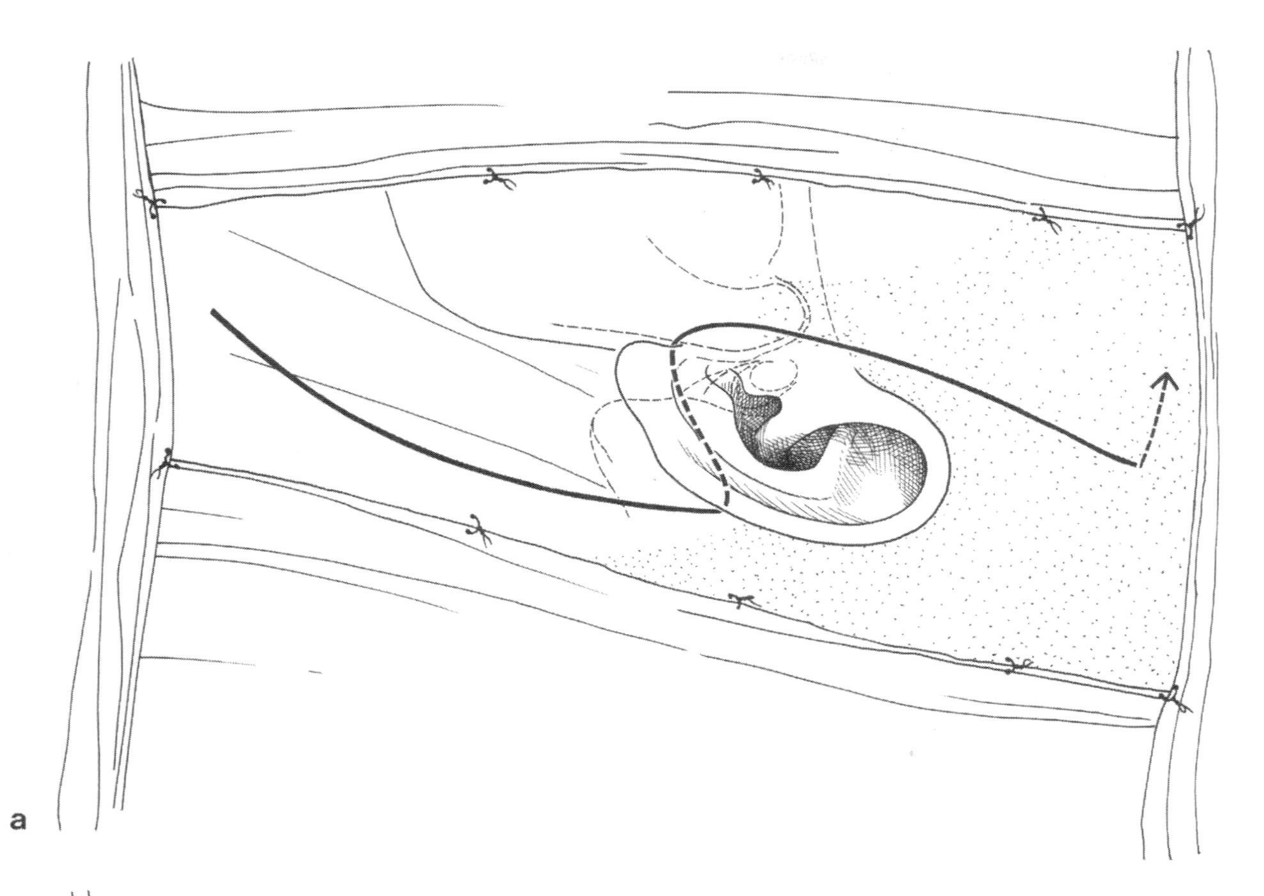
a

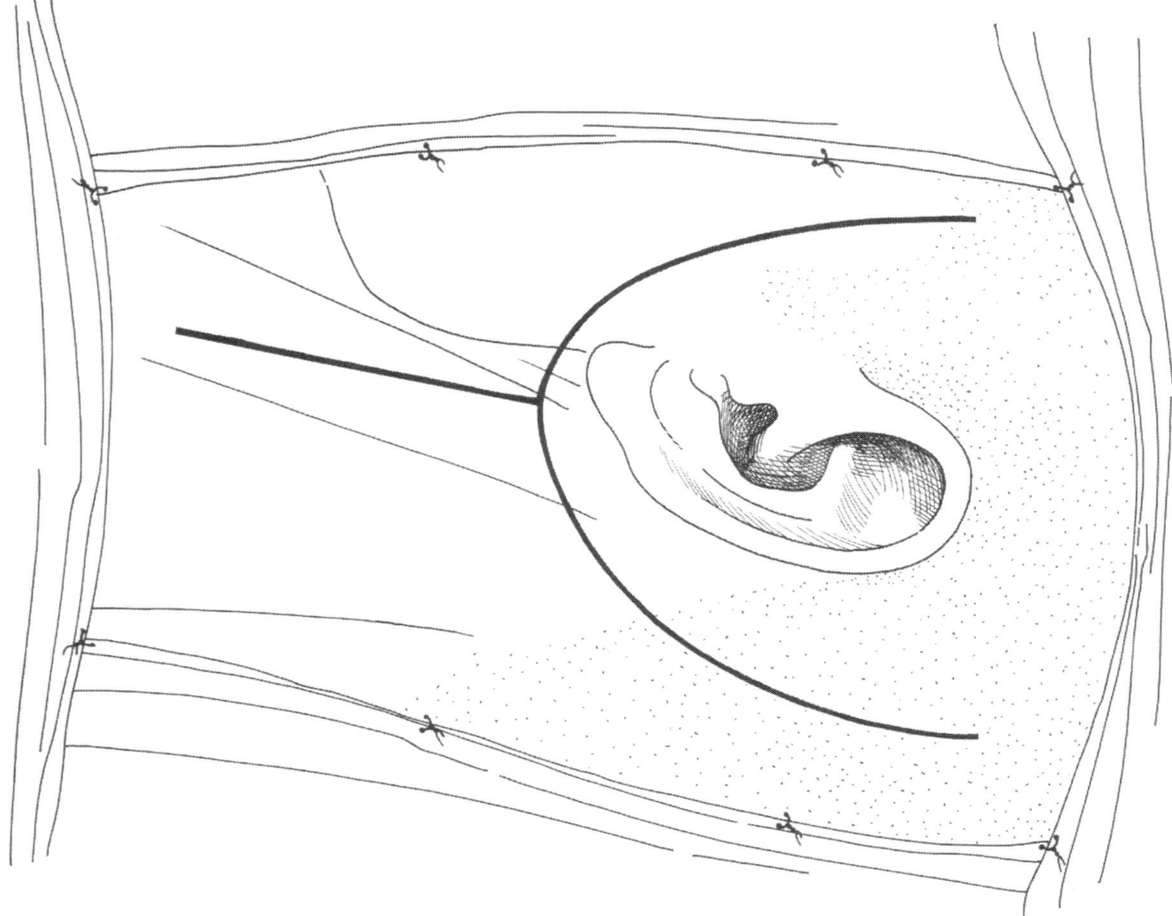
b

towards the bulb, where it is sectioned leaving the internal wall in place, so allowing removal of the tumor in one piece. The inferior petrosal sinus is occluded with muscle. The posterior belly of the digastric is swung into the cavity. After hemostasis, the tympanic membrane is retored in place, the skin of the meatus replaced, the meatal caliber restored by packing with tulle gras and the skin planes sutured. At any time, if the size of the tumor is greater than expected, the access can be extended and transformed into a proper type A infratemporal route.

The "Extended" Infratemporal Routes

Fisch [16] suggests extending the opening forward by dividing the zygomatic arch, which is turned downwards with the temporalis muscle detached from the squama, and with the masseter, the attachment of which to the zygoma is preserved. Also, the bony resection on the pyramid is extended to the glenoid fossa, while the ramus of the mandible is dislocated forward or, if need be, partly resected. In depth, drilling is continued along the carotid canal as far as the clivus (type B) or even upward to the cavernous sinus (type C). For good exposure of the region, one must divide the mandibular nerve and if necessary the maxillary nerve. These only exceptionally used routes allow access to the clivus or the parasellar region, but we have no personal experience of them.

The Extended Transcochlear Route

This combination of a transcochlear and an infratemporal route is in fact an adjusted petrectomy [6], which widely and simultaneously exposes the CPA, the infratemporal regions and the clivus, while the petrous resection is taken as far as the entry of the internal carotid into the cavernous sinus (Fig. 104).

Superficial Stages

Skin Incision

We have long used a Y-shaped incision as advised by Conley [9] (Fig. 105 b). It spreads upwards, widely circumscribing the outer ear and mobilizes a wide scalp flap including the ear. The lower limb of the Y follows the anterior border of the sternocleidomastoid, stopping at about the level of the cricoid cartilage. This incision facilitates the entire subsequent parotid stage, but there is a risk of ischemia of the upper flap, especially when preop-

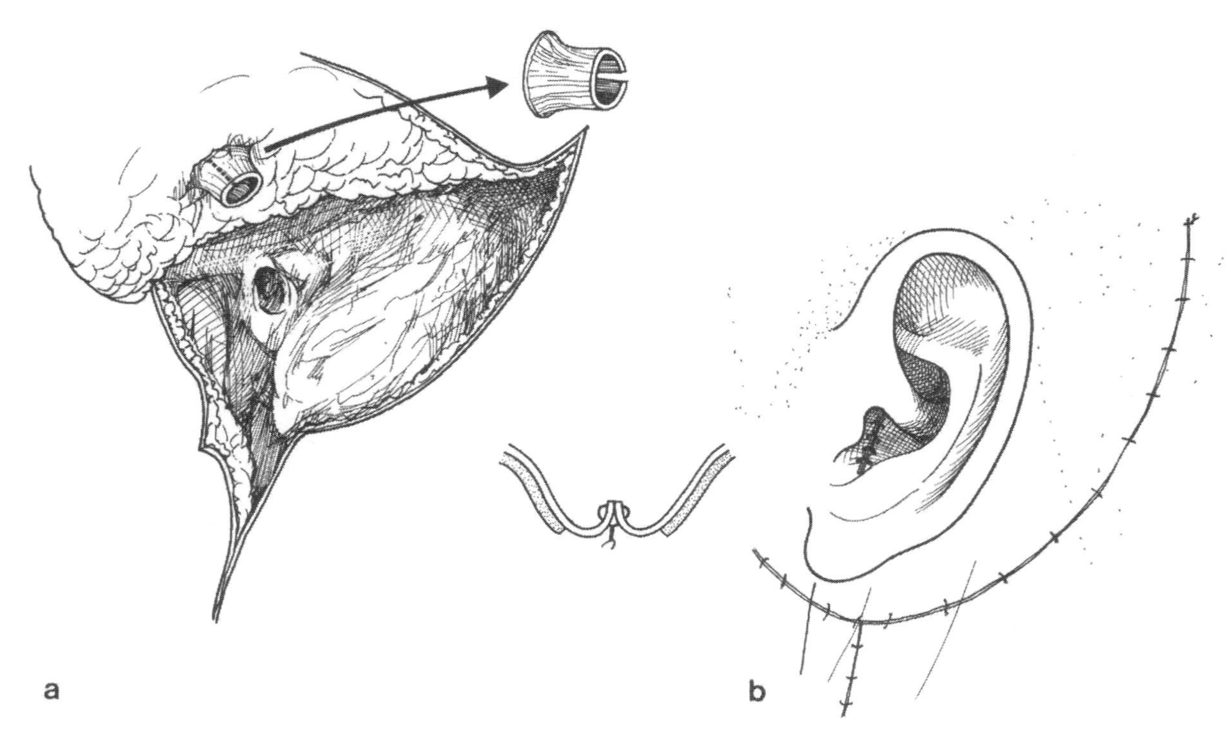

Fig. 106 a, b. Left extended transcochlear route. **a** Excision of cartilaginous meatus. **b** Closure of external orifice of EAM

erative embolization has been performed. This is why we recommend in these cases a straighter incision (Fig. 105 a), beginning very high up in the parietal region, vertically crossing the entire temporal region to pass in front of the tragus, then curving just under the ear-lobe to reach the mastoid at the tip of which it curves again to descend along the anterior border of the sternocleidomastoid.

The Superficial Cephalic Stage

Whatever incision is used, the subcutaneous planes are stripped and the EAM is divided at its osteocartilaginous junction. After resecting the cartilaginous meatus, the cutaneous orifice of the meatus is initially sutured (Fig. 106). One has then exposed all the mastoid field, the temporalis and the bony external acoustic meatus, the zygoma and the temporomandibular joint. The temporalis muscle is detached in its posterior part and turned forward.

The Cervical Stage

The cervicotomy exposes the origin of the sternocleidomastoid muscle, which is detached from the mastoid but left adherent to the skin and subcutaneous planes to preserve the maximum bloodsupply to these layers. The digastric is then detached and reflected downwards to expose the whole of the upper cervical region. One must then identify the facial nerve, dissect its branches and free the posterior border of the parotid, giving a good exposure of the jugulo-carotid axis and the nerves (Fig. 107). Each of these vascular and neural structures is marked by a tie, if possible silastic. The branches of the external carotid are dissected and ligated, beginning with the occipital and ascending to the bifurcation. In the case of a tumor of the foramen lacerum the ascending pharyngeal artery is particularly large and its ligature often considerably reduces the tumoral blood-supply.

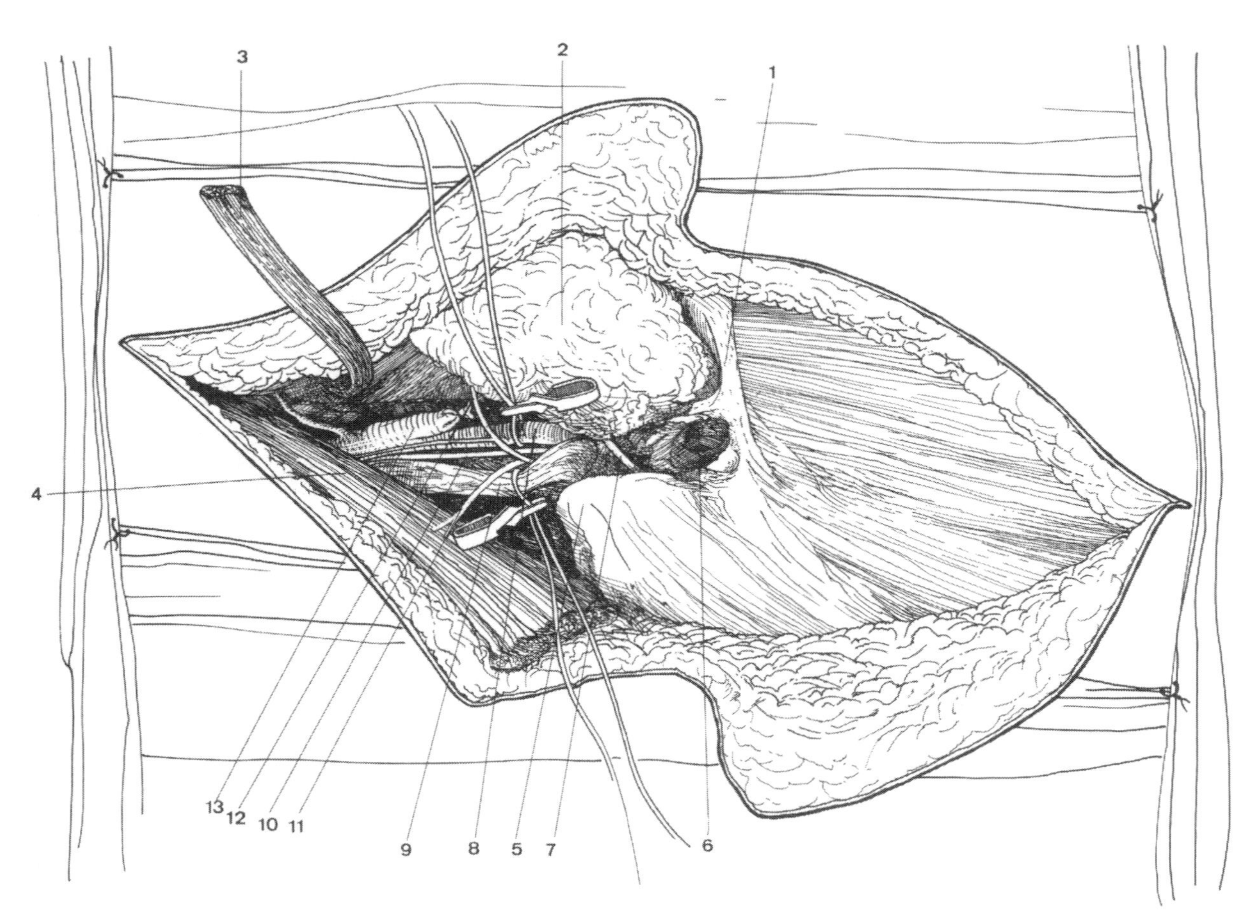

Fig. 107. Extended left transcochlear route. Cervical stage. Zygomatic arch *(1)*, parotid *(2)*, digastric muscle *(3)*, sternocleidomastoid muscle *(4)*, mastoid process *(5)*, EAM *(6)*, retroparotid facial nerve *(7)*, internal carotid artery identified by a ligature *(8)*, internal jugular vein similarly identified *(9)*, vagus nerve *(10)*, accessory nerve *(11)*, great hypoglossal nerve *(12)*, external carotid artery ligatured *(13)*

The Bony Otologic Stages

It is necessary to progressively resect virtually the entire petrous pyramid in successive stages (Fig. 108).

The Extended Translabyrinthine Stage

The petrectomy begins with the conventional ETL approach: the initial extralabyrinthine stage, followed by the labyrinthine stage with skeletonization of the 2nd and 3rd parts of the facial canal, opening of the fundus of the IAM and resection of the posterior three-fourths of the periphery of the meatus. All the structures of the meatus can then be identified and the facial nerve located at its entry into the facial canal.

Rerouting of the Facial Nerve

The EAM is now resected, at the same time that one opens the 2nd and 3rd parts of the facial canal, then its first part. The bony resection also involves the posterior part of the zygoma. The entire course of the facial nerve can then be cleared up to the stylomastoid foramen, and the nerve can be safely rerouted after section of the petrosal nerve. The temporomandibular joint is dislocated forward after division of the joint capsule, so widening the approach.

Transcochlear Stage

After infraction of the cochlea, drilling is continued toward the petrous apex, opening the carotid canal and progressively exposing the dura internal to the aperture of the IAM from the inferior to the superior petrosal sinus, and inwards as far as the clivus. The bony stage is completed by resecting with the

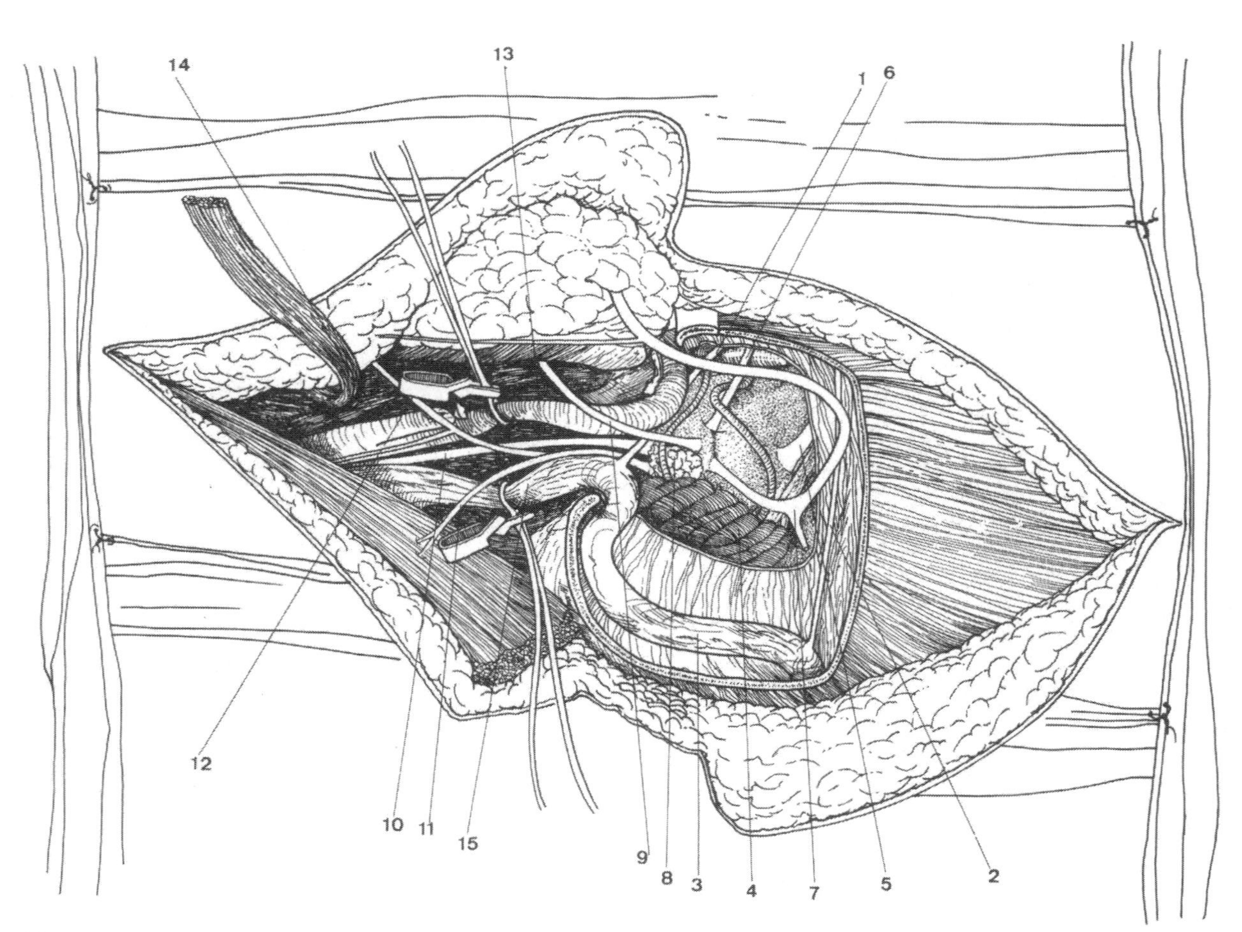

Fig. 108. Extended left transcochlear route. Operative field at end of approach after total petrectomy. Inferior petrosal sinus *(1)*, superior petrosal sinus *(2)*, sigmoid sinus *(3)*, cerebellum *(4)*, trigeminal nerve *(5)*, trochlear nerve *(6)*, rerouted facial nerve *(7)*, choroid plexus *(8)*, glossopharyngeal nerve *(9)*, vagus nerve *(10)*, accessory nerve *(11)*, great hypoglossal nerve *(12)*, internal carotid artery *(13)*, external carotid artery ligatured *(14)*, internal jugular vein *(15)*

beaked nibblers the lateral and anterior margin of the foramen lacerum to give good exposure of the bulb and the emergence of the mixed nerves, which bend like a clothes-horse over the posterior margin of the foramen. When this stage is completed, one has exposed (Fig. 108) the internal carotid artery from the thyroid cartilage to the skull base, and the continuity of the sigmoid sinus, jugular bulb and internal jugular vein. The mixed nerves and the XIIth have been dissected. The facial nerve is rerouted upwards.

The (Neurosurgical) Tumoral Stages

Opening the Dura Mater

This is unnecessary when the tumor is purely extradural. On the other hand, for certain jugular glomus tumors (Fisch's type D) or a neurinoma of the mixed nerves, the dura must be incised on either side of the IAM and halfway between the two petrosal sinuses. This incision is extended towards the apex of the petrous under direct vision, taking care not to damage the abducent nerve. A line of splitting is then taken downward towards the foramen lacerum, relaxing the dural diaphragm around the tumor.

Excision of the Tumor

When there is a tumor of the glomus jugulare, it is helpful, before starting the excision proper, either to ligature the sigmoid sinus by means of a suture passed via two small dural incisions made on either side of the sinus or to pack it with a long strip of Surgicel crammed in distally from an incision made at the outer aspect of the sinus. After this procedure the tumor collapses somewhat and becomes much less hemorrhagic. All the afferents derived from the external carotid, especially the ascending pharyngeal, have already been coagulated during the cervical stage. Sometimes there are meningeal afferents running on the dura which may be coagulated without having to open the posterior fossa. There remain the branches derived from the internal carotid in its intrapetrous course, which may be coagulated flush with the artery in the carotid canal, now opened. Excision of the cervical extension of the growth is now done. The simplest method is to strip the internal jugular vein. After section of the vein between two ligatures below the lower pole of the tumor, one works progressively upwards, pulling on the proximal end of the vein with a thread and cautiously dissecting the upper cervical internal carotid and the mixed nerves. Eventually, one is able to

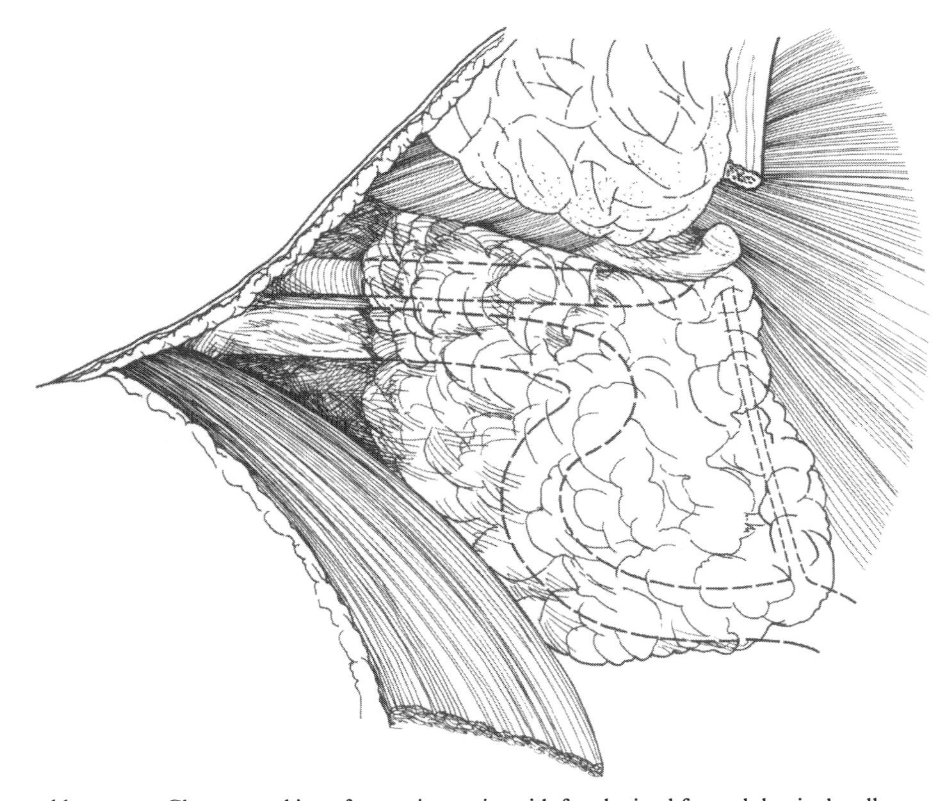

Fig. 109. Extended left transcochlear route. Closure, packing of operative cavity with fat obtained from abdominal wall

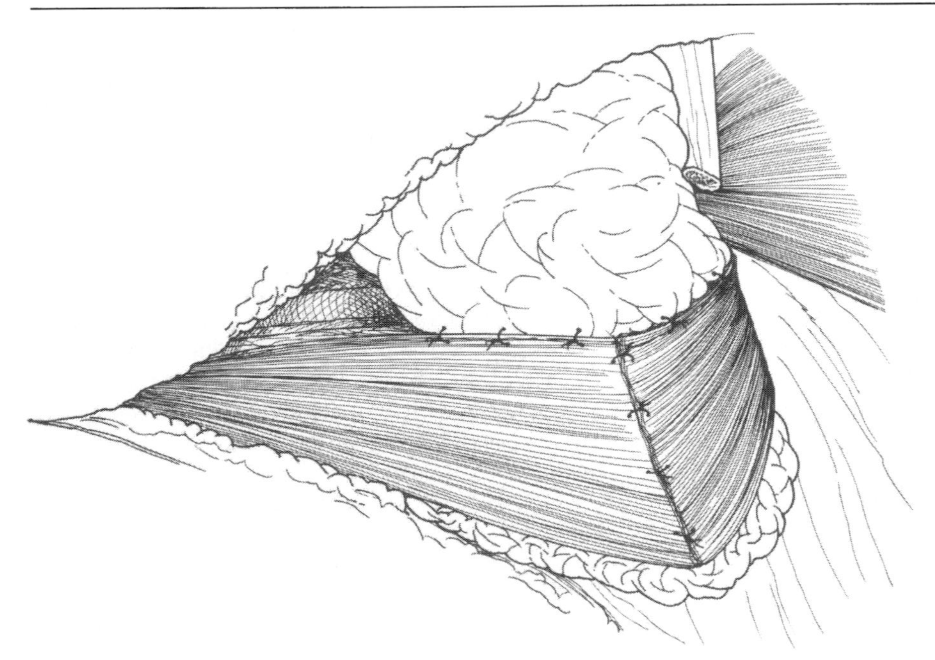

Fig. 110. Extended left transcochlear route. Closure of subcutaneous planes with a flap of temporalis muscle sutured to the digastric and sternocleidomastoid muscles, then suture of parotid gland at this muscular level

"pediculize" the tumor on the bulb. The wall of the vein is then sectioned in such a way as to leave the internal wall in place, which allows removal of the tumor in one piece without having to dissect the nerves at the external orifice of the lacerate canal. The gaping inferior petrosal sinus often bleeds profusely. Hemostasis is made by occluding it with a new strip of Surgicel. If there is an intracranial extension, it is then resected by repeated coagulation and fragmentation. Dissection of the structures of the CPA other than the mixed nerves is usually effected without much difficulty, provided due respect is paid to the arachnoid layers.

In cases of neurinoma of the mixed nerves, the tumoral excision is much simpler for this type of tumor is much less vascularized and only very slightly adherent to the adjacent anatomic structures, particularly the vessels. However, the tumoral growth arising in the angle must be properly evacuated and then removed, and this calls for section of the nerve which is its source and the dissection and safeguarding of the other nerves of the region.

Closure

This must be scrupulous, to avoid escape of CSF. The meatus of the EAM has been occluded at the start of the operation. The auditory tube must be occluded with a fragment of muscle. If a breach has been made in the dura, this is closed by means of a plasty using epicranium or fascia lata. It is vain to attempt to close this breach in water-tight fashion and the best course is to pack the excisional cavity with a large fragment of subcutaneous fat removed from the abdominal wall (Fig. 109). The musculo-aponeurotic plane is then reconstituted. To do this, a large muscle flap is freed at the posterior part of the temporalis muscle. This flap is turned over downwards and sutured to the sternocleidomastoid and the digastric. The posterior border of this muscle flap is sutured to the epicranium and the nuchal muscles. At the anterior border the parotid is restored in place (Fig. 110). Hemostasis must be perfect if the skin is to be closed without drainage to limit the risk of a CSF fistula.

References

1. Bergogne-Berezin E (1985) Diffusion des antibiotiques. In: L'infection en milieu chirurgical. JWPU Anesthésie-Réanimation. Pitié Salpétriére. Arnette éditeur: 121-150
2. Boulard G (1986) Effets de l'isoflurane sur débit et métabolisme cérébral, dynamique du L. C. R. et pression intracrânienne. Applications en neuroanesthésie. In: Journées Méditérranéennes d'Anesthésie-Réanimation Marseille Arnette éditeur: 41-53

3. Brackmann DE (1982) Trans-labyrinthine removal of acoustic neurinomas. In: Neurological surgery of the ear and skull base. Brackmann DE Edit. Raven Press Pub (New York): 235-241
4. Brodersen BR, Barky N (1979) Acoustic Tumor Surgery: Anesthetic Considerations. In: Acoustic Tumors. Vol. II: Management. House WF, Luetje CM University Park Press Publ: 3-14
5. Cannoni M, Pech A, Pellet W, Zanaret M (1985) Techniques et indications des pétrectomies réglées. Intérêt de l'association d'une voie transcochleaire et infra-temporale. Ann Oto-Laryng (Paris) 102: 31-45
6. Cannoni M, Pellet W, Zanaret M, Scavennec C, Collignon G (1985) Les voies d'abord du rocher. Encycl Méd Chir (Paris France) O. R. L. 20052 A 10 28 p
7. Clerc P, Batisse R (1954) Abord des organes intra-pétreux par voie endocranienne. Ann Oto-Laryng 71: 20
8. Cohen NH (1984) Fluid management. In: Newfield P and Cottrell JE: Handbook of Neuroanesthesia: Clinical and physiologic essentials. Little Brown Ed. 154-170
9. Conley J (1970) Concepts in head and neck surgery. Cancer of the ear. G Thieme Verlag Edit (Stuttgart): 69-79
10. Dabadie P, Erny P, Destribats B (1986) Hémodilution et oxygénation tissulaire. Ann Fr Anesth-Réanim 5: 204-210
11. Delgado TE, Buchheit WA, Rosenholtz HR, Chrissian S (1979) Intraoperative monitoring of facial muscle evoked responses obtained by intracranial stimulation of the facial nerve: a more accurate technique for facial nerve dissection. Neurosurg 4: 418-421
12. Dureux JB, Gerara P (1985) Critères de choix d'un antibiotique. In: l'infection en milieu chirurgical. JEPU Anesthésie-Réanimation. Pitié-Salpêtrière. Arnette Edit 151-174
13. Farrior JB (1984) Infra-temporal approach to skull base for glomus tumors: anatomic considerations. Ann Otol Rhinol Laryngol 93: 616-622
14. Fisch U (1970) Transtemporal surgery of the internal auditory canal. Report of 92 cases. Technique, indications and results. Adv Otorhino-laryngol 17: 203-240
15. Fish U (1977) Vestibular neurectomy. In: Silverstein H, Norrel H (edit) Neurological surgery of the ear. Aesculapius Pub (Birmingham - Alab) 144-149
16. Fish U (1978) Infra-temporal fossa approach to tumors of the temporal bone and base of the skull. J Laryngol 92: 949-967
17. Fish U, Fagan P, Valavanis A (1984) The infra-temporal fossa approach for the lateral skull base. In otolaryngologie clinics of North america 17/3: 513-552
18. Fischer C, Ibanez V, Mauguiere F (1985) Monitorage peropératoire des potentiels évoqués auditifs précoces. Presse Med (Paris): 14-37, 1914-1918
19. Gantz BJ, Fisch U (1983) Modified transotic approach to the cerebellopontine angle. Arch Otolaryngol 109: 252-256
20. Garcia Ibanez E, Garcia Ibanez JL (1980) Middle fossa vestibular neurectomy: a report of 373 cases. Otolaryngol Head Neck surg 88: 486-490
21. Granthil C, Léna P, Colavolpe C (1982) Le sevrage de la ventilation artificielle en période post-opératoire chez l'adulte. Ann Fr Anesth-Réanim 1: 617-628
22. Granthil C, Duflot JC, Bouffier C, Latière M, Colavolpe C (1986) Place des épreuves fonctionnelles repiratoires dans l'analyse du risque respiratoire en période de sevrage post-opératoire. In: Mises au point en anesthésie-réanimation. Kremlin Bicêtre. Arnette Edit 1: 89-98
23. Grundy BL, Jannetta PJ, Procopio PT, Lina A, Boston JR, Doyle E (1982) Intraoperative monitoring of brain stem auditory evoked potentials. J Neurosurg 57: 674-681
24. Hitselberger WE, Pulec JL (1972) Trigeminal nerve posterior root retrolabyrinthine selective section. Arch Otol Laryngol 96: 412-415
25. House WF (1961) Surgical exposure of the internal auditory canal and its contents through the middle cranial fossa. Laryngoscope 71: 1363-1385
26. House WF (1964) Monograph of trans-temporal removal of acoustic neuromas. Arch Otol laryngol 80: 587-756
27. House WF (1968) Monograph II. Acoustic neuroma. Arch Otolaryngol 88: 575-715
28. House WF, Hitselberger WE (1976) The trans-cochlear approach to the skull base. Arch Otolaryngol 102: 334-342
29. House WF, Luetje CM (1979) Acoustic tumors. Vol II: Management. University Park Press (Baltimore) 1: 239
30. Jenkins HA, Fisch U (1980) The trans-otic approach to resection of difficult acoustic tumors in the cerebello-pontine angle. Amer J Otol 70-76
31. Le Corff G, Elias A, Bouvier JL, Serradimigni A (1986) Prévention de la thrombose veineuse post-opératoire in: Journées Méditerranéennes d'Anesthésie-Réanimation, Marseille. Arnette Edit
32. Levine RA (1980) Monitoring auditory evoked potentials during acoustic neuroma surgery. In: Neurological surgery of the ear. Silverstein H, Norrel H Edit Aesculapius Publ (Birmingham Alab) 2: 285-293
33. Mangez JF, Roux P, Alibert F, Colas G, Winckler C (1984) Problèmes per et post-opératoires posés par la chirurgie des neurinomes de l'acoustique. In: Résumés des communications. Forum de clubs d'anesthésie et réanimation chirurgicale. Cinquantenaire de la S. F. A. R. Septembre 1984. CHU Pitié Salpêtrière Paris
34. Marcillon M, Marcotte C, Merot S, Blin F, Lereau L, Moisan L, Maestracci P (1986) Le cathétérisme de l'artère radiale en anesthésie-réanimation. Ann Fr Anesth-Réanim 5: 48-57
35. Møller AR, Jannetta PJ (1983) Monitoring auditory functions during cranial nerve microvascular decompression operations by direct recording from the eighth nerve. J Neurosurg 59: 493-499
36. Ojemann RG, Levine RA, Montgomery WM, Mc Gaffigan P (1984) Use of intra-operative auditory evoked potentials to preserve hearing in unilateral acoustic neuroma removal. J Neurosurg 61: 938-948
37. Paillas JE (1975) Système nerveux central. Nerfs craniens. In: Nouveau traité de technique chirurgicale. Tome VI. Masson Edit. 9
38. Panse R (1904) Ein gliom des Akusticus. Arch Ohrenh 251-255
39. Pellet W, Cannoni M, Lavielle J, Lehmann G, Vacherat S, Mouren P (1973) Volumineuse tumeur du glomus jugulaire. Voie d'abord combinée oto-neuro-chirurgicale. Neurochirurgie 19: 567-579
40. Portmann M, Bébéar JP, Lacaze JL (1972) Neurectomie de la VIIIème paire cranienne par voie de la fosse cérébrale moyenne. Cah O. R. L. (Montpellier) 7: 787-793
41. Prass RL, Lüders H (1986) Acoustic (loudspeaker) facial electromyography monitoring. Part I: evoked electromyographic activity during acoustic neuroma resection. Neurosurgery 19: 392-400
42. Raudzens PA, Shetter AG (1982) Intraoperative monitoring of brainstem auditory evoked potentials. J Neurosurg 57: 341-348
43. Richmond IL, Mahla M (1985) Use of antidromic record-

ing to monitor facial nerve function intraoperatively. Neurosurgery 16: 4358-462

44. Ricksten SE, Medegard A, Curelaru I, Gustavsson B, Linder LE (1986) Estimation of central venous pressure by measurement of proximal axillary venous pressure using "half-way" catheter. Acta Anesthesiol Scand 30: 13-17
45. Roux FX, Cioloca E, Constans JP (1985) Bases physiques et techniques de l'hémostase tumorale avec un laser CO_2 en Neuro-chirurgie 31: 412-415
46. Roux FX, Oswald AM, Merienne L, Chodkiewicz JP (1985) Neurochirurgie au Laser CO_2. Bilan de 61 interventions. Neuro-chirurgie 31: 416-420
47. Scherpereel B, Rousseaux P, Bernard MH, Guyot JF (1979) Le non rasage pré-opératoire en neurochirurgie. Neurochirurgie 25: 244-245
48. Sekhar LN, Estonillo R (1986) Transtemporal approach to the skull base: an anatomical study. Neurosurgery 19: 799-808
49. Sekiya T, Iwabushi T, Kamata S, Ishida T (1985) Deterioration of auditory evoked potentials during cerebellopontine angle manipulations. J Neurosurg 63: 598-607
50. Serradimigni A, Chiche G, Elias A (1983) La maladie thrombo-embolique post-opératoire. Encyclopédie Med Chir (Paris) Anesthésie Réanimation 36827 A 10, 10
51. Shapiro HM (1986) Anesthesia effects upon cerebral blood flow, cerebral metabolism, electroencephalogram and evoked potentials. In: Miller RD Anesthesia. Vol 2. 2ème édition. Churchill Livingstone Edit 1249-1288
52. Soulier MJ, Fraysse B, Lazorthes Y, Da Motta M (1985) Potentiels évoqués auditifs per-opératoires dans la chirurgie de la fosse postérieure. Neuro-chirurgie 31: 519-526
53. Sterkers JM (1973) Technique simplifiée de découverte du conduit auditif interne par voie temporale. Ann Otol Laryngol (Paris) 90: 232-325
54. Sterkers JM (1976) Microchirurgie du conduit auditif interne et des régions avoisinantes. Monographie de chirurgie O. R. L. et cervicofaciale. Arnette Edit. Paris
55. Sterkers JM, Batisse R, Gandon J, Cannoni M, Vaneecloo FM (1984) Les voies d'abord du rocher. Arnette Edit. Paris 1: 195
56. Wigand ME, Rettinger G, Haid T, Berg M (1985) Die Ausräumung von Oktavusneurinomen des Kleinhirnbrückenwinkels mit transtemporalem Zugang über die mittlere Schädelgrube. HNO 33: 11-16

Applications

W. Pellet, M. Cannoni, and A. Pech

In Collaboration with
S. Malca, J. M. Thomassin, J. M. Triglia,
S. Valenzuela, and M. Zanaret

It is not difficult to have one idea. What is difficult is to have every idea.
(Alain)

The petrous or peripetrous topography makes it possible to distinguish, among the intracranial tumors, a quite particular group which confronts neurosurgeons with altogether specific problems inherent in their localization, their possible extension and, for some, their singularly "vascular" character. Although comprising tumors of very disparate nature in terms of their histology, site of origin and size, the very preferential location at the posterior aspect of the petrous gives this particular group a reputation for operative difficulties. These arise from the proximity of the brain-stem, the small size and awkward access of the osteodural dihedron in which they develop and the complexity of their anatomic relations, whether neurovascular, sinusoidal or otologic. This reputation is worsened by the existence of intrapetrous extensions pertinent to the intrapetrous origin of the tumor, or, on the contrary, by its invasive potential, always facilitated by the cellular structure of the petrous, and also sometimes by extradural extensions, especially through the posterior lacerate canal or through the cavities of the ear. The classical neurosurgical approaches do not afford proper exposure of these extensions, so that tumoral eradication appears a matter of chance or even frankly impossible. Finally, the particularly "vascular" nature of some of these tumors, especially those of the glomus jugulare, involves the risk of very profuse operative bleeding, the more formidable because their blood-supply comes from multiple peri- and also intrapetrous sources whose flow cannot be interrupted without complete exposure of their intrapetrous course. All this means that the otoneurosurgical routes we have just described offer neurosurgeons the solution to problems which, by themselves, they would be incapable of mastering. In fact, drilling of the petrous provides a simultaneous solution to almost every problem: direct access to the tumor, exposure of the nerves, in particular of the facial nerve, in a healthy area, complete clearance of all the intrapetrous and extracranial extensions, exposure of all the vascular pedicles, all with the assurance of a perfect view of the cerebellopontine angle and, if need be, its borders (foramen of Pacchioni, foramen magnum, clivus).

It is true that these otoneurosurgical routes of access, which usually traverse the labyrinth, as a rule sacrifice hearing on the side of the approach. This problem will be discussed later, but we may stress the following points here and now: (1) the preservation of hearing really arises for only a small proportion of cases (at most 10%), the true position being that, at the time of operation, the tumor is such that it is usually bound to cause definitve deterioration of the ear involved (2) that this problem must be discussed in a global context taking into account the prognosis for life, the perfection of the excision and the preservation of the facial nerve, and (3) that the suprapetrous route offers the possibility of safeguarding both hearing and the facial nerve in those cases where this problem really arises. This does not mean that, in our view, the classical neurologic approaches - whether suboccipital, subtemporal or transtentorial - should be abandoned. Quite the contrary, for we employ these whenever the tumor remains strictly intradural, such as a cholesteatoma of the angle, or when its nature and site suggest that still functioning facial and cochlear nerves may be safeguarded without great difficulty, a practicable objective for certain trigeminal neurinomas or petrosal meningiomas. The indications can be discussed in relation to the type of operation, whether suprapetrous, transpetrous or extended transpetrous, in terms of the region whose exposure is desired: the IAM alone, the CPA or the posterior lacerate canal, or even in terms of the size of the tumor. We think it more practical to base the discussion on the histologic nature of the lesion, considering that the diagnosis is usually made before operation thanks to the information provided by modern paraclinical investigations, and that in any case similar gross appearances usually call for similar surgical solutions.

Therefore, with special regard to the frequency of histologic types, we shall discuss successively the indications for the neurinomas, the glomus tumors and finally for the other tumors, among which the meningiomas certainly have pride of place.

The Neurinomas

In our experience [142], as in the majority of reported series [46, 90, 105, 189], in 9 cases out of 10 the tumor approached is a neurinoma. Certain authors, such as Sterkers et al. [172] or Vaneecloo et al. [185] report even higher figures (96% for the first, 94% for the second), which may be explained by the more otologic nature of their practice. In our experience and that of others, 95% of these neurinomas are so-called "acoustic" neurinomas. Thus this anatomic-clinical type forms the heart of the matter and a great part of the discussion will be devoted to it. The neurinomas arising from the other nerves of the region, which are much rarer, will be discussed separately, since their special topography calls for appropriate solutions which cannot be provided by the approaches usually employed for neurinomas of the VIIth nerve. When we are dealing with a neurinoma of the facial or mixed nerves, including the very rare neurinomas of the hypoglossal, the solution seems preeminently otoneurosurgical. On the other hand, our attitude to trigeminal neurinomas is more often neurosurgical.

Neurinomas of the VIIIth Nerve

These form the prototype indication, and it is their context that we generally discuss the essential problems raised by surgery of tumors of the cerebellopontine angle, a discussion worth developing here to give some support to the consistently otoneurosurgical attitude we have adopted.

The Problems

These are numerous. Some arise in endeavoring to attain the objectives universally regarded as imperative: not to damage the brain-stem, to remove the entire tumor, to respect the facial nerve. Another has arisen more recently: to preserve hearing if at all possible. The letter results from the difficulty in the early discovery of these tumors while they are still of small size.

The Imperatives

Not to damage the brain-stem, so as to guarantee the vital prognosis. Several factors may be involved in producing such lesions.

• *Direct trauma.* This is primarily the outcome of too sharp surgery, which is why the "blunt finger dissection" of the early authors, efficient but hasty, has long been excluded from the surgical repertoire. There is no need to stress the dominant role of optical magnification in the decisive improvement in operative results acquired in the last twenty years. Nor need we stress the importance of the most careful procedure throughout the dissection, which should be as anatomic as possible, using when possible, the plane which separates the arachnoid envelope from its tumoral content, the best method of safeguarding the neurovascular structures situated outside this covering. We feel that the very anterior approach afforded by the extended translabyrinthine route particularly facilitates the exposure of this arachnoid layer from the aperture of the IAM to the ring of reflection on the tumor (Fig. 28). It is unnecessary to revert to the value of tumoral evacuation, which, by simple reduction in size, allows stripping of the periphery from its parenchymatous bed, or to the usefulness of the vacuum rotary dissector of Hurban House, its intermittent suction is the most effective method we know for performing this manœuver of evacuation and stripping.

On the other hand, one must stress the danger of the retractor pressing against the cerebellum, even if, as is usual, it is covered by a protective swab. First, there is the purely mechanical effect, acting directly on the neural parenchyma and, in certain cases, a distortion of the cerebellar peduncle and brain-stem secondary to hemisheric retraction. A vascular ischemic effect is probably added to this tissular effect, resulting in cell damage manifested rapidly by the flattening and then the disappearance of electrical activity, a phenomenon long familiar. Numoto and Donaghy [43, 130] have shown that this suppressive effect is proportional to the pressure exerted, and that the cortical "extinction" sets in when the pressure exerted on the cortex reaches 35 cm of water. Bennet et al. [12] emphasize the vascular mechanism of the phenomenon by showing that the suppressive effect of evoked proprioceptive potentials received at the cortex depends on the differential existing between the perfusion pressure and the pressure exerted on the cortex. This ischemic effect is accompanied by damage to the capillary walls [96], intraluminal obstruction [153] and probably disturbances of the blood-brain barrier [1, 2]. In view of the anatomic arrangements, it may be imagined that these mechanical and vascular consequences immediately subjacent to the retractor blade, have more diffuse regional effects throughout the posterior fossa, a closed compartment whose compliance is reduced because of the occupying mass. It seems to us that the very anterior approach of the ETL route, initially exposing the external pole of the tumor, avoids any major traction, the only role of the blade placed on the dura being to obliterate the sigmoid sinus and barely depress the cerebellar hemisphere. In any case, this procedure is automatically limited by the fact that the retractor is placed on the dura mater and not directly on the cerebellum, and that the impaction is restricted by adhesions of the sinus to its bony groove. It must also be realized that the retractor has even more diffuse effects of a general nature by indirectly producing variations in blood-pressure and heart rhythm. Laha et al. [109] have clearly shown the frequency of systemic disturbances provoked by placing a retractor on the cerebellum, in particular a marked fall in arterial pressure in 60% of cases and bradycardia in 46%. These authors have also stressed the aggravation of these problems when the patient is operated in the sitting position, a maneuver which produces a fall in blood-pressure in half the cases. Hence the usefulness of operative positions which diminish this seated position and the value of the otoneurosurgical routes, which are performed with the patient in dorsal decubitus.

• *The vascular risk.* Besides the role of the retractor in compressing the cerebellar vascular bed and of more general pressure variations, the pontobulbar consequences of lesions of the anteroinferior cerebellar artery have been known since the work of Atkinson [5]. During his autopsies, this author had noted the existence of a softening of the lateral tegmen of the pons, always with an interruption of this artery by a wound or clip. We should stress that in all these cases Atkinson also noted that the caliber of the postero-inferior cerebellar artery was always very fine, suggesting that the latter had been unable to take over the territory of its interrupted neighbor. In fact, there seemed to exist a sort of anatomic predisposition produced by the relative atresia of one artery, whereas under normal conditions the ordinary diameter allowed perfect compensation by means of the anastomoses. This explains why interruption of the AICA does not always have the same effects and why some authors have maintained that it could be interrupted without ill-effects. However this may be, the softening of the brain-stem is a frequent cause of death together with more general causes (infarction, embolism, infection). House [90] believes that 30% of the deaths encountered are due to an "AICA syndrome". Many authors, like Olivercrona [132], believe that softening of the brain-stem is the origin of 75% of deaths, but that this is not solely determined by obstruction of the AICA, other factors being often associated (direct trauma, ischemia under the retractor, fall in systemic pressure, anesthetic problems, disorders of ventilation). Whatever the truth, the studies of Atkinson [5] have had the merit of underlining the importance of respect for the vessels of the posterior fossa; and "intra-arachnoidal" dissection, with coagulation of those vessels only contained in the sheath is without doubt the best means of attaining this objective.

• *Air embolism.* This is a potential risk whenever the patient is operated in the sitting position, or even half-sitting, and the recent technical methods of detecting it peroperatively have shown that it is probably commoner than usually thought. Albin et al. [2] found such a complication in 20% of their operations in the sitting position, while experimentally Laha et al. [109] detected an air embolism in 53% of their dogs operated in the sitting position. Of course, the gravity of this accident depends on the amount of air aspirated. Sometimes minimal and unrecognized, it may also produce failure of heart pumping. The severity depends very much on the possibilities of passage of air from the right atrium to the left-sided cavities through a still-patent foramen ovale, a possibility encountered in 35% of cases according to Hagen at al. [76]. The bubbles may then lodge anywhere, but with particularly serious consequences when they obstruct the coronary arteries or the cerebral arteries. In certain cases (massive embolism, myocardial infarction) the accident is immediately obvious during operation, but it may pass unperceived and, by the ischemic lesions it produces, give rise to various postoperative cardiopulmonary, cerebral or parenchymatous complications. Numerous expedients have been suggested, either to reduce the risk (pneumatic head tourniquet, intermittent cervical tourniquet, positive pressure ventilation, anti-G suit) or to immediately detect the air (constant auscultation, Doppler monitoring, capnography in particular) or to evacuate it (a Swann-Ganz catheter in the pulmonary artery). Recently, Lechat et al. [111] have suggested the preoperative detection by contrast ultrasound of a patent foramen ovale, a useful method for preventing the risk of cardiac or neurologic complications but one that is not entirely reliable since it does not necessarily detect the partial defects of impermeability which may open up in the intracavitary hemodynamic conditions created by embolism. In practice, as the seated position is the chief factor, it is logical to avoid it as far as possible. Therefore many surgeons use less upright positions: the position of Mount, ventral, lateral or even dorsal decubitus, the use of a Gardner's stirrup allowing good exposure of the suboccipital region. The ETL route and its position in dorsal decubitus find support here.

Removal of the entire tumor to prevent the risk of recurrence, which is very common in cases of incomplete resection as shown by German [62], who reviewed the patients in whom Cushing had performed only a simple intracapsular evacuation and found a 5 year recurrence rate of 56%, a figure quite similar to that of Olivecrona [132] who regretted his 40% of recurrences when he had been unable to do a complete resection. Such recurrence has a bad prognostic reputation, as evidenced by the 50% postoperative mortality reported by Pennybacker [143] or the 41.5% of Horrax and Poppen [86]. It is certain that present-day operative conditions have radically modified the data of the problem. It is also certain that the experience each surgeon acquires helps to improve his results. The best proof of this is the analysis of successive series published by House [87, 88, 90]. His percentage of incomplete excisions fell from 30% to 21% and finally to 6.8%. Despite his unique experience, it has to be stated that he may plan a partial excision, usually for reasons of advanced age or other defects (0.5% of cases), or settle on leaving some fragments (4.2%), usually because of the occurrence during operation of autonomic disorders or disturbances of arterial pressure. Most authors now aim at a total resection rate of around 95% of cases, a success rate which explains why recurrences have become much more rare. However, it is difficult to form a precise idea since few authors divulge their figures in this context. What is certain is that the risk is very different depending on whether the residual fragment is large, or if it is a matter of micro-residues which seem most likely to degenerate *in situ*, especially if it has been possible to coagulate them. House [91] emphasizes that the risk of recurrence from these microfragments should be less likely since they are in appearance "devascularized"; and in this context he believes that the fragments abandoned on the brain-stem are very likely to be less "viable" than those which have been in the IAM close to the afferents which supplied the tumor. We feel it important to stress that the quality of exposure of the IAM is very certainly an important consideration under the circumstances, and that the technique of trepanation of the IAM via the posterior fossa offers an undeniably restricted exposure of the fundus of the meatus, especially when one has decided to try to preserve hearing. Now, we have noted in the course of our experience that very few acoustic neurinomas (about 20%) do not reach to the fundus of the meatus, whereas (and this must be stressed) if 63% do reach the fundus of the meatus, 17% ascend into the first part of the facial canal, from which only the ETL route can dislodge them. In our view, this is a major argument when selecting a route of approach. It is also a reason for remaining very cautious about any affirmations that may be made as to the certainty of the completeness of excision, especially since it is known that it takes at least 5 to 6 years for a recurrence to become clinically obvious after leaving a somewhat large fragment, while the time required for a microscopic fragment to give rise eventually to obvious recurrence is simply unknown. The future may provide information on this subject.

Safeguarding the Facial Nerve. Now that microsurgery and progress in anesthesia and intensive care have radically transformed the vital prognosis, surgical preoccupations have turned to the functional prognosis. Hence the safeguarding of the facial nerve has gradually become a priority.

It is a very classical concept that the facial nerve is, paradoxically, often spared before operation. In our experience (Triglia [182]), only 14% of our patients had facial symptoms, and in the great majority of cases (95%) this amounted to only a very partial deficit, limited at the most to a positive Souques sign or a minor facal asymmetry. Every author agrees on this point and, in view of the cosmetic, psychologic and especially ophthalmologic consequences of postoperative paralysis, they take every step to prevent it. Rougerie and Guyot [155] were the first in France to raise this problem, stressing from the outset that the difficulties to be resolved during operation resulted from imprecise location of the facial nerve, its tedious dissection and the existence of growth within the canal.

Imprecision in location is due primarily to the variability in position of the nerve in relation to the tumor. The latter, arising in the IAM from the (mainly inferior) vestibular nerve, has every chance of pushing the facial nerve in front of it and flattening it against the anterior wall of the meatus and the anterior rim of its aperture. Koos [106] found this in

50% of cases, but the random nature of tumoral growth, possibly related to blood-supply, explain why the facial nerve may be found in a very different position, often antero-inferior (21% of cases) but also superior, at the ceiling of the meatus, in 11%, on the meatal floor in 11% and even in a posterior position in 7%. While these variations are not very serious as long as the nerve remains relatively sheltered in the meatus during the drilling, it is certain that the nerve is particularly exposed in a posterior situation during trepanation, the more so since the surgeon does not expect to find it there and drills without special caution. In general, the position in the meatus determines the subsequent position of the nerve on the tumoral convexity in the angle. Most often, the nerve is slung over the internal or infero-internal pole, but it may be found anywhere on the convexity including the outer pole, to the extent that the latter must be approached only with great caution and always after assessment of the approximate path of the nerve. We have also at times encountered a nerve which suddenly curves to make a complete change of direction by virtually 90°. This is a quite dangerous situation, especially if the nerve is sought through the tumor. But in fact, rather than the variability in position of the nerve, it is its spreading out and especially the separation of its fibers which makes its identification difficult. Koos [106] states that in two out of three cases the nerve trunk is unidentifiable, spread out over the tumoral convexity and very difficult to distinguish from the "capsule", so that it may be damaged merely in searching for it. It only a third of the cases is it well-marked, collected in an identifiable bundle, the most favorable conditions for locating it. Many methods have been recommended in attempting to locate it, the simplest being to keep a hand on the cheek or to monitor the face with a rear-view mirror under the drapes to detect the least twitch. Currently, peroperative monitoring with a loudspeaker appears the method of choice. In practice, whatever method, however sophisticated, is used to detect a noxious influence on the facial fibers, it must be recognized that the warning comes after the event, the damage having already been done. This is why peroperative electrical stimulation seems to provide supplementary information since it permits immediate establishment of the direct relation between the point stimulated and the motor or sound response. The stimulator is to the neurosurgeon what his white stick is to a blind man, but one cannot help thinking that "blind" surgery is undoubtedly more hazardous than when it is clear-sighted.

In fact, the facial nerve has two fixed points, one at its emergence from the brain-stem, just medial to the vestibulocochlear nerve and the choroid plexus, the other at the fundus of the IAM where it enters the facial canal. It seems logical to try to identify it as these sites and it is undisputed that the most constant landmarks are to be found at the fundus. In practice, as soon as the tumor is sufficiently bulky to come into contact with the brain-stem or, worse, to deform it, the usual landmarks - choroid plexus, nerve emergence, pontomedullary groove and vein - may be unidentifiable or even displaced to the margins of the depression in which the neurinoma is embedded. Identification is thus awkward or even dangerous. On the contrary, the bony nature of the landmarks in the fundus of the meatus, the falciform crest, Bill's bar and the entry to the facial canal, guarantees their fixity. Only those cases where there is invasion of the first part raise more difficult problems, but drilling of this first part, admittedly not easy for the neophyte, allows exposure of the facial nerve distal to the tumor and thus of the plane of cleavage between the nerve and the neurinoma. Theoretically, trepanation of the meatus should offer the same advantage but the angle of approach, the depth of the trench that must be excavated to reach the fundus of the meatus and the impossibility of extension if the integrity of the labyrinth is to be respected ensure that access via the posterior fossa is necessarily more restricted. As for exposure of the facial nerve by this method, except in the 20% of cases where the tumor does not reach the fundus of the meatus, it is much more difficult, because vision is very restricted or even impossible. This compels one to pull somewhat blindly on the end of the tumor, which increases the risk for the nerve and also the danger of leaving a fragment of tumor behind. It is doubtless to avoid these difficulties that some authors, like Malis [119] in the USA or Fischer [81] in France, advise seeking the facial nerve at the convexity of the tumor after having cut the neurinoma in two. On reflection, this method is very sensible if one recalls that the tumor is clothed in an arachnoid layer and that the facial nerve, once past the adhesions at the meatal aperture and the zone of reflection of the arachnoid layer, lies free in the ponto-cerebellar cistern, outside the arachnoid covering. It is certainly the case that the nerve is most readily detached at this site. However, it seems to us that in the absence of precise landmarks and in the knowledge that the position of the nerve on the tumoral convexity is variable, and because section of the tumoral substance rapidly mars the normal appearance of the tissues, this technique cannot be very easy; and, as it does not deal with the problem of excision in the meatus, it does not constitute a decisive argument for choosing the suboccipital route rather than the extended translabyrinthine route with its several advantages.

Dissection of the facial nerve is difficult. Even when it is suitably isolated and identified, the nerve - stretched, flattened and its fibers separated - is very difficult to follow from end to end. To begin with, it is difficult to find the plane of cleavage. Where it emerges from the brain-stem, unless the tumor is small enough (stage I or II) to leave the origins of the nerves free, one must start by dividing the cochlear and vestibular fibers which serve as a form of attachment to the brain-stem. In view of the proximity of the nerves at their emergence, section of these fibers involves the risk of also cutting the emerging facial fibers, the more so since there often exists at this site a pontomedullary vein, enlarged because it drains the tumor, whose coagulation or, worse, bleeding result in local readjustments which subsequently greatly hamper identification of the normal structures. It is true that tumoral evacuation facilitates the maneuver, but as soon as the tumor is just a little smaller the profuse vascularization at the approaches to the pons and the pontomedullary groove considerably hampers the dissection. We have seen that transection of the tumor makes it possible to locate the facial nerve more distally on a segment where it is more easily cleavable, but this method, however attractive, does not seem very easy and certainly calls for surgical experience. In the meatus cleavage is relatively easy, even though the nerve is directly in contact with the tumoral substance. With the suboccipital approach and trepanation of the meatus, the great problem is to find this plane of cleavage by first traversing the tumoral tissue or by pulling on the end of the intracanalicular portion of the growth. In view of the flattening of the nerve, its approach cross-wise and also its variable position, this is a maneuver with numerous hazards. On the contrary, the ETL approach which begins by opening the fundus of the meatus and exposes the entry to the facial canal, permits immediate discovery of the plane of cleavage. Further, the direct view of the latter and the access to the nerve along its axis and not cross-wise greatly simplify the maneuver provided one is sufficiently meticulous and adequate instrumentation is available. Dissection is also complicated because there almost always exists, even in the case of small tumors, a zone of tight adhesions in the vicinity of the aperture of the IAM. Several reasons may be invoked to explain their existence at this site. First, this is the constriction zone through which the tumor becomes embedded in the angle. Next, it is the zone of reflexion of the arachnoid sheath (Fig. 28) and the adhesion of the two layers probably gives rise to a dense fibrosis. It is also the zone of attachment of the tumor to the dura mater. Here, it receives its blood-supply from the meningeal branches derived, at the inferior border, from the meningeal branches of the ascending pharyngeal artery and, at the upper border, from the last branches of the superior petrosal artery which arises from the middle meningeal just above the spinous foramen. Finally, and perhaps most important, it is here that the tumor usually obtains one or even two or three branches derived from the arteries of the facial nerve itself. This region of the aperture is always the most tricky to cross, especially as the sudden expansion in size in the CPA sharply bends the already spread-out nerve ribers over the edge of the orifice. Finally, if the cleavage plane is difficult to find and follow and if the tight adhesions are troublesome to separate, it is certain that the dissociation of the fibers spread out over the tumor will seriously complicate the maneuver and that flattening of the nerve will be more marked if the tumor is bulky. If, as we believe, the ETL route is the method offering the surgeon the greatest chance of safeguarding the continuity of the facial nerve, it follows that this technique is so much the more indicated when the tumor is a large one.

The Problem of Hearing

Since the surgeon is capable of respecting the continuity of the facial nerve, it would seem logical for him also to try to preserve hearing. However, it must be stressed, before any discussion on this point, that while the patient usually presents for operation with a normal or near-normal facial nerve, this does not apply to hearing, disturbance of which is generally the earliest and often the most marked manifestation. It is obvious that, in principle, the ETL approach leaves this problem out of account. It is proper, therefore, to consider the matter seriously before using it. Many questions may be raised, and we discuss them under three headings:

Audiometric Problems

• *The contralateral ear.* To begin by considering the problem of the ear opposite to the tumor may appear paradoxical, and yet it is certainly the essential preoccupation since it is the state of this ear that determines all subsequent considerations.

- In the great majority of cases it is normal and, though some authors have reported very exceptional secondary changes [30, 72], it will probably remain so. In these conditions, the loss of one ear may be acceptable as a minor handicap for the most part, except perhaps for certain patients who may possibly find their professional activities somewhat hampered as a result (teachers, lawyers, sales staff). House [91] notes that many patients operated by the ETL route believe they can hear with both ears after operation. We have made the same observation in a fair number of our patients, some even maintaining that they heard better than before, although our procedure had destroyed their labyrinth. It may be that the preoperative hearing in these patients, although not useful, was associated with distortion of the sound message giving rise to poor understanding. The suppression of this distortion would then give them an impression of improvement. Under these conditions, the preservation of hearing might be regarded as rather a "second best", especially as the functional value of the postoperative auditory function is often very debatable. This is why, in such cases when the opposite ear is healthy, it is worth discussing the utility of preserving hearing.

- The problem is quite different when the opposite ear is affected. In that case, safeguarding the least possibility of hearing is imperative, and to succeed in this many would advise therapeutic decisions that would seem unacceptable in other circumstances: partial resection, even abstention from surgery with regular monitoring, the delay being put to good use for teaching the patient sign language. True, the severity of these cases must be affected by the cause of the contralateral involvement. We know that presbyacusia or hypoacusia due to exposure to noise only rarely progress to complete deafness. It might be that a chronic otitis could be definitively eradicated after appropriate treatment and the hypoacusia stabilized. On the other hand, when it is a case of von Recklinghausen's disease the problem is very difficult or even insoluble, total deafness being the usual outcome for most of these patients once they present auditory localization.

Table 1. Distribution of tonal audiometric levels related to size of tumor

auditory loss		25 dbs	50 dbs	80%
stage I	6%	37.5%	37.5%	19%
II	3.6%	25%	35.7%	35.7%
III	6.8%	34.5%	20.7%	38%
IV	3.5%	31.5%	28%	40%

Table 2. Distribution of stage I and II neurinomas with hearing-loss of less than 50 dbs in which an attempt to preserve hearing may be considered

Stage I: (11%)	43.3%	4.7%
Stage II: (23%)	28.6%	6.5%

• *The tumoral ear.* This is affected in the very great majority of cases. In the light of our experience with 232 acoustic neurinomas operated since 1973: 11% stage I (intracanalicular), 23% stage II (less than 2.5 cm), 27% stage III (between 2.5 and 3.5 cm) and 39% stage IV (over 3.5 cm), we can produce figures perfectly consistent with those generally published (Table 1):

- Hypoacusia exists in over 95% of cases and only 4.5% of patients with an acoustic neurinoma can be regarded as having normal hearing
- A third of the patients (31.5%) still have useful hearing, the auditory loss at pure tone audiometry being less than 50 dbs
- There is a predominance of total deafness with the large tumors (40% of stage IV), but the small tumors sometimes cause as much damage (19% for stage I). This means that the auditory involvement is not directly proportional to the size of the tumor, and this is confirmed by the average level of auditory loss which is more or less comparable whatever the stage of the neurinoma (61 dbs for stage I, 73 dbs for stage II, 71 dbs for stage III and 74 dbs for stage IV).

As for speech audiometry, the data in the literature show that 30% of acoustic neurinomas are accompanied by correct discrimination (60% or more of words presented), which corresponds on the basis of the generally accepted classification to good discrimination as against poor discrimination (between 30 and 60%), very poor (0 to 30%) or nil discrimination.

• *What hearing should be preserved?* For efficient fuction in bicochlear hearing, the sole system capable of providing a spatial sense of sound, the interauricular difference must not exceed 25 decibels. It cannot be denied that it is possible to preserve such bicochlear hearing in the rare cases where it still exists. The improvement of damaged hearing so as to reestablish bicochlear audition constitutes a real success obtained by some [33, 51, 68, 91, 161, 177] but in actually exceptional cases. Between 25 and 50 decibels of hearing loss, a much commoner situation, the ear is still useful because it allows the patient to still seize some scraps of conversation, but the usefulness of this ear is limited. Beyond 50 decibels loss, the subject uses only the sound ear and, even with its residual hearing capacity, the affected ear may be considered useless. As pointed out by Clemis [31], at this stage the discussion is of only academic significance. One must also stress the risk of simultaneously preserving some very unpleasant auditory disturbances, always more difficult to bear when some hearing persists and which may even cause the patient to regret the preservation of his cochlear nerve (as we have noted in one case). We sum up by saying that it is good to try to preserve hearing, but not any kind of hearing.

Anatomic Problems

The integrity at the end of the operation of the entire cochleovestibular apparatus (cochlear nerve, internal ear and blood-supply) is the condition *sine qua non* of success. But respect for this integrity may not always be compatible with the demands of tumoral surgery if this is to effect complete excision.

• *The cochleovestibular apparatus.* The cochlear nerve, classically flattened out by the tumor like the facial nerve, must be progressively dissected. Not to mention the technical difficulties, this dissection may come up against various impossibilities arising from the fact that the natural course of the neurinomas is towards invasion and destruction of the cochlear nerve.

According to Ylikoski [190], initially the neurinoma invades only its nerve of origin, usually the inferior vestibular nerve, the other nerves of the meatus being perfectly identifiable (stage I of Ylikoski). At a more advanced stage, the tumor infiltrates the cochlear nerve, which may appear grossly unaffected but is found to be invaded at histologic examination: this is stage II. Later, the nerve will be totally destroyed: stage III. Neely [129], studying ten apparently unaffected cochlear nerves removed during operations for neurinomas, confirms this microscopic infiltration even though these were stage I or II tumors. It is therefore possible that the dissection of a cochlear nerve the continuity of which is apparently respected, may in fact consist in freeing a bundle of nerve fibers stuck together by tumor, indicative of an incomplete excision. This ought considerably to limit the indications for preservation of hearing once the tumor, having reached a certain size, has invaded the CPA, even if the cochlear nerve appears quite separate to the surgeon. It should also be noted that the auditory state does not allow any prejudgement of the state of the cochlear nerve. Dandy [35] had already shown, in 1934, that an incomplete but major section of the cochlear nerve could produce only a minor loss of hearing for high-pitched sounds. Neff [130] confirmed this idea, and Schuknecht [160] showed that interruption of 75% of the fibers might produce no loss of the auditory threshold for pure tones provided the organ of Corti remained intact. Therefore, the fact that preoperative hearing is preserved does not mean that the surgeon has a greater chance of finding an unblemished nerve. Of course, this is not an argument for rejecting attempts at preservation, but it is an important reason for moderating the hopes placed in these attempts.

The generally accepted separation of the cochlear and vestibular components is not always effective; Rasmussen [150], in his now classic study, clearly showed that a quarter of the vestibular fibers were in fact mixed with cochlear fibers. This intermingling of fibers at once suggests that the plane of surgical cleavage may not correctly respect the cochlear trunk. Further, this interpenetration of cochlear and vestibular fibers can only promote propagation of the tumor from the vestibular into the cochlear nerve. This probably accounts for the very early infiltration of the cochlear nerve, even at stage I, as reported by Neely [129].

The internal ear must be scrupulously respected. As is known, the least infraction of the bony housing of the vestibule or cochlea is immediately manifested by definitive cophosis. This is one of the difficulties in trephining the IAM for better access without infringing the bounds of the bony labyrinth. We very much doubt the possibility, reported by Palva [136], of preserving hearing despite opening the labyrinth by immediately occluding the breach. On the other hand, it may happen that former disease has produced a very progressive obstruction of the vestibule while respecting the cochlea, to the extent that in the absence of any communication an infraction of the vestibule may have no effect on the cochlea. Such a case has been reported recently by Palva and Johnsson [136].

It is perhaps less wellknown that obstruction of the endolymphatic sac or duct may also lead to progressive cophosis in the mid-term. Kimura [103] has shown this experimentally very clearly; and this probably accounts for certain cases of progressive deafness developing after operation even though the surgeon had perfectly respected the acousticofacial bundle, the inner ear and vessels, but had coagulated more of less widely the dura mater clothing the petrous external to and below the IAM. This is one more snag to beware of when incising, rasping and coagulating the dura in order to burr the posterior face of the meatus, another obstacle to good exposure of the meatus.

• *The blood-supply of the internal ear*, especially of the cochlea and cochlear nerve, largely conditions auditory function. Perlman et al. [145] have shown that five minutes' obstruction of the labyrinthine artery produces degenerative lesions of the cochlea and vestibule. The ciliated external cells are the most sensitive. The vascular stria and the spiral ligament are less affected. With prolonged obstruction, atrophy of the entire organ of Corti and the vestibule develops with fibrosis and ossification of the perilymphatic spaces. These facts are well illustrated by the very fine histopathologic study of Belal, Lynthicum and House [11]. The problem for the surgeon, therefore, is to respect the labyrinthine blood-supply, and this is particularly difficult in view of the fact that there are several small branches and not just one large labyrinthine artery which would be easier to dissect. It is then that the surgeon may be tempted to abandon some fragments on the nerve. House [91] remarks, as we have already stressed, that, while the tumoral periphery adjacent to the nerve trunk seems to receive the last blood-supply and constitutes a "necrotic" zone where it is perhaps not too serious to leave some fragments, the zone situated in the internal acoustic meatus is particularly well vascularized and constitutes an "active" zone, capable of being the origin of a recurrence should the surgeon ever leave in place any still well-supplied tumoral fragments.

• *The internal acoustic meatus*. The precise measurements made by Geurkink [63] and more recently by Domb and Chole [42] are well-known. They have shown that trephining of the posterior wall of the meatus, if it is to respect the vestibule, can only expose 80% of the length of this meatus. Taking account of the oblique external view afforded by the suboccipital approach, it is never possible to expose the fundus of the meatus and the entry of the different nerves into their respective channels. In view of the frequency with which the tumor reaches the fundus of the meatus (63%) and sometimes even ascends into the first part of the facial canal (17%), it may be imagined that exposure of the nerves may be difficult by this posterior route and that the risk of leaving a fragment is considerable.

This is why trephining the roof by the suprapetrous route seems more indicated, since it allows one to reach the fundus of the meatus without opening the labyrinth.

Tactical Problems. The attempt to preserve hearing does not relieve the surgeon of his obligation to respect the three great principles of the surgery of acoustic neurinoma, as previously set out (see under "Imperatives").

The preservation of hearing should be a "plus" and should in no case add to the risks of this surgery. In proposing this "plus" to the patient, one must be able to guarantee him chances equivalent to those offered to other patients.

It is indisputable that the handicap of a postoperative facial palsy is far more difficult for the patient to bear than that of unilateral deafness which, in any case, existed to a greater or lesser degree even before operation. It would also be regrettable if a recurrence should curtail or compromise the benefits of an operation that had preserved hearing. In view of the anatomic data supplied by Neely [128], we do not think it prudent to routinely try to preserve the cochlear nerve, which really seems, except in certain privileged cases, to be "a nest for schwannomas". Without denying the importance of the preservation of hearing, it therefore seems necessary to keep it in its proper place and to stress that it comes well after the other imperatives imposed by this surgery.

The Problem of Early Diagnosis

The increasingly early detection of acoustic neurinomas may be expected from the routine investigation of the least vertigo or instability, especially of any unilateral hypoacusis, above all from modern paraclinical techniques of investigation. In theory, such early diagnosis should resolve every problem, simplify operative risks and lessen the risk to life, facilitate the dissection of the facial nerve and the preservation of hearing. In fact, these hopes have remained in great part vain. All authors agree that the incidence of the intracanalicular forms (stage I) is not increasing and that, even with well-directed and vigilant teams [53, 88, 90, 169], it is usually less than 10%. Even if the field is extended to tumors of 1,5 mm (small stage II), as Glascock does [68], the figure is no higher than 17%. However, the picture is not altogether negative since the diagnosis of neurinoma of the VIIIth nerve is made much sooner than formerly; but this has much more influence on the clinical than on the anatomic condition. The tumor discovered is in practice as large as formerly (60% stage IV – over 3 cm – in Fischer's series [53] and 39% in our own), but the patients are no longer in the disturbing "neurosurgical" condition that used to cloud the vital prognosis from the outset. This certainly constitutes progress since the patients present for operation in much better condition, but this very disparity between the clinical and pathologic states often makes it very difficult for the surgeon to explain to the patient why the operation must be regarded as very serious despite the mildness of his symptoms. It is often difficult to appease the patient's incredulity with a valid explanation. It is probable that the "vascular" factor plays an important part in the sometimes quite rapid development of symptoms (vertigo, unsteadiness, sometimes deafness), either because a long compressed labyrinthine or cochlear artery has finally occluded or because changes within the tumor (necrosis and edema, lipid degeneration, hemorrhage) have produced a rapid increase in size. But how is one to account for the prolonged quiescence of the majority of cases, since, even nowadays, in our experience [142] two out of three tumors in the angle exceed 2.5 cm in diameter (stages III (23%) and IV (39%)), and conversely how is one to explain the early expression of the intracanalicular forms? Events suggest that these latter represent a particular anatomical-clinical type. Our experience has led us to note certain facts (Triglia [182]).

- The intracanalicular neurinomas are mainly the type A neurinomas of Antoni (62% of cases), whereas the same histologic type is observed in only 31% of large type IV neurinomas (over 3,5 cm in the angle). It might be thought that the Antoni type A tumors are younger tumors of more compact texture, whereas the type B lesions are altered and older tumors. If such is the case, why do the large type B tumors not become clinically manifest when they are younger and of type A? It could be objected that the course is not similar in the small and large tumors; but then, how can it be explained that we have observed exactly the same number of each of these two histologic types (41% A, 41% B and 18% mixed A and B) whereas the large tumors predominate in our series?
- The intracanalicular neurinomas expand the IAM more often (94% of cases) than do the large neurinomas (stage IV: 67%). This seems rather surprising at first sight, but may demonstrate that the growth of the former remains strictly within the canal, not because they have been rapidly discovered but because they have not succeeded in extending into the angle as the neurinomas usually do.
- Hearing loss is almost the same for the intracanalicular neurinomas (61 dbs average loss) as for the large neurinomas of stage IV (74 dbs average loss). This proves that it is the hearing loss that leads to discovery of the tumor, which is almost self-evident, but seems to confirm that the local evolution differs as between an intracanalicular neurinoma and a large tumor. In other words, the diagnosis is only apparently earlier in cases of small tumors: the tumor becomes clinically manifest when neural damage has resulted. These reflections lead us to seriously doubt the cogency of the attitude of those who believe that the discovery of a small tumor proves that the conditions are more propitious for the preservation of hearing. Our view is in agreement with the conclusions drawn by Neely [128] from the pathologic study of operative specimens from all small intracanalicular tumors. Even in these apparently more favorable cases, he found it impossible to predict the anatomic condition of the cochlear nerve from preoperative audiometric examinations. Only the absence of obvious infiltration, the value of which has already been discussed, and the ease of dissection yield any hope of a favorable outcome.

Indications

Acoustic neurinoma calls for a surgical decision.

The Otoneurosurgical Option

This seems justified after reviewing the various problems that have been discussed above.

Reasons. These can be resumed as follows: acoustic neurinoma poses a predominantly neurosurgical problem to the otologist who discovers it, while it confronts the neurosurgeon desirous of removing it with otologic obstacles that he cannot overcome on his own. This situation impels these two specialists to collaborate and this creates the conditions favorable to resolution of most of the problems that arise. Let us review the main ones: surgery in dorsal decubitus, a very anterior direct approach to tumors of the cerebellopontine angle, minimal risk of injury to the brain-stem and cerebellum, initial exposure of the facial nerve in a healthy area, easier dissection of the nerve, complete excision of the intracanalicular growth, a particular approach to the problem of preservation of hearing. The value of otoneurosurgical collaboration is not confined to these several technical or tactical advantages. In the light of its experience, and thanks to the sharing and comparison of the concepts appropriate to each of its components, the otoneurosurgical team acquires new capacities which enable it to attack different problems, sometimes more complex such as excision of tumors of the glomus jugulare. This is surely of decisive importance.

Constraints. As we have already stressed, these result from the efforts each party must make to acquire an understanding of the fundamentals and technical capacities proper to each of these two specialities, both in the context of the problems to be dealt with and in the collaborative spirit which must be exhibited to maintain the necessary collaboration. In practice, this particular attitude must be able to promote the blossoming of a new specialty whose specific organization will grant it decisive scope.

Special Considerations. It is quite clear that the transpetrous approach cannot be decided on lightly or in routine manner. The bone structure traversed contains functionally important zones the safeguarding or destruction of which must be planned. Again, the position of the sigmoid sinus is open to major variations which either facilitate or complicate the drilling stage. This is why it is so important to define preoperatively not only the nature, size and extent of the tumor of the angle to be approached but also the exact state of the IAM, whether dilated or burst open, the position of the bulb in relation to it and also the position of the first bend of the sinus in relation to the EAM. Therefore, a precise neuroradiologic investigation must be made available for each case. We do not at all share the opinion expressed by some neuroradiologists that, on the basis of the undeniable advances made in the field of magnetic resonance imaging, this examination by itself suffices for preoperative assessment. It may well be adequate for those responsible for demonstrating the existence of the tumor, but other information is required for those responsible for operating on it. Therefore, we always insist on a good tomographic study of both petrous bones, frontal and lateral, so as to properly define the course of the sinus, the position of its first bend, the height of the bulb, the width of the intersinuso-facial space and the diameter and conformation of the IAM, all very useful information to have in mind during drilling of the pyramid (Fig. 111). CT scan of the bony window provides much the same information. It shows particularly well the bulb behind the IAM in major prolapses (Fig. 112) but, short of reconstructions in the coronal plane which are not always available, it does not allow precise evaluation of the position of the bulb when this does not reach the floor of the IAM. This is why we remain faithful to the standard tomograms. It may be that progress in scanning and, no doubt, in our own interpretation of its images may one day allow us to limit our radiologic studies to this single investigation. We hardly any longer perform arteriography for acoustic neurinomas, reserving this method for cases when there is a doubt as to the probable histologic nature of the tumor of

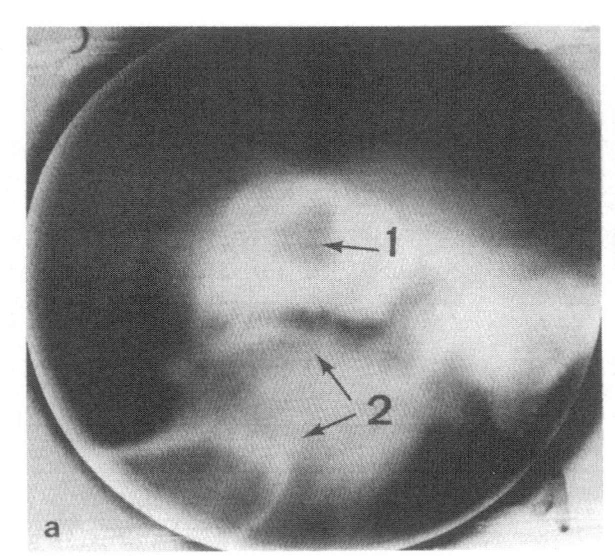

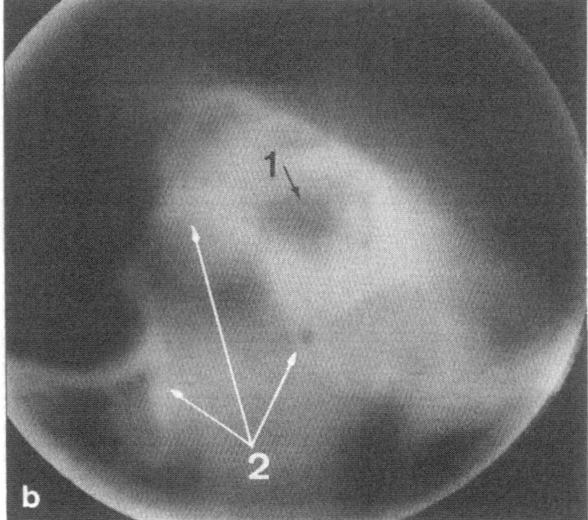

Fig. 111 a, b. Tomogram of petrous, lateral view: IAM *(1)*, jugular bulb *(2)*. **a** The bulb remains below the floor of the IAM. **b** The bulb reaches the floor of the IAM

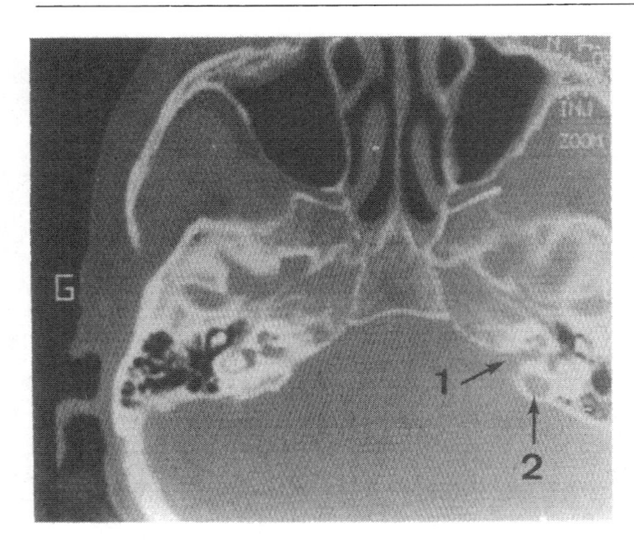

Fig. 112. CT scan. Prolapse of bulb. IAM *(1)*, jugular bulb *(2)*, with dome situated behind IAM

the CPA. Obviously, complete audiometric assessment is essential to decide in the light of all the available information whether to sacrifice or preserve the labyrinth.

However, in view of the average size of the tumors on which we are led to operate (two-thirds stages III and IV, only 11% stage I and 23% stage II) and the major hearing loss noted in the majority of cases (two-thirds of the patients have auditory loss of over 50 decibels), the indications seems fairly simple.

Routine Indications

The Extended Translabyrinthine Route. This is the approach currently most used: 94% of our operations. Provided the otologist is prepared to take his bony resection to the limit, especially superficially but also around the IAM, so as to give the term "extended" its full value, this route allows resection of every tumor, even the largest and including extensions towards the foramen of Pacchioni (tentorial foramen), the clivus and the foramen magnum. As the evacuation of the central part of the tumor proceeds and its size is reduced, the existence of the arachnoid envelope allows spontaneous stripping of these extensions, so that they can be drawn towards the access route and safely fragmented under direct vision. The arterial pedicles reach the tumor in the vicinity of the aperture of the IAM and, when there are branches arising from the cerebellar arteries, these are always distal and in our experience do not give rise to serious vascular risks during operation. Paradoxically, it is the most lateral pole of the tumor that has caused us most difficulty in certain very large tumors. This is because, not unusually, there exist here tumoral arterial afferents or draining veins of quite large size which tether the tumor in its cortical cerebellar bed. As resection of this external pole has to be done first, and since the tumoral mass still in place blocks this outer pole under the cerebellum in which it is embedded, it is sometimes impossible to draw it, even when well evacuated, towards the access route for fragmentation. It is in such cases that we have sometimes been compelled to open the dura mater behind the sigmoid sinus to resect the outer pole by the suboccipital route before continuing the ablation of the neurinoma by the translabyrinthine route. But this course is really exceptional and, as House [88] said long ago, the translabyrinthine route seems to us the most reliable for removal of the largest acoustic neurinomas. We are altogether opposed to the opinio of Glassock et al. [69] who argue that the translabyrinthine approach does not ensure a perfect view of the confines of the CPA in cases of acoustic neurinoma. Of course, it is not the same when the tumor is adherent to the dura medial to the aperture of the IAM or at the internal orifice of the posterior lacerate canal or even at one of the nerves at the boundaries of the CPA (trigeminal or mixed nerves). This will be discussed later. This security has doubtless appeared decisive to such authors as Sterkers [172], who, long skilled in the suboccipital routes (we prefer this term, already used by Dandy [35] to that of retrosigmoid), have become fervent advocates of the extended translabyrinthine approach. This route is chosen whenever hearing is no longer functional or when the chances of preserving hearing, in view of the size of the tumor, are quite uncertain. Naturally, this decision is to be made only if the opposite ear is intact.

The Suprapetrous Route. Though this is often used by otologists, particularly for vestibular neurectomy, it is indicated much more rarely for cases of acoustic neurinoma: 5% of our operations. Its essential indication is the preservation of hearing, which is why the hearing status constitutes the pivot of the debate. We believe that in the majority of cases the gravity of the problems to be solved render any concern for hearing superfluous. Moreover, when this problem arises, all the anatomic, technical and tactical arguments we have previously debated lead us to regard the suboccipital approach as inadequate, except in the 20% of cases where the neurinoma does not reach the fundus of the meatus. Knowing how difficult is still is to verify this situation, it seems logical to eliminate this approach and to use only the suprapetrous approach when this problem arises. The problem then is to define the selection criteria, which are audiometric and anatomic.

In audiometric terms, Glassock [68], starting from the fact that hearing is really bicochlear (or, as some would say, binaural) only when the interauricular difference does not exceed 25 decibels, proposed in 1978 that the maximum hearing loss beyond which there is no point in preserving hearing should be fixed at 25 decibels, and that discrimination in speech audiometry should not be less than 80%. Experience has taught us, exceptional as it may be, that operation may produce not only preservation of hearing but even a genuine improvement. It therefore seems legitimate to reduce the lower limit of maximal hearing loss in pure tone audiometry. In 1979, House [90] suggested fixing the limit at 40 decibels. It now seems, and Glassock agrees [69], that one may go as low as 50 decibels. Tator et al. [177] suggest reducing the lower limit of loss of discrimination in speech audiometry to 60%. All the same, it must be realized that at this level the ear is not of great functional value, and that the benefits of preservation are, all said and done, only very relative. One must also hope, although there direct cause and effect relationship, that this outcome is not spoilt by the existence of often very unpleasant auditory phenomena.

In terms of the morbid anatomy, most authors accept as a reasonable limit of size the tumor of 2.5 cm [31, 33, 46, 78, 136, 169] and Glassock [69] puts it at 2 cm. It must be made clear that all are referring to overall size and not to the size in the CPA, which, allowing for the normal dimensions of the IAM, reduces the average diameter considered reasonably compatible with attempts at preservation of hearing at 1.5 cm in the angle. This relates mostly to stage II tumors. Thus, the problem arises only for a third of our patients, tumors being too bulky in the other two-thirds; but as these patients must also satisfy the minimum criteria demanded by audiometry, the real candidates for preservation of hearing actually constitute a much lesser number (Table II), around 11% of the total number of neurinomas. So far, using only the suprapetrous route, we have operated only stage I tumors, which explains the much lower figure (5%) reported by us. Our criteria are very limiting. However, the publications of Wigand [187] on extension of the suprapetrous route with, if need be, section of the superior petrosal sinus and the tentorium cerebelli to give better exposure of the angle, offer new possibilities though our experience is still inadequate to allow a valid opinion. It is probable that this approach will extend the indications to stage II neurinomas satisfying the criteria. This discussion cannot be closed without stressing that this suprapetrous approach, while offering the hope of safeguarding the cochlear nerve, adds a potential risk to the facial nerve, given its habitual superior position in the IAM. Charachon et al. [26] stress this on the basis of their experience.

The Combined Approach. This combines the extended translabyrinthine route with opening of the dura mater behind the sigmoid sinus. In view of the variations of the torcula and unless precise angiographic studies of the latter are available, we do not recommend ligature and section of the lateral sinus as has been proposed. These combined routes represent about 5% of our ETL procedures, but this figure should be adjusted since we used this method for our first 9 patients before adapting the ETL route. We now limit this approach, as we have explained earlier, to those cases in which the particularly large size of the tumor blocks its external pole, even after evacuation, in the lateral recess of the posterior fossa. The only problem in simplifying the extension stage is to plan a somewhat larger skin incision, going slightly beyond the asterion posteriorly, so that, if need be, one can expose the dura of the posterior fossa rather more merely by slightly enlarging the occipital resection with the nibblers. This is a possibility to be borne in mind whenever confronted with a very large tumor. However, it is not obligatory to have recourse to it. Leaving out our first 9 cases, this technique was required only 6 times, i. e., in 3% altogether of our operations for acoustic neurinoma.

Particular Indications

Certain circumstances justify some changes in our admittedly routine attitude to acoustic neurinomas.

Preoperative Hydrocephalus. This eventuality is far less common than formerly. As we have seen, earlier diagnosis has not so much modified the anatomic conditions at the time of operation as transformed the patients' clinical condition. Patients with raised intracranial pressure, which represented 50% of the cases reported by Pertuiset [146] in 1970, have regularly decreased in number along the years: 12.5% of our series before 1979, 11% in 1980, 6% from 1981 and just 2% since 1985.

It is logical in these cases to construct a shunt, and we prefer this to be definitive rather than external from the outset to minimize the risk of infection. Clearly, we base this decision on the clinical status, particularly astasia-abasia and intellectual disorders of frontal nature, and on the scanographic image of ventricular dilatation and deviation or sometimes blockade of the fourth ventricle. However, where scanography is concerned, we feel that this procedure is indicated more by the "active" nature of the condition, especially with transependymal resorption and obliteration of the cortical sulci, than by the ventricular dilatation in itself, unless of course this is major. It should be followed, within 8 to 10 days if possible, by excision of the tumor. In view of the duration of this latter procedure, we take care to place the pump on the side opposite the tumor and very low down, at the level of the upper part of the sternocleidomastoid, so that prolonged pressure should not produce a sore. It is the case that unishunt pediatric valves are particularly practical and suitable for these cases. In accord with the precepts of our teachers, when we first began using the ETL route only we thought it necessary in the case of a large tumor to open the foramen magnum and resect the posterior arch of the atlas. As advised by House [88] in 1968, at one time we began the operation by carrying out this procedure in ventral decubitus by a midline posterior approach, before turning the patient into dorsal decubitus and continuing with a classic extended translabyrinthine approach. Later, the clinical state of patients with a large tumor appeared satisfactory enough for us not to fear a postoperative problem with infection and we ceased to practise this maneuver without having to regret any complications. Therefore we no longer contemplate its performance and are content with simple shunting of the patients if necessary.

Bilateral Neurinomas. These are something of a nightmare for those concerned with acoustic neurinomas. Fortunately, they are quite rare: 6 cases (2.7%) in our series, 4% for Hitselberger [82], 5.5% for Hughes et al. [95], 7.4% for Yasargil et al. [188], 8% for Sterkers et al. [170] and 8.5% for Fischer et al. [53]. Few authors really deal with this problem, many being content simply with reporting the incidence of their cases. It must be said that the problem is a distressing one, too difficult for anyone to deal with very well.

It is distressing for several reasons. First, it may be a manifestation of von Recklinghausen's disease, known for its dissemination and confusing course. As is known, these bilateral neurinomas which may form part of a classical form of this disease, associated with the characteristic skin lesions, often manifest as a central form which may then be combined with other intracranial or intraspinal tumors, particularly neurinomas or meningiomas, the skin signs then being minor or even altogether absent. There even exist apparently isolated forms without associated manifestations, so that every acoustic neurinoma should initially be considered as such and, as reported by Sterkers et al. [170], one may fail to check the opposite side and then have the unpleasant experience of noting the development of contralateral deafness secondary to the first operation. Then there is a form in young subjects, the deafness classically developing during the second decade, often bilaterally, which immediately raises the most distressing problems as to the social and professional future of these adolescents. Finally, there is a hereditary disorder in which the younger relatives and descendants must be checked for the earliest stages of the disorder, when only the most early diagnosis and treatment offer any hope of preventing the total deafness that threatens all these patients on the more or less long term. It is very difficult indeed to have to inform the families of these apparently unaffected individuals that the risks to the facial nerve and hearing are not, by definition, nonexistent and that, in any case, one cannot exclude the subsequent development of one or more other tumors.

It is a difficult problem, for whatever decision is made about treatment the longterm outcome is uncertain. We find it irritating that most authors have nothing to say on this subject, and particularly regret the current silence of Hitselberger. In 1968 [82] this author had recommended a very aggressive attitude to multiple tumors, with altogether encouraging immediate results. However, he does not even touch on the subject in the otherwise very exhaustive work of the team of the Ear Research Institute of Los Angeles, published in 1979 [90].

No-one knows exactly what to do and many solutions come up for considerations, including those that would be unacceptable under other circumstances. In this context, the decision to abstain or to perform partial resection advised in certain cases by a surgeon like Yasargil [188] are instructive and somewhat mitigate our own remorse after some of our abstentions from treatment. There is an admittedly great risk of precipitating the onset of total deafness, whereas perhaps, in view of the sometimes very slow evolution of cases that are simply monitored (30 years in a case of Sterkers [170] and over 10 years in 3 others), it might have been possible to facilitate rehabilitation of the patient by teaching him sign language, easier while he is still capable of hearing, and a job more suited to his handicap. From the reported experience, we conclude:

- That one must try to detect these tumors as soon as possible, as soon as the diagnosis of von Recklinghausen's disease arises, and to operate on them if they are still in the intracanalicular stage or, in any case, less then 2 cm. Hitselberger [82] has thus been able to report bilateral preservation of the facial and cochlear nerves
- That the tumor must be removed if the ear is totally deaf, the translabyrinthine route then seeming the best to save the facial nerve
- That one must try to preserve all the residual audition in either ear, either by the extended suprapetrous route, or by a combined suprapetrous and suboccipital route to empty the meatus
- That the larger tumor must be removed first when there are signs of increased intracranial pressure, but that there is a risk that this procedure may cause sudden detrioration of the other ear. If those still has its hearing, one may temporise
- That the tumor on the side of the worst ear should be tackled first, which may produce improved hearing on the opposite side, an improvement to the greatest benefit of one of our patients, whose case we summarize:

Miss BRE, 21 years old. Stage IV neurinoma of the left side with a pure tone curve showing virtual cophosis (90 dbs loss) and a stage III neurinoma on the right side where the pure tone curve showed a moderate loss of 60 dbs. Operation by ETL on the left side on 10 May 1985. Subsequently, pure tone audiometry revealed a restoration of near-normal hearing on the right side (10 dbs loss).

- That one must know how to prepare the patient psychologically, and his home circle, and especially to direct his reeducation and social rehabilitation with the handicap awaiting him
- That one must always suspect the development of another intracranial or intraspinal tumor and therefore follow up the patients
- That one must wait a long time before considering the immediate postoperative result as definitive, since it is not impossible for hearing to deteriorate secondarily because of cochlear sclerosis or recurrence.

Our attitude is guided by these several concepts. It cannot be defined very clearly and largely depends on the size of the tumor. Difficult as it may be, we shall sum up our position:

- A cophotic ear is operated by the ETL route

- Problems arise when auditory function persists and whatever the degree of its involvement. It is impossible to consider a translabyrinthine approach since it is essential to respect the least morsel of hearing. Excision of the tumor is to be attempted by the extended suprapetrous route, supplemented if necessary by the suboccipital route. The size of the tumors and the degree of hearing loss at each ear should be assessed before coming to a decision

- If the diagnosis is made when both tumors are intracanalicular:

- Operation is by the extended suprapetrous route on the side of the better ear with total excision
- If hearing is preserved, one should wait at least a year to be sure of success before operating on the second side
- If hearing is only partly preserved, and of course if it is lost, operation is not performed and the tumor is monitored *in situ*. Operation is reserved for any developing neurologic complication.

- If the diagnosis is made when there is at least a large tumor:

- If there is cophosis on one side, total excision of the tumor by the translabyrinthine route, then monitoring of the opposite side, which is not operated except for neurologic complications and then attempting to preserve the residual hearing (suprapetrous and then suboccipital approach, or partial resection by a suboccipital route)
- If some hearing persists, operation on the side of the larger tumor by the suprapetrous and then suboccipital route, with monitoring of the opposite side
- If hearing on the side of the large tumor is useful, abstention and monitoring. If any complications, excision by the suprapetrous then the suboccipital route.

This situation is not too dissimilar from those fortunately rare cases where there is only one tumor but the contralateral ear is not altogether healthy, either because it is the site of previous permanent disease (such as posttraumatic disorder) or of a rather stagnant disorder (such as presbyacousia), or because it has a lesion capable of getting worse (such as chronic otitis).

- In the case of a permanent affection, and if the hearing is very useful (less than 25 dbs loss on the nontumoral side), the tumor is treated as if there were no problem, i.e., an ETL approach is made unless the tumoral ear satisfies the criteria for an attempt at preservation of hearing. If the hearing on the nontumoral side is more affected, the tumoral side is monitored with intervention by the extended suprapetrous and then the suboccipital route if any neurologic complication develops.

– In the case of a disorder capable of deteriorating, and unless there is an imperative reason to deal with the tumoral side, the nontumoral ear is treated first and stabilized if possible before tackling the tumor. The relevant pathology is often a chronic otitis and then it is not unusual for the chronic infection to be bilateral, clearly calling for bilateral treatment of the otitis before doing anything else. The following case is a good example of this situation:

Mrs. DEL, 51 years old, operated 10 years previously for bilateral cholesteatoma with preservation of good bone conduction. For 1 year, progressive deterioration of hearing on the right side, the left remaining normal. ENT assessment demostrated bilateral recurrence of the cholesteatoma and dilatation of the IAM on the right side. Scan: tumor of the right CPA, stage IV. Vestibulometry: right arreflexia. First stage: two operations for excision of the cholesteatomas by bilateral Palva technique. Two months later: right transcochlear approach to eliminate the cholesteatomatous pyramid and resect the tumor of the CPA. Its size and especially the adhesions to the posterior face of the petrous, under the aperture of the meatus, necessitated a combined suboccipital approach. Postoperative course straightforward. Very transient right facial palsy. Bony transmission intact on the left side.

Abstention or Abandonment of Surgery. Surgical abstention may be defined as the decision not to operate on a patient who, in absolute terms, could be operated. In our view, this measure differs from a contraindication, which is imposed on the surgeon for one or more reasons usually of a general nature (associated pathology, visceral disorders, infection, etc.), the leading part in making the decision being usually taken by the anesthetist/ intensive care specialist. By abandonment, we refer to cases where, during operation, for various reasons, it is decided to go no further with the excision and to abandon quite a large portion of the tumor. These two decisions are not necessarily absolute and may subsequently be revised and further operation contemplated.

It is reasonable to bracket these two situations because basically they derive from the same process, which is the assessment of tumoral development. Nowadays, the CT scan and MRI afford the possibility of easily and precisely monitoring the development of a tumor *in situ* that in certain circumstances it may be possible to elect for this guarded surveillance. Recently, some authors have adopted these tactics under certain conditions and it is useful to analyse their experience. Sylverstein et al. [175] followed up 7 patients aged over 65 years and with small neurinomas (less than 2,5 cm) for a period of from 1 to 6 years. By serial scanning they were able to show that in 80% of the cases tumoral development was slow, the average annual increase in diameter in the angle, measured from the aperture of the IAM, being 0.2 cm. Clark et al. [29], in a series of 6 patients followed over 2 years, with small neurinomas (less than 1.5 cm), found much the same figure for annual growth (0.15 cm) but noted that this could be as much as 0.4 cm in some cases. Thus, these two reports seem to provide useful information as to the possible development of certain neurinomas, though it must be realized that these are simply reports, that there is no general rule, and that development is sometimes disturbing, as in the case reported personally by Glasscock to Clark et al. [29] of one of his patients, aged 80, the diameter of whose neurinoma doubled in only 3 months (from 1.5 to 3 cms). However, on the basis of the above remarks, and provided it is possible to keep the patients under close surveillance, it is sometimes possible to decide on abstention. Three sets of circumstances seem to justify this attitude:

• *The bilateral tumor:* we have already raised this problem and stated our opinion

• *The elderly subject.* It may be imagined that the discovery of a virtually asymptomatic tumor (simple auditory disturbances or unilateral deafness or even minor vertigo) might raise this problem, and would authorize abstention while waiting for the appearance of more disabling features (unsteadiness, headache, etc.). But it would then have to be decided at what age one would feel justified in so doing. Sylverstein et al. [175] have chosen 65 years. This seems still too young for many cases, especially if the patient is a woman (two cases out of three) with her longer life-span and known tendency to more rapid growth of a neurinoma, as reported by Kasantikul et al. [101]. The dilemma would be to subsequently regret having to operate on a patient under less satisfactory conditions and thus having spoiled the chances of a good result, which are practically the same as average in elderly subjects provided the tumor is not too large. Personally, we have temporized with patients over 65 years of age only if there was some associated systemic disease (history of myocardial infarction, Parkinson's disease, etc.).

• *The small tumor:* this is a particularly thorny problem. It is almost certain that there is no great difference in the risk to the facial nerve or the risk to life when one operates on a tumor of 1 or 2 cm diameter in the angle, but the situation changes entirely when hearing is at stake. In our opinion, abstention can be considered only if the problem of hearing is already determined by a major deterioration at the time of diagnosis, the more so when the patient is elderly or else, as we have already stated, when some hearing persists but the other ear is profoundly damaged or cophotic because of a bilateral tumor.

• *The problem of abandonment.* There is an initial difference, depending on whether one abandons micro-fragments or a sizeable fragment; also, it seems, on the site where the fragment is abandoned. This has already been discussed in dealing with the question of total excision. This cannot be an habitual tactic, not even a "quite common" tactic as seems to be recommended by Sheptak and Jannetta [165]. Strategic withdrawal is only to be envisaged in certain quite special circumstances:

- The development of peroperative problems. As stressed by Hitselberger and House [84], this usually relates to the onset of disorders of cardiac rhythm or arterial pressure which the anesthetist is unable to correct quickly. We have experienced this on three occasions (1,2% of our operations), each time for stage IV tumors.

Mr. BON. A 64-year-old man, obese and atheromatous, complaining of severe headaches, intellectual impairment, astasia-abasia and progressive urinary incontinence for 2 months. Also, very old right deafness. The paraclinical investigations revealed a large right acoustic neurinoma and considerable triventricular hydrocephalus with signs of transependymal resorption. A ventriculo-peritoneal shunt very markedly improved the patient's intellectual performance and balance. Two weeks later, translabyrinthine approach: a very "red", pseudomeningiomatous tumor. Partial resection. Operation discontinued after 8 hours at the anesthetist's request because of the appearance of disorders of rhythm and inversion of the T wave. Convalescence straightforward. Discharged a month later, walking with a stick. This patient was reoperated 8 years later with success.

Mrs. VIR, aged 34: deafness and vertigo for 3 years, then headache and maxillary neuralgia and hypesthesia of the right half of the face. On examination, involvement of right V and VII, left pyramidal syndrome and right cerebellar syndrome. Investigation showed a very large neurinoma of the right VIII. Ventriculo-peritoneal shunt, followed after 5 days by a translabyrinthine approach revealing a very large tumor adherent to the brain-stem of which only three-quarters had been resected after 20 hours' operating. This patient underwent complete excision of the residual fragment 4 years later.

Mrs. MAR, aged 36. Intermittent left facial neuralgia slowly developing for 12 years. For 10 years, left hypoacusia. Headache for 2 years and unsteadiness on walking and left-sided awkwardness for the last 10 months. Examination: astasia-abasia, left facial palsy, left cerebellar syndrome, left cophosis. Left corneal reflex weak but present. The paraclinical study showed a stage IV acoustic neurinoma. Ventriculo-peritoneal shunt and after 4 days excision of the tumor by the ETL route. A fragment of tumor very adherent to the protuberance was left in place. Postoperative facial paralysis. Lost to follow-up for 11 years. Reviewed in 1987: persistent facial paralysis despite a transfacial anastomosis done elsewhere. CT scan: very minimal comma-shaped hyperdensity in the left CPA.

Hitselberger and House [84] believe that one should wait 6 months and then reoperate. They state that the operation is much less risky that it seemed to the older generation of surgeons. They believe that during the lapse of time between the two operations the tumor becomes "disengaged" from the brain-stem while the circulation is re-established in the latter. Our last case was lost to follow-up. In the first two, we preferred to temporise and wait for neurologic signs to reappear; we had no occasion to regret this decision and were astonished by the simplicity of reoperation. This confirms the impression of Hitselberger and House [84] and would seem to show that the ETL route had a great deal to do with it, probably because it provides a direct access to the tumor, without any aggressive manipulation of the cerebellum or brain-stem.

- A serious neurologic picture in an elderly subject. Here, the conditions are particularly unfavorable. But here again, a good reduction of tumoral size can markedly reduce the compression of the brain-stem and cerebellum, lower the intracranial pressure and, by improving the overall cerebral circulation, allow restoration of nerve-cell function. We feel that the ETL route is particularly indicated in these cases, especially if the tumor is large, because of the comfort afforded the patient and surgeon and the minimal aggressiveness it inflicts in procedure. Everyone who uses this approach stresses this sentiment.

- Finally, certain bilateral tumors sometimes justify deliberate partial resection. We have already mentioned this possibility. We return to it only to stress that, if the resection is intended to be incomplete, it must nevertheless be large enough to be effective. The problem in these cases is the preservation of hearing; and it must be admitted that one does not always fully apprehend the limit not to be exceeded if there is to be no risk of compromising the blood-supply of the inner ear and also, doubtless, of the cochlear nerve, which is probably the determining factor for the success or failure of such conservatism.

Results

Ideally, in a patient with an acoustic neurinoma whose only handicap, now usually the case, is a hypoacousia or some vertigo, one hopes to succeed in removing the entire tumor without causing additional neurologic defect, in particular without producing facial paralysis, while preserving hearing as it is or, better still, restoring it. But these last aims are almost utopian and cannot reasonably be expected except for very special cases marked by quite small size of the tumor, the fact that it is not too embedded in the fundus of the IAM, the integrity of the cochlear nerve and doubtless also by the perfection of the operative procedure. This is why we feel it obligatory to make a reasonable choice and - except in the favorable case that presents only about once in ten - sacrifice hearing or rather what is left of it for the greater benefit of the facial nerve and brain-stem. The arguments already advanced explain our choice of otoneurosurgical techniques. Our experience is based on 232 acoustic neurinomas between 1973 and the end of 1986. Apart from the very first case, operated by the translabyrinthine and transtentorial route as then advised by Morrisson and King [104] and the 9 subsequent cases operated in two stages, a translabyrinthine stage to dissect the facial nerve and isolate the tumor and then a suboccipital stage to remove the latter, all our remaining cases since were operated either by the translabyrinthine route (220 cases, 6 with a suboccipital extension at the same stage) or by the suprapetrous route (12 cases).

Imperatives

Survival. We have to report 11 deaths in our 232 patients (4.7%). This figure is much the same as that of Fischer et al. [53]: 6.3%. For us, as for Fischer, however, this is an overall mortality which does not precisely reflect the current risk which can only decrease as experience progresses. Unfortunately, in view of the potential gravity of this form of surgery, it does not seem that one can honestly hope for the total elimination of this risk. From 1973 to 1980 we had 7 deaths in 68 operations (10.3%), and then only 2 deaths for the following 62 operations in 1981 and 1982 (3.2%). At the end of 1985 [142] we were delighted to have had not a single death in our last 50 operations. However, in 1983, 2 new deaths in 38 operations sharply brought us back to reality. It is true that over the period 1983-1986 our mortality amounted to 2.2%, but this gave the impression of being a bottom level on which it would be very difficult to improve. It is clear that the risk increases with the size of the tumor: 6 deaths in our 90 stage IV cases (6.9%), 4 for the 118 stages II and III (3.7%). It should be noted that a risk is attached to even the smallest tumors and we had one death, due to a Mendelson's syndrome at waking, in our 24 stage I cases. At first glance, this complication was not the result of any direct traumatism to the neuraxis, but it does show that any operation, of whatever magnitude, carries an intrinsic risk. This risk is primarily the direct outcome of the operative procedure, as has been verified 7 times in our experience. On 3 occasions, at the start of our practice, there was probably a softening of the brain-stem, now attributed to a softening in the territory of the AICA, the classic syndrome of Atkinson [5]. The disorders of consciousness followed by cardiorespiratory arrest, in the absence of a hematoma of the angle in the scan, seem to certify this diagnosis which thus accounts for 27% of our deaths, a figure close to the 30% reported by House [90] among the causes of death in his series. These 3 cases occurred at the beginning of our practice when we had more difficulty in dissecting and respecting the neurovascular structures, especially at the aperture of the IAM. On 3 occasions, there was a hematoma as demonstrated by the scan. In 3 cases, deterioration was rapid in the 24 hours following operation and the patients died despite reoperation, once rapidly in 48 hours and twice much later: in one case with gastrointestinal hemorrhage and stress ulceration as visualized by fiberoscopy, in the other case by progressive decline in a patient who developed bilateral pneumopathy and renal and hepatic failure. Finally, one patient died at the 4th postoperative day from a fulminating Shigella meningitis whose abrupt onset made us think in terms of an accidental infection rather than of the CSF fistula which developed, but which might also have been a manifestation of the CSF hypertension resulting from the meningitis. We have never had the problem of air embolism. 4 patients died from the indirect effects of the operative procedure on the nervous system. Apart from the Mendelson's syndrome mentioned above, there were one myocardial infarction and two pictures of massive pulmonary embolism occurring at the 3rd postoperative day when the patients were doing very well. It is likely that these last two cases were the outcome of phlebitis secondary to the prolonged immobility and the chilling of the patients responsible for peroperative stasis in the lower limbs. This is why we now routinely use elastic stockings and a heated mattress and blanket.

Total Excision. This was achievable in 96.6% of cases. We had to abandon some tumoral fragments in 8 of our 232 operations (3.4%). In 5 cases the residues were microscopic, abandoned because of dense adhesions, usually to the facial nerve in the region of the aperture. Carefully coagulated *in situ*, they carried virtually no possibility of being the origin of a recurrence and we believe that many would consider these cases as total excisions, which would increase our total excisions to 98.7%. In only 3 cases was the fragment macroscopic (1.3%). In each case this was the residue of a bulky tumor of stage IV. Twice, we were compelled to abandon a fragment on the brain-stem because of dense adhesions impassable after hours of operating. In the third case the anesthetist asked for operation to be discontinued in an obese patient aged 64 exhibiting anomalies of repolarization and in whom the excision was rendered tedious by the pseudo-meningiomatous nature of the tumor. This last patient, who had a stage IV tumor responsible for a virtually bedridden state with astasia-abasia, recovered very well postoperatively, as did our two other cases, both young women who manifested intracranial hypertension before operation with a clinical otoneurosurgical syndrome. The man and one of the two women emerged without facial paralysis, while the other young woman was left with a very severe facial paralysis (our stage II, [House's] stage V). Of these, only the first young woman was subjected to regular clinical and scanographic follow-up. We had to reoperate on the first two patients, without facial paralysis, during 1986 because of reappearance of headaches and disturbed balance while the scan revealed an increase is the size of the residual fragment. The translabyrinthine approach, very simple in both cases, this time allowed complete excision of the tumor, 8 years in one case and 4 years in the other after the first operation. Convalescence was trouble-free in both cases, with regression of the symptoms motivating the reoperation. Unfortunately, while the man recovered very quickly from a postoperative facial palsy, the young woman, after a suggestion of recovery, was left with a severe paralysis (marked stage II) which justified a hypoglosso-facial graft at the beginning of 1987. The ease of the reoperation and the possibility of complete excision were unexpectedly gratifying and led us to think that a repeat operation by the translabyrinthine route and with the aid of microsurgery is an easily performed undertaking. In 1987, we also reviewed the third patient who had been lost of follow-up. Apart from his facial palsy, his clinical status was satisfactory and the scan showed the quiescence of the residual fragment. She declined the hypoglosso-facial graft that we suggested. None of the other 5 patients has shown any recurrence to our knowledge, just as with the 224 patients in whom we consider the excision to have been complete.

The Facial Nerve. Overall, we have been able to respect the continuity of the facial nerve in 207 of our 232 patients, i. e., in 89% of cases. Naturally, these results relate to the difficulties encountered in dissection of the nerve, which in turn obviously depends on the anatomic state of the nerve on the tumor. The size of the latter plays an essential part. 17 of the 25 facial sections that we were unable to avoid (68%) occurred during the excision of large stage IV tumors, 18.8% of cases, whereas with the smaller stage II and III tumors this occurred only 6 times (5% of cases). It should be stressed that on 2 occasions we divided the facial nerve when operating on stage I tumors. This was at the beginning of our series, once during a suprapetrous exposure, once during a translabyrinthine exposure when the small but very firm tumor suddenly turned around the end of the sucker during the dissection. With practice and also with more suitable equipment, these accidents should not recur. It is true that practice makes perfect. Among the first 130 cases dealt with in the thesis of Triglia [182], the continuity of the facial nerve could be respected in 109 cases (84%) and 98 times in the 102 following cases (96%). These macroscopic anatomic results are not uniform and of variable quality, depending on whether the dissection was more or less easy. Thus it is that between the virtually normal and the cut nerve there exist every intermediate macroscopic forms and obviously every classic degree of involvement of the nerve fibers, from neurapraxia to neurotmesis via axonotmesis. This explains why one must always make a distinction between the anatomic and the functional results.

• *Assessment of the facial defect.* Before considering the functional aspect of the results, it is important to define a rather more objective method of assessing facial function. House [94] proposed a rather complicated classification in our opinion, consisting of six grades, but with the merit of showing that one must take into account the rest tonus, the movements, particularly palpebral occlusion, the existence of a hemispasm. In the light of these three parameters, it seems possible to distinguish, apart from normal facial mobility, three degrees of facial involvement. In the first stage (slight involvement) the rest tonus is normal, ensuring correct facial symmetry, and complete palpebral occlusion is possible. A minor or weak type I and a marked type I can be distinguished according as whether or not there is a moderate hemispasm slightly interfering with expression. In stage II (moderate involvement), complete palpebral occlusion is impossible. Usually, there is a facial hemispasm and synkinesias which interfere with expression, the more so when involvement is minor and rest tonus normal. In stage III (complete involvement) the rest tonus is abolished, palpebral occlusion is impossible and there is no hemispasm (Table 3).

Table 3. Assessment of facial deficit

	Eyelid closure	Hemispasm	Rest tonus
Stage I	+	0 (minor) ± (marked)	+
II	–	+	+ (minor) – (marked)
III	0	0	0

• *Functional results.* These vary with each team, first in relation to the experience gained, and also of course in relation to the average size of the operated tumors and lastly in relation to the elapsed period since operation. The series are quite difficult to compare, since some teams operate only on small neurinomas, others on tumors of every size and yet others mainly on large tumors. Thus, in the long term Fisch [46] had 4.5% of defibitive facial paralyses but operated on only 5% of large tumors. Koos [105] had 7% of paralyses, but with 11% of large tumors, Thomsen [180] and Tarlov [176] 10% with 18%, Yasargil [188] 17/5% with 35% and King [124] 56% of paralyses but with 80% of large tumors. Moreover, the routes of access used by these different teams were also different.

In the present state of our experience, based on 232 tumors, of which 39% were stage IV, 27% stage III, 23% stage II and 11% stage I, our immediate postoperative findings in 207 patients benefiting from anatomic preservation was that there were 72% with facial paralysis and 28% with normal or barely impaired facial innervation (minor stage I). In 60% of the paralysed cases the involvement was partial (marked stage I or stage II) and 40% total. A year after operation, whereas the patients who had a virtually normal facal musculature all had normal expression, 94% of the patients with immediate postoperative paralysis had recovered completely (50%) or almost completely (mainly minor stage I but also a marked stage I with hemispasm in 1 patient in 6), while 6% of patients were left with marked sequelae (usually minor stage II). The recovery occurred in the first 6 months in two-thirds of the cases. Of course, the functional state was better when recovery was more rapid. The marked stage I or minor stage II cases with hemispasm predominated when recovery was retarded.

Of the 25 divisions of the facial nerve, 19 underwent repair, either immediate (10 cases) or secondary, between 15 days and 3 months (9 cases). A hypoglossal-facial anastomosis was done in 17 cases (8 immediate and 9 delayed) and end-to-end fixation with biologic glue in the I.A.M. in 2 cases. The results were very satisfactory, with good recovery of resting tonus, speech and even smiling, but the more marked asymmetry reappeared in laughing or energetic expression. Although the longterm results of the facial paralysis are now very diminished, it remains the main preoperative problem for every patient. It is proper in every case to discuss the probabilities of success and the chances of recovery, while clearly stating that it is impossible to make any promises except, in the worst case, useful function by means of hypoglossal-facial anastomosis.

• *Postoperative care.* Apart from the fortunate cases where facial mobility is normal or near-normal after operation, every patient must benefit from surveillance directed mainly towards the ubiquitous ophthalmologic problem.

The minimum course, for practically every operated case, is the two-hourly instillation of an eye-drop (methyl-cellulose) and, at night, the application of a strip of hypoallergic adhesive plaster transversely across the upper lid from the root of the nose to the zygoma, without ointment which would loosen the plaster. Regular ophthalmologic surveillance, especially if there is the least sign of eye trouble, is essential.

If palpebral occlusion is very impaired, it is our practice, as advised by Zlotnik et al. [196] to inject 10 cm^3 of air into the areolar spaces of the upper lid. We use a fine needle, not siliconized, to prevent escape of air from a gaping puncture-site. This iatrogenic subcutaneous emphysema inflates the upper lid for 3–5 days and keeps it closed. The injection can be repeated several times without problems.

If the facial paralysis is complete and especially if signs of keratitis appear, the best course is to perform a tarsorrhaphy. As a rule, this will be done at the end of the operation when there has been anatomic section of the facial nerve. It should be a lateral tarsorrhaphy, leaving the medial two-thirds free, which is the least handicap to vision. Under local anesthesia the mucosa of the outer third of the margins of both eyelids is excised with a scalpel and the raw surfaces apposed with a U stitch passed through two small plastic tubes, the stitches being removed after 15 days to 3 weeks. The tarsorrhaphy is undone simply with a knife after recovery of facial function.

Muscle reeducation must be given in every facial paralysis and continued until complete recovery. It should be both passive (massage) and active (movements of the different muscle groups of the 3–4 times daily in front of a mirror).

In theory, the operative outcome can be predicted from the operative findings:

- The facial nerve is intact, not traumatized and responds normally to stimulation. The postoperative course is usually straightforward, sometimes without facial paralysis, usually with a partial paralysis which regresses completely and permanently in a few days or weeks.
- The facial nerve is intact, operative trauma moderate, the response to stimulation good. There is usually a more or less severe postoperative facial paralysis. In most cases, recovery without sequelae occurs in a few weeks. But it may take several months, and then always has sequelae: lacrimation, synkinesias, localized deficit, etc.
- The facial nerve is probably not interrupted, but very traumatized. The postoperative facial paralysis is nearly always immediate and complete. The presence of a response to peroperative stimulation allows hope for recovery, but this cannot be assumed in advance, nor its timing or quality.
  The tarsorrhaphy must be continued as long as palpebral function is inadequate. The situation becomes complicated when the paralysis remains complete or severe after 12 months. While very delayed recovery is possible, even after more than two years as every author reports, it is always of poor quality: poor facial symmetry at rest, inadequate eyelid closure, synkinesias, contractures, etc. If total facial paralysis persists 12 months after operation without electromyographic evidence of regeneration, it seems reasonable to consider a nerve repair.
- The facial nerve is known to be interrupted. Various situations may exist:
  - When there remains a good length of facial nerve after its emergence from the brain-stem, end-to-end suture can be attempted after rerouting of the mastoid portion. In view of the absence of an epineurium on the proximal end of the nerve, one has to be satisfied with a single suture made with monofilament 7/0, catching the epineurium of the distal end and passing through the proximal end. The fragility of the neural structures is such that the coaptation cannot hold unless the suture is made without tension. The use of a guide-groove, as suggested by Fischer [49], appears a useful technique.
  - When the proximal end is very short, although it is often difficult to identify with certainty, it is possible after rerouting the distal end to interpose between the two nerve-ends a fragment of the superficial cervical plexus, sutured as previously described. Brackmann and Hitselberger [16] report 9 cases so repaired, with 8 successes. These authors prefer this solution to hypoglossal-facial anastomosis. We outselves [21] prefer to perform such an anastomosis during the original operation whenever this is possible, combined with a tarsorrhaphy at the outset, but the anastomosis may have to be done secondarily if there have been operative problems. When it is done at the end of operation, a simple vertical preauricular and cervical access route is adequate. The facial trunk is identified and sectioned at the stylomastoid foramen without doing a parotidectomy. The hypoglossal nerve is divided as near as possible to the tongue. To perform the anastomosis without tension, the hypoglossal is brought under the posterior belly of the digastric in the parotid region after division of its descending branch, flush with the main nerve. Suture is end-to-end with 7 to 9 stitches of 9/0 silk, catching only the epineurium of the two nerve-ends. The longterm results seem much the same, whatever the level of anastomosis with the facial nerve. Passive exercises are begun after operation and up to the 7th month. Active remedial exercises (movements of the different muscle groups of the face together with tongue movements several times a day in front of a mirror) begin when recovery of the facial muscles starts, usually around the 3rd or 4th month, continuing to the 12th month. Recovery is always of good quality, with facial symmetry at rest, near-normal expression and complete eyelid closure. However, facial expression and ordinary movements (smiling, eyelid closure) are all the better when accompanied by tongue movement, and this calls for a major effort by the patient during the retraining so that these tongue movements become automatic. The frontalis is never reinnervated, nor is it with any of the other procedures. The only fault of this method is that expression is not voluntary; the facial paralysis reappears and distorts the face in any marked emotional situation, particularly open laughter.

Hearing

The arguments developed at length earlier explain our very conservative attitude in this matter. Among the 232 neurinomas we have had to operate on, there were only 12 in which it seemed proper to attempt to preserve hearing. These patients were operated by the suprapetrous route and hearing was preserved in only 6 cases. In 3 of these 6 cases hearing was much the same as preoperatively, with slight improvement in conversational frequencies; in the other 3 cases it was rather less good at the same frequencies (Fig. 113). These results are more

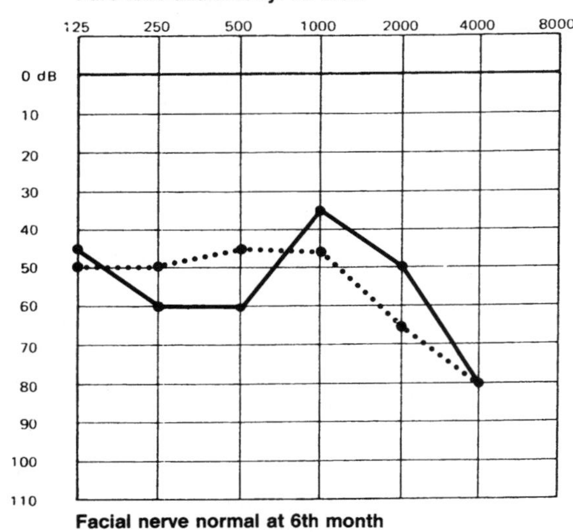

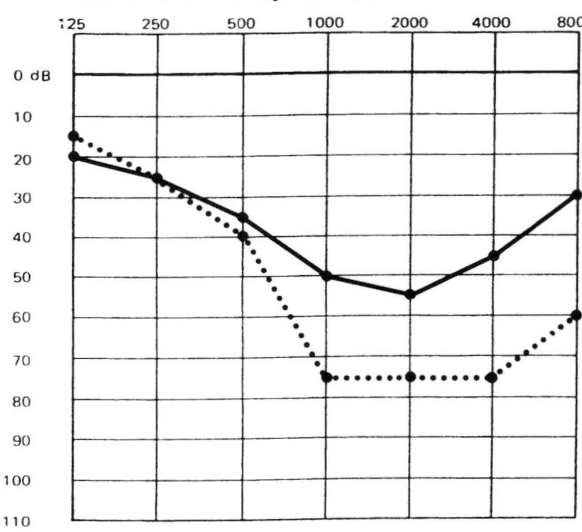

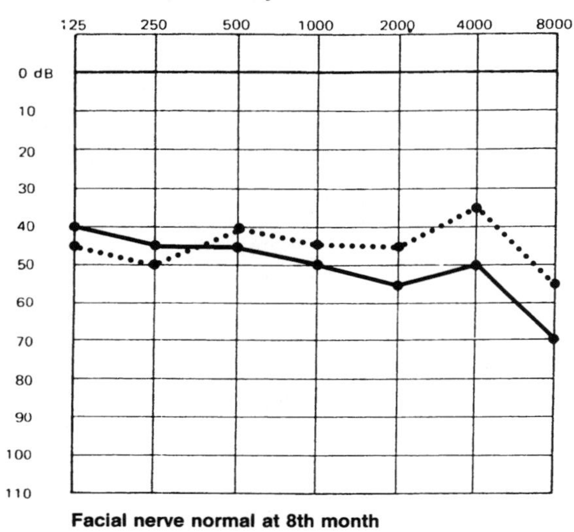

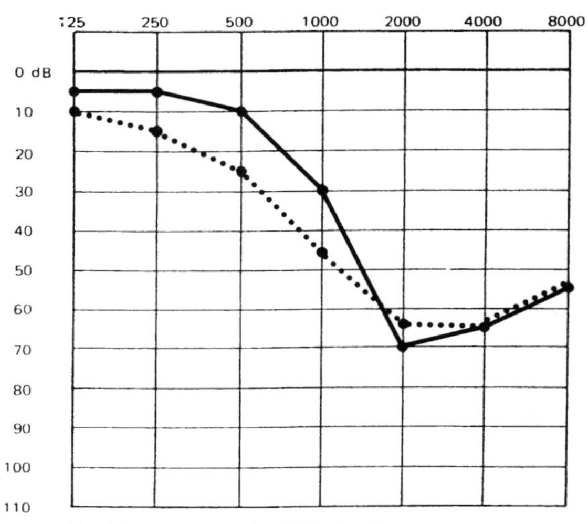

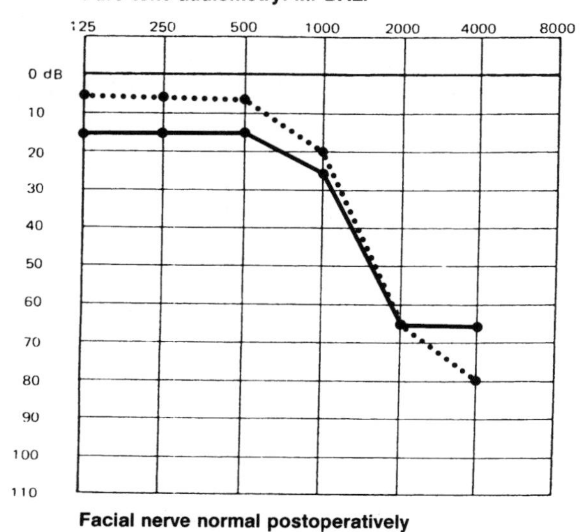

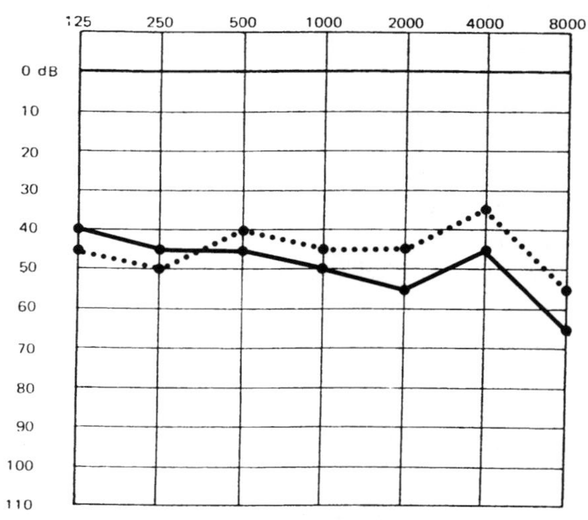

Fig. 113. Tonal audiometry

or less comparable with those reported by many authors [17, 33, 68, 78, 90, 135, 161, 169, 177). Obviously, our selection criteria are very restrictive; in particular, the strict limitation to stage I neurinomas is debatable, especially as the possibilities of extension offered by the technique of extended suprapetrous access proposed by Wigand [187] might influence us to select cases with larger tumors. But it should be stressed that Wigand [187] admits to having been able to perform a complete resection in only 51 of his 63 operated cases for various reasons. It is true that this relates to an experience that is still new, and that he reports that among his last 40 cases only one was the subject of a partial resection due to inadequate exposure of the tumor, but this 19% of incomplete resections seems to us altogether prohibitive, even if he has been able to preserve the integrity of the acoustic nerve in 44 cases and to preserve hearing in 28 of these cases. In our view, other aims have priority. Nor can one overlook the functional value of the hearing so preserved; it seems important to stress once more that it is pointless to preserve traces of hearing of no functional value, except of course in cases with bilateral tumors. Many of the cases reported as successful seem to us unimportant or almost useless, because the auditory loss is well above the threshold of 25 decibels of interauricular difference or because the intelligibility is such or has such a threshold that we are convinced the ear is functionless.

Complications

These are quite rare. Three types can be distinguished:

Neurologic Complications. Apart from facial paralysis, dealt with elsewhere, or rare and always transient disorders of swallowing due to injury of the IXth or Xth nerves, we have had 4 patients with persistent neurologic complications due to a different process in each case.

- Atkinson's syndrome [5]. The patient, who survived 3 months' coma, was left with a neurologic picture combining quadriplegia more marked on the left side, complicated by osteoarthropathies, facial paralysis and left facial anesthesia. Her neuropsychologic state, however, was compatible with a very good relational life, evidence of the integrity of her cerebral cortex. She was lost to follow-up.
- Postoperative hematoma. Reoperated as an emergency, this 41-year-old woman survived 3 weeks' coma complicated by bilateral pneumopathy. She regained a normal interior and exterior life, has no facial paralysis, but complains of rightsided awkwardness, some moderate disturbance of balance, disorders of speech of a "bulbar" nature and moderate disorders of swallowing.
- Bithalamic unexplained hemorrhage in a woman aged 72, with a stage IV neurinoma, who had a successful initial shunt. After a trouble-free operation, she exhibited a state of mutism which lasted 4 months. She then recovered a hint of bilateral motor function and an elementary phasic activity allowing her to answer questions by simple words (yes, no, good morning). She was discharged to a rehabilitation center and has never been followed.
- Brain-stem contusion in a woman aged 37 with a large tumor, who had suffered a cerebral vascular accident. The neurinoma was revealed by a scan made to display the brain-stem. She was left with disturbed balance, a cerebellar syndrome, facial paralysis and anesthesia and disordered swallowing due to palato-pharyngeal paralysis. Nevertheless, she regained a full interior and exterior life, managing her home and driving her car.

Cerebrospinal Fluid Fistula. This was a common complication at the initiation of the translabyrinthine route. Shea and Robertson [164] reported an incidence of 30%, Glassock et al. [66] 25%, House [88] 20%. In practice, the incidence has been considerably reduced by careful packing of the auditory tube and the tympanic cavity with muscle fragments, occlusion of the posterior tympanotomy and the mastoid cells with bone-dust and/or a ceramic, combined with biologic glue, and obstruction of the route of access with fat taken from the abdomen. Our initial 23% incidence [141] fell to 10% [Triglia, 182] and is now 7.7% for our series overall and only 4.9% for the 102 cases operated since 1984. Apart from forgetting to occlude a cell or the vestibular window when the stapes has been mobilized, we now feel that such a fistula can occur only if an increase in CSF pressure develops, and this can occur in only three circumstances:

- Postoperative meningeal hemorrhage
- Meningitis or meningeal "reaction"
- Disturbance of the CSF circulation.

Reoperation does not therefore seem a matter of urgency. First, a meningitis must be excluded by doing a lumbar puncture, which will at least have a decompressive effect, and then a temporary reduction of intracranial pressure will be sought by combining several minor measures: semi-seated position, administration of Diamox to reduce the secretion of CSF and of a laxative to prevent straining at stool, a mildly compressive dressing, daily lumbar puncture. If meningitis is confirmed, systemic and intrathecal antibiotic treatment usually produces cure within a week. If the discharge persists, a peritoneal shunt will usually dry it up. In two cases only we have had to reopen the operation wound because a cerebrospinal rhinorrhea on the side of the operation seemed to indicate that there was an open gap in the middle ear. The problem was immediately solved by the occlusion of an aircell in one case and of the vestibular window in the other. Finally, the problem of fistula now seems to be so much under control as to be considered negligible, and of no more importance during translabyrinthine surgery then during suboccipital approaches [79].

Obviously, other problems may arise. They result in some cases from the lengthy immobilization, intrinsic to the operation and the secondary awakening we usually employ. This is the source of some cases of phlebitis of the lower limbs, 3 or 4 of which were complicated by pulmonary embolism, one massive with sudden death and two others requiring the placement of a clip on the inferior vena cava. This is a problem of every prolonged operation.

Other Neurinomas

Neurinomas of the Facial Nerve

This is a rare site, since an exhaustive study of the English-language and French literature made recently by Chobaut [27] collected only 150 cases, to which we may add the 3 cases of Murata et al. [125], 1 case of Jung et al. [99] and the 12 cases of Sterkers et al. [174] reported since. We ourselves have observed 12 cases.

The Problems

Anatomic. In theory, these tumors may be situated on each of the parts of the facial nerve, so that, like Murata et al. [125], one may distinguish intracranial, intratemporal and extracranial forms, these last constituting cervicofacial tumors outside the frame of our study. There is no doubt that the neurinomas developing in one of the three portions of the facial canal, mostly in the first two portions according to Fisch and Rüttner [45] but usually in the third according to Fuentes and Uziel [55] and in our own experience (3 cases), as well as those found on one of the branches of the nerve (greater petrosal nerve, stapedial nerve [150], chorda tympani [55]) are primary tumors of the facial nerve. On the other hand, there is some ambiguity about those involving the facial nerve within the canal and in the CPA. When the tumor is quite small and only involves the facial nerve, leaving the cochlear and vestibular nerves intact, or, as in case 2 of Sterkers et al. [174], spreads "like a string of beads" on the geniculate ganglion, there can be no doubt. We have seen 4 cases of this type. On the other hand, when the tumor is quite bulky and ensheaths the entire acoustico-facial bundle (4 cases in our series), one may wonder whether this is not in fact a "classic" vestibular neurinoma which has prematurely invaded the cochlear and facial nerves. It is the habitual respect for the facial nerve in the "classic" forms which led us to exclude from this group the 4 cases that we encountered. Our peroperative diagnosis was based on the fact that the facial nerve manifestly disappeared within the tumor and not at its periphery, which might have suggested the extreme splaying of the facial nerve by a neurinoma developing on the vestibular nerve.

Diagnosis. In view of the extreme rarity of clinical facial nerve involvement by the generality of acoustic neurinomas, the existence of a facial paresis, and certainly of a paralysis or hemispasm are of course the best signs suggesting the diagnosis (5 cases in our series). The gradual development is a further supplementary argument but, as we have noted once and Iwagana et al. [97] twice, it may develop very quickly and lead to a misdiagnosis of paralysis due to cold until this is finally corrected by the failure of recovery. Not uncommonly, the clinical picture consists entirely of cochlear or vestibular signs and is suggestive in every aspect of an acoustic neurinoma. If the paraclinical investigations fail to correct the diagnosis, then there is great risk that this will be an "unpleasant operative surprise" as stressed by Charachon et al. [25], sometimes even a surprise in terms of its morbid anatomy, as when the tumor develops from the geniculate ganglion or the petrosal nerve and resembles a temporal tumor. Attention should already have been drawn to it by pure tone audiometry, which shows a transmission or mixed deafness whereas there is a manifest tumor of the angle. But it is mainly the deformations of the facial canal in the tomograms and the bone-window scan that must carry conviction, as demonstrated by our case 7:

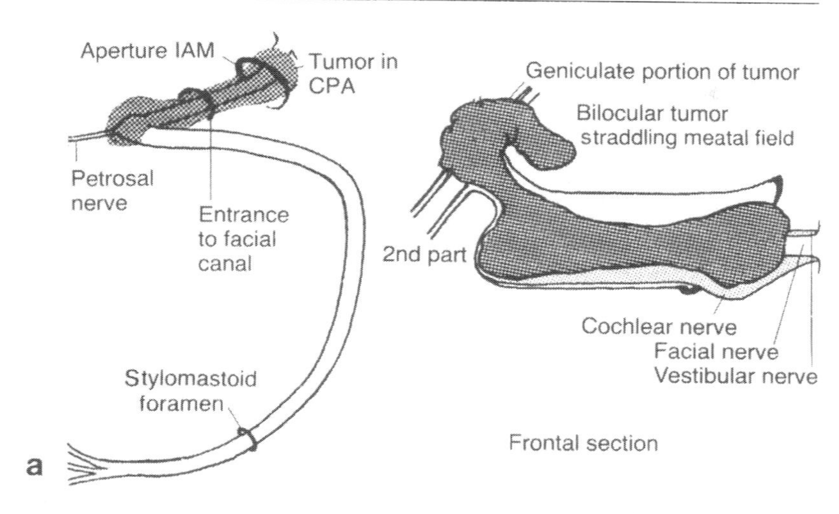

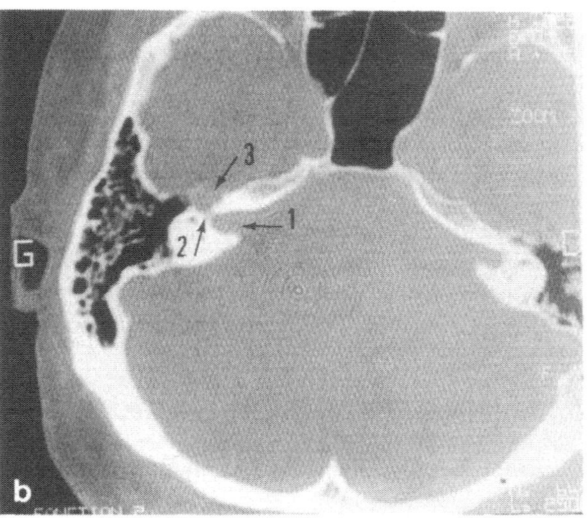

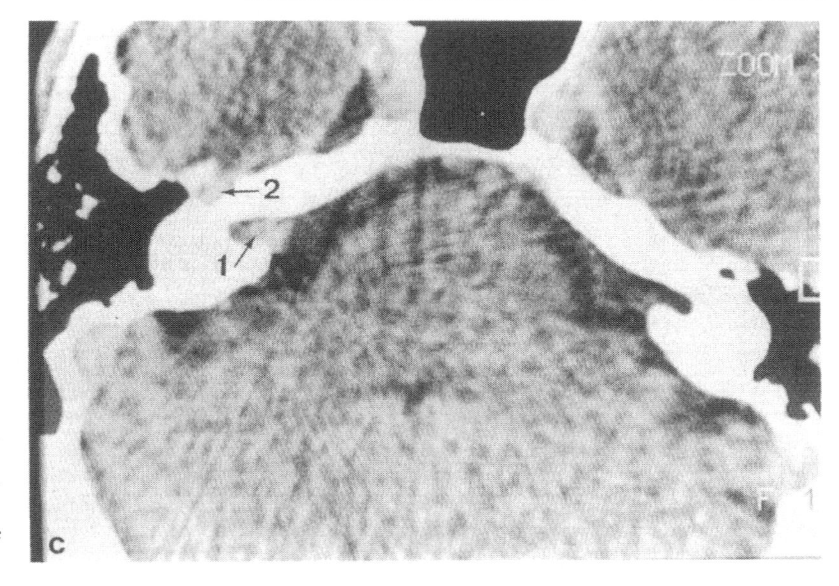

Fig. 114a-c. Neurinoma of intrapetrous facial nerve. **a** Diagram of operative findings. General arrangement of tumor on facial nerve. **b** CT scan in bony window. Expansion of IAM *(1)*, widening of first part of facial canal *(2)*, erosion of compartment of geniculate ganglion *(3)*. **c** CT scan in parenchymatous window. Intracanalicular window *(1)*, tumor in compartment of geniculate ganglion *(2)*

Mr. SAS, aged 51. 1 year's "discomfort" in the left eye due to absence of secretion. 6 months progressive left facial paralysis. 3 months: hebetude and unsteadiness in walking. Tonal audiometry: bilateral hypoacousia for high notes not exceeding 40 dbs loss. Vestibulometry: marked left vestibular hyperreflexia to caloric testing. Evoked auditory potentials: slowing of conduction on left side. Tomography: enlargement of first part of facial canal in frontal view and Pöschl's view (Fig. 114). Scan: intracanalicular tumor with slight extension into the angle in parenchymatous window and erosion of the first part and of the geniculate ganglion compartment in the bony window (Fig. 114). Diagnosis was made by Mme. Dr. Castan-Tabouriech on these findings. The EMG confirmed longstanding progressive denervation of the facial nerve. 29/10/84 operation by suprapetrous route: there was a neurinoma of the geniculate ganglion extending into the CPA (Fig. 114). A hypoglossal-facial anastomosis was done immediately and at 9 months facial mobility was very satisfactory (House type III).

Finally, we may concur with Sterkers et al. [174] that, excluding extracranial forms which present as parotid tumors, three pathologic types can be distinguished: the facial neurinoma developing in the angle and IAM which usually simulates an acoustic neurinoma; the neurinoma developing in the vicinity of the geniculate ganglion, which is more likely to cause gradual or sometimes sudden facial paralysis; and finally, the tympanomastoid neurinoma which produces a transmission deafness associated with a tumor visible behind the tympanum or sometimes embedded in the EAM.

Treatment. The chief problem is of course the repair of the facial nerve. Some authors have succeeded in performing facio-facial grafts with interposition of a graft taken from the cervical plexus and placed in either the IAM [174] or the facial canal [97, 125, 162, 172]. Many proceed to a hypoglossal-facial anastomosis, a technique we employ with satisfactory results [21]. It is exceptional to be able to preserve the continuity of the facial nerve as have Sterkers [174, case 1] or Fuentes and Uziel [55] whose case I was a neurinoma of the nerve of Wrisberg and case II a neurinoma of the chorda tympani.

Another problem is that of hearing. In the privileged cases of very small tumors, within or at the facial canal, hearing may be preserved and even improved when the expansion into the tympanic cavity produces a blockage of the chain of ossicles and a transmission deafness (the cases of Sterkers et al. [174], Iwagana et al. [97] for example). The proximity of the cochleovestibular cavities and the first two parts of the facial canal explains how the integrity of these cavities is greatly threatened by tumoral growth, and why the surgeon must avoid burring the bone too zealously around the intrapetrous tumoral extension to dislodge it for fear of opening the labyrinth.

Finally, one must bear in mind the extensive aspects of the cases reported by Fisch and Rüttner [45] and especially the "beaded" appearance reported by Sterkers et al. [174] in their case 1. The concern for total excision should lead to a review, even during the operation of all the radiographic records in the search for such extensions if one discovers a facial neurinoma when the diagnosis has not previously arisen. Again, the tumoral resection should be wide enough to divide the nerve in a healthy region.

Indications

These can be strict only if it has been possible to make a precise anatomic assessment before operation. At a first glance, the need to monitor the whole or part of the facial canal contraindicates the suboccipital route, as confirmed by Fuentes and Uziel [55] in their case 1.

The Suprapetrous Route. In theory this is the best approach, and it does properly expose the tumor in the IAM, in the first part of the canal, at the compartment for the geniculate ganglion and even, if need be, in the second part of the canal. Unless it is large, an extension into the angle should be perfectly controlled via an extended suprapetrous approach as advised by Wigand [187]. Thus, the suprapetrous approach should be capable of dealing with the majority of forms in the IAM and at the geniculate ganglion, especially when hearing is still useful.

The ETL Route. This ist the route of election for tumors arising in the angle and the IAM, especially when they are bulky and present as tumors of the angle with predominantly vestibular signs, the more so when hearing is impaired.

The Retrolabyrinthine Route. This allows access to tumors arising from the third part or, after a wide posterior tympanotomy, those involving exclusively, or also, the second part of the nerve.

Some Special Cases

Though it would seem logical to remove every neurinoma diagnosed, in certain circumstances it may be more sensible or more profitable for the patient to know when to refrain. This was our attitude to our case 9, with obvious von Recklinghausen's disease, exhibiting a moderate transmission hearing loss of 15 to 20 dbs but with a positive stapedial reflex and normal evoked auditory potentials. The diagnosis of a probable neurofibroma of the facial nerve despite the absence of any facial paresis was made by tomography, which demonstrated the existence of a dilatation of the second and third parts of the canal. Since the patient's symptoms were benign, we decided to temporise while keeping a careful watch on progress.

Our attitude may be compared with that of Sterkers et al. [174] towards their case 5, though this was somewhat more active. This patient had an isolated progressive facial paralysis associated with enlargement of the compartment of the geniculate ganglion. The suprapetrous approach revealed a "hypertrophic" facial nerve at the genu and all along the second part. All that was done was a simple opening of the canal to free the zone of constriction, which resulted in an "80% improvement" in facial mobility in 10 months.

Neurinomas of the Mixed Nerves

This again is very rare site since in an exhaustive review of the literature [Malca 117] we found only 102 reported cases, to which we were able to add 3 unpublished cases (2 from the neurosurgical clinic of our teacher, JE Paillas and one communicated by Guégan of Rennes) and 2 personal cases.

The Problems

Morbid Anatomy. Tumors arising at the posterior lacerate canal may develop in bilocular form, with an intracranial and an infratemporal portion. Only the neurosurgeon can access the intracranial portion, but he cannot reach the infratemporal extension. Conversely, a cervical approach cannot reach the intracranial tumor. We have described previously the narrowness of the space giving access under the bony arch to the posterior retroparotid region where the tumor develops. The key to this region is the resection, by methodical burring, of the petrous pyramid, thus opening the posterior fossa and the styloid region widely and simultaneously. This is the principle of the extended transcochlear route.

Kaye et al. [102] distinguish the essentially intracranial forms (type A), the intraosseous forms (type B) which may possibly exhibit an extension into the CPA, and the strictly cervical forms (type C). Our experience suggests the recognition of a type D, combining the three preceding types because there exists the bilocular shape mentioned above. It is obvious that a precise anatomic diagnosis is essential in deciding on which technique to adopt to approach each of the neurinomas of the mixed nerves diagnosed (Fig. 115).

A neurinoma of the mixed nerves which arises in the cerebellopontine angle progressively fills the cerebello-medullary cistern. Our limited experience does not allow us to state whether a tumor arising in the posterior lacerate canal can push ahead of it and become ensheathed by an arachnoid layer which separates it from the neurovascular struc-

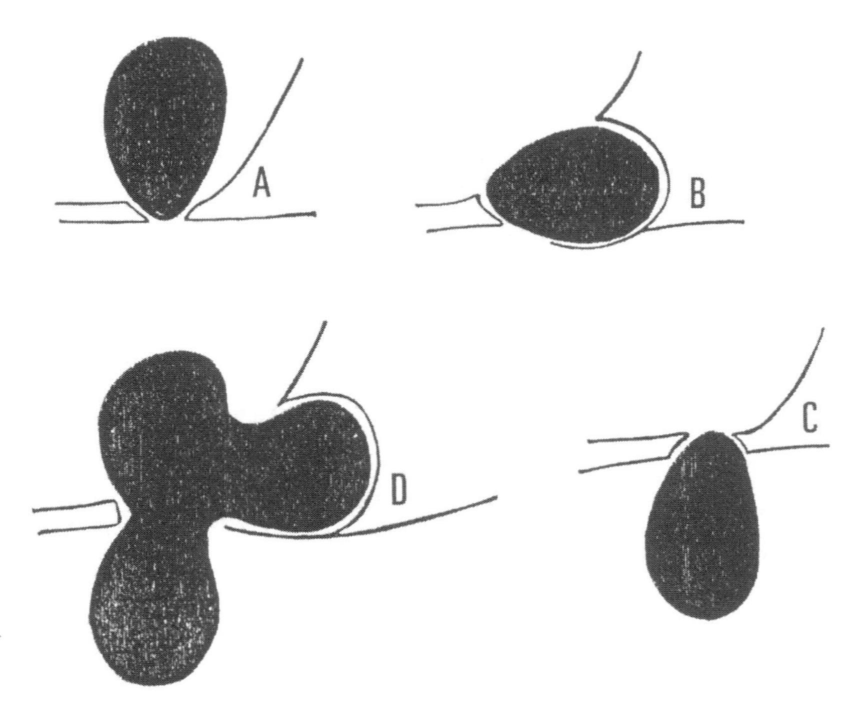

Fig. 115 A–D. Neurinomas of mixed nerves: structural types. **A** strictly intracranial. **B** intraosseous. **C** strictly cervical. **D** bilocular

tures contained in this cistern, as is the case with acoustic neurinomas. On the other hand, we have very clearly observed that the acousticofacial bundle and the AICA were perfectly separated from the tumor by an arachnoid sheath rendering their dissection very easy provided one remained in the right plane. This is cardinal for safeguarding these structures during operation.

Diagnosis. It is quite obvious that an adequate approach depends on preoperative recognition of the origin of the tumor on the mixed nerves and its extensions. Yet it is clear that the diagnosis is most often erroneous at the time of operation. In the 107 cases we collected, the preoperative diagnosis was accurate in only 13.5%. In half the cases, the diagnosis made was of a tumor of the angle, usually an acoustic neurinoma. In a third of the cases, the situation at the posterior lacerate canal was identified but the diagnosis was frequently that of a tumor of the glomus jugulare. It seems likely that nowadays the scan will prevent the more important errors formerly made (tumor of the brain-stem, of the cerebellum, of the cisterna magna, of the foramen magnum). It must be said that the symptomatology is often deceptive. While, in half the cases, attention may be attracted by signs of involvement of the mixed nerves (especially dysphonia, dysphagia, spinal amyotrophy, hypogusia), in two-thirds of the cases there are signs of audio-vestibular involvement (hypoacousia, vertigo) or facial or even trigeminal signs promoting diagnostic error. It should

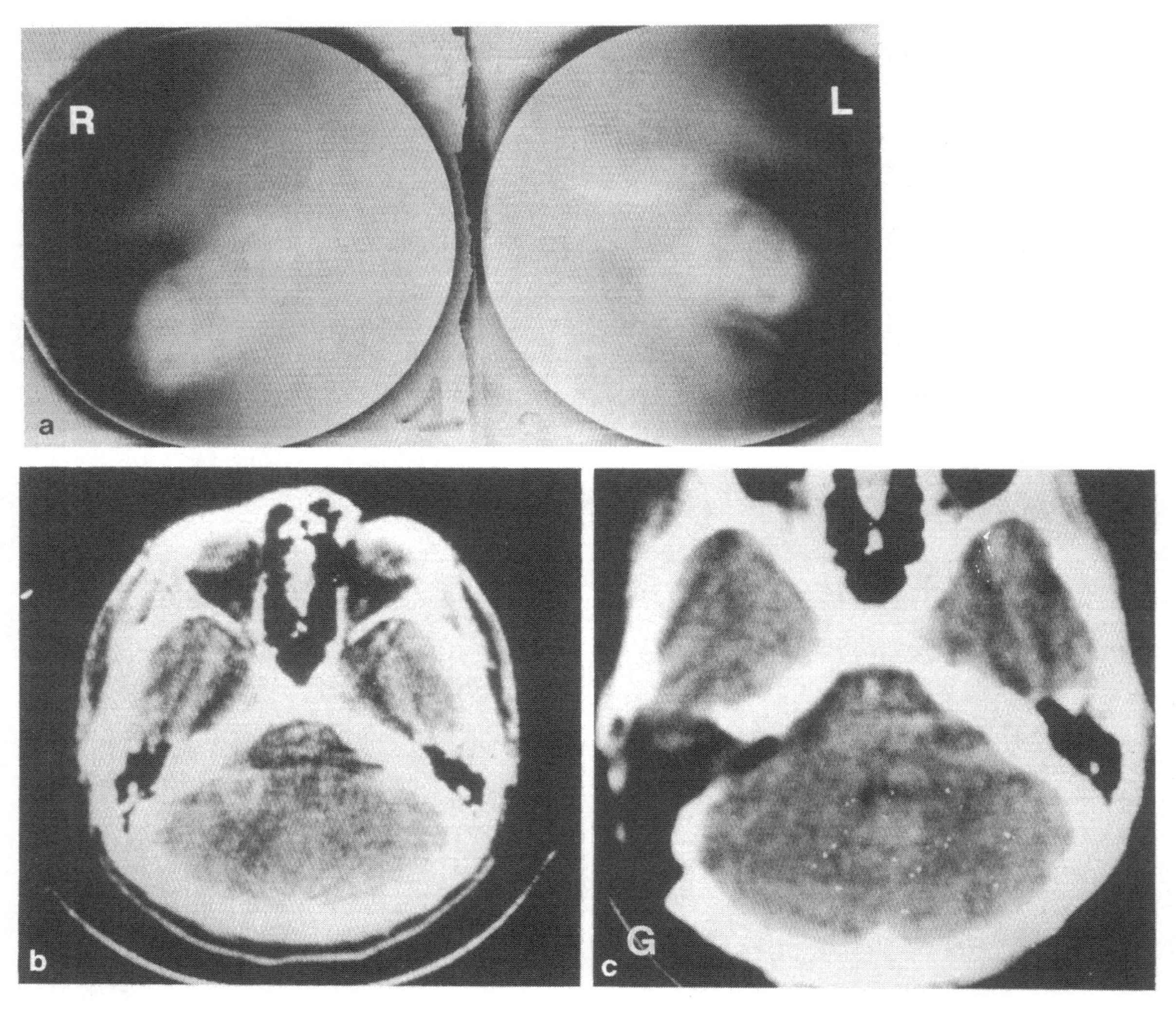

Fig. 116a–c. Neurinoma of mixed nerves (case BON. 1st operation). **a** Horizontal tomogram of both IAMs. The left is enlarged. **b** CT scan 1980: tumor in left cerebellopontine angle. **c** Postoperative scan: note the bony resection of the left translabyrinthine approach

also be stressed that in one case in four the symptoms are strictly otologic.

The neuroradiologic studies, therefore, if properly requested and performed, are decisive. The regular dilatation of the posterior lacerate canal is very suggestive (Fig. 117 A). Low section with the scanner in the bony window (Fig. 117) also show the deformation and filling of the lacerate canal by the tumor. Even better, successive sections well show all the extensions. It is likely that MRI, of which we have no relevant experience, many be even more useful.

Treatment. Neurinoma of the mixed nerves, a well-encapsulated and not very hemorrhagic tumor, poses no great problem apart from requiring perfect exposure for its total extirpation. The best way of illustrating the problem is by the following case history:

Mr. BON, aged 30. April 1980, sudden onset of left deafness accompanied by loss of balance, left earache and headache. Treatment on vascular lines diminished the disturbance of balance and the headaches. Neurologic examination revealed only horizontal nystagmus to the right, moderate deviation of blindfold walking to the left and left hypoacousia. Pure tone audiometry: auditory loss of 70–80 dbs on the left – stapedial reflex at 110 dbs on the left. Vestibulometry: left vestibular arreflexia. Evoked auditory potentials: slowing of conduction II–V on the left. Petrous tomography: enlargement of the left I.A.M. in horizontal section (Fig. 114a) but not in frontal section. Scan: lesion occupying the left CPA, stage III (Fig. 116b) 4/6/1980, translabyrinthine approach. No tumor in the meatus. The VII–VIII bundle, well-sheltered in its arachnoid sheath, elevated by the tumor of the angle. This was quite adherent to the posterior foramen lacerum where dissection of the mixed nerves gave rise to disorders of cardiac rhythm. Morbid anatomy: mixed A and B neurinoma. Recovery uneventful, no facial paralysis. 3 months later resumed work. Postoperative scan: no tumor in the angle (Fig. 114c). 6 months later, gradual development of dysphonia and atrophy of left half of tongue. 16 months postoperatively, complained of dysphonia and slight dysphagia. Scapu-

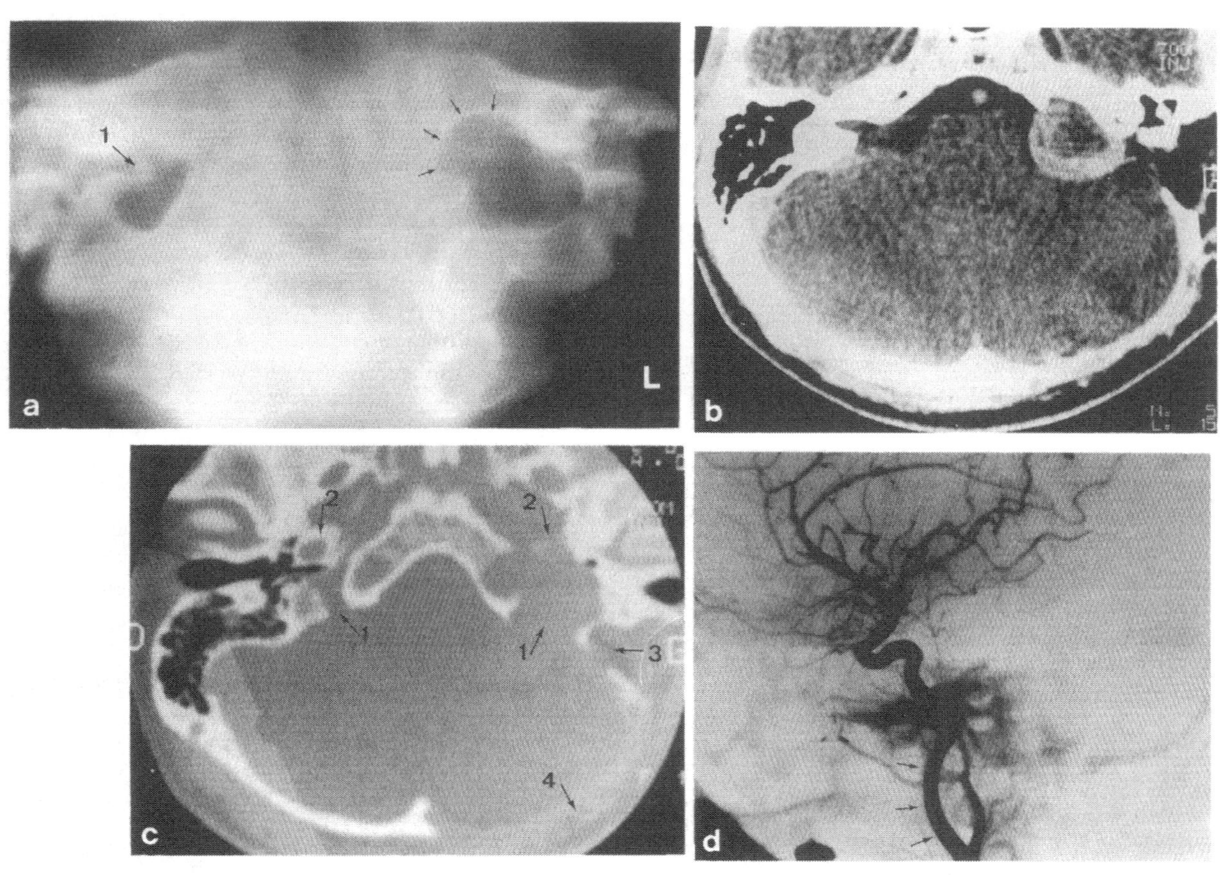

Fig. 117 a–d. Neurinoma of mixed nerves (case BON. 3rd operation). **a** Tomogram of posterior lacerate canals: enlargement of left canal with sharp borders, disappearance of jugular spine of temporal (clearly visible on right) *(1)*. The arrows show the expansion of the pars nervosa. **b** CT scan 1983 (parenchymatous window): tumor in left CPA. **c** Scan 1983 (bony window): enlargement of lacerate canal *(1)* on left. Note erosion of carotid canal *(2)* on left, bony resections of anterior translabyrinthine route of approach *(3)* and suboccipital approach *(4)*. **d** Carotid angiogram: note the forward displacement of the cervical internal carotid

lar pain. Left spinal amyotrophy, also affecting left half of tongue. Tomography of posterior lacerate foramina: sharp-edged enlargement of left foramen lacerum (Fig. 116a). Brain-scan: expansive lesion low in the left CPA. 5/2/1982, suboccipital approach: neurinoma embedded in the left posterior lacerate canal. Excision could not be pursued beyond the canal. Subsequent disappearance of scapular pain but persistence of other symptoms, dysphagia worse. 3 months later, return of scapular pain and at 6 months left facial hemispasm, dysphagia and dysphonia improved after injection of Teflon into the left vocal cord. At 11th month, development of headache and nausea. 15 months: scan showed tumor filling the left posterior lacerate canal (Fig. 117b and c). Angiography: forward displacement of left internal carotid (Fig. 115d). Absence of drainage by the left jugular bulb. 24/3/1983, extended transcochlear approach with complete excision of the tumor which consisted of 3 portions: jugular canal, intracranial and infratemporal. Recovery complicated by an acinetobacter meningitis, rapidly cured by systemic and intrathecal treatment. Left facial paralysis. Discharged at the 30th day. At 6 months, movement of face and soft palate normal, tongue normal. Left pharyngo-laryngeal paralysis but swallowing perfect. Had returned to work at the 3rd month. Regular follow-up since and is doing very well.

Indications

Provided the diagnosis has been made, these depend essentially on the extensions of the tumor and the state of the posterior lacerate canal. If this is enlarged, a limited approach to one of its orifices will be inadequate.

Extended Transcochlear Approach. This should be considered whenever the investigation has demonstrated the presence of a type D bilocular tumor. However, sacrifice of hearing by this approach should be discussed, especially since the protection of the acousticofacial bundle in its arachnoid sheath gives hope of easy dissection, particularly of the facial nerve. The sacrifice of hearing for the benefit of the facial nerve which forms part of the importance of the ETL route is less evident here, for the rerouting of the facial nerve involves much more risk to this nerve than does the ETL route. If the tumor is not too large in the CPA, it would seem more judicious to undertake an infratemporal approach followed by a suboccipital approach as advised by Fisch [47].

Infratemporal Route. In theory, this is the ideal approach for type C tumors and probably also those combining the intracanalicular type B and the infratemporal type C. We have noted its value, when combined with a suboccipital approach, for the bilocular forms without much loss of hearing.

Suboccipital Route. Although this is not an otoneurosurgical route, it is useful to discuss it here. In fact, as we have just seen, it is an indispensable complement to certain infratemporal routes. It may also be the best route when dealing with a purely intracranial type A neurinoma. In fact, the particular position of this tumor arising in the cerebello-medullary cistern (Fig. 31) and the absence of intrapetrous extensions invalidate all the arguments we have advanced for neurinomas of the VIIIth nerve and the ETL route. Provided the procedure is restricted to the cerebello-medullary cistern, the risks to the acoustico-facial bundle should be reduced as much as possible. However, it must be stressed that the suboccipital route, used alone, seems inadequate once the intracranial tumor has crossed the posterior lacerate canal. For bilocular tumors of small size, Hakuba et al. [77] recommend a suboccipital route advanced to the mastoid so as to expose the inner border of the sigmoid sinus. They propose a first stage to approach the intracranial tumor by the usual transdural route and to resect it flush with the internal orifice of the posterior lacerate canal. In a second stage, they would proceed by an extradural route, extending the occipital and temporal craniotomy in the direction of the posterior foramen lacerum and opening its external circumference. They state that it is then possible to fragment and extract the intraosseous portion and then the cervical portion. They merely advise that one should "take care not to damage the facial nerve", but without specifying how it is possible to expose the foramen lacerum widely without opening the third part of the facial canal. It may well be that, when the intersinuso-facial space is wide, this maneuver can be performed without too many problems; but even if it exposes the outer border of the lacerate canal, the space available must be restricted and would only allow quite safe dissection for the intracanalicular growth, and that only provided it was not too big. In any case, as neurinoma of the mixed nerves is a tumor of the pars nervosa, there will always be the bulb and the origin of the internal jugular vein interposed between the surgeon and the tumor, and this venous flow would seem a major obstacle, especially at the bottom of the hole these authors use. As for removing the cervical part of the growth from here, this seems particularly hazardous. At first glance, this technique seems practicable only if the intraosseous extension of the tumor is minimal and the intersinuso-facial space particularly wide. And even under these conditions we very much doubt the real possibilities of this techniques. All in all, we feel that the combination of an infratemporal route and then a suboccipital route offers much more assurance of efficacy and security.

Postoperative Course

It is natural that the dysphonia and dysphagia should be increased after operation. This is not constant but common and requires that special care should be taken in every case to avoid false passages and the bronchopulmonary infection that would inevitably follow. Having experience of such problems, we can only advise anyone who may be faced with a neurinoma of the mixed nerves to make it a matter of routine to pass a stomach-tube during operation and not to remove it until one is certain that the patient can swallow without difficulty; and even then, these patients must have their pulmonary status carefully supervised with an eye for the least congestion or the least rise of temperature. When there is only a glossopharyngeal paralysis, the various problems of swallowing after operation with difficulty in swallowing solids without a false passage usually pass off rapidly provided the patient helps by taking a mouthful of water. The tube can be removed after 3–4 days. When there is a paralysis of the Xth nerve, and of course when the IXth and Xth are paralysed together, the problems are much more serious and long-lasting. There is combined dysphonia and dysphagia, with false passages after the least mouthful of water or the least salivation, and pulmonary problems are often very quick to appear. All oral feeding must be forbidden and gastric feeding resorted to in the semi-sitting position, avoiding large meals which might promote reflux. If there is frank paralysis of one vocal cord, there should be no hesitation in giving an injection of Teflon which corrects the gaping of the glottis and markedly reduces the risk of false passages, while improving the dysphonia. The patient can then resume taking semi-solid or solid food, still taking precautions with drinks. It often takes 3 to 4 months for this basic and vital activity to become more or less automatic once more.

Neurinomas of the XIIth Nerve

This is a very rare tumor, never encountered by ourselves. The majority of the reported cases relate to intracranial forms. Ulsø et al. [183] in 1981, including their own case, were able to list only 17 cases from the literature to which can be added the more recent case of Dolan et al. [41]. This is a tumor which develops in the CPA, but whose initial symptom in the 18 known cases has always been a gradually developing hemiatrophy of the tongue. Ulsø et al. [183] stress the usual delay in diagnosis, the very wide errors and the importance of the scan in displaying the tumor. The suboccipital approach has seemed most suitable in view of the site of the tumor, but in the great majority of the 16 operated cases the subsequent course was complicated by false passages and by severe pulmonary infection, which caused death in a third of the cases. This seems only natural, since the hypoglossal nerve is associated with the mixed nerves in the cerebellomedullary cistern and these latter must be in actual contact with the tumor. Despite microsurgical dissection, operative traumatism is the rule and, even if the paralysis regresses, it is responsible for immediate postoperative difficulties which call for routine precautions including tracheotomy and feeding by stomach-tube. The case of Dolan et al. [41] is interesting since it had progressively invaded and destroyed the infralabyrinthine and perifacial regions, thereby justifying an infratemporal and then a suboccipital approach to remove the intracranial fragment inaccessible at the first operation. Given its cervical course, it is not surprising that some parapharyngeal tumors prove to be neurinomas of the XIIth nerve. We have encountered only 6 cases, all of which raised the problems of a pharyngeal tumor. This has led us to carefully study, among other things, the outlines of the condylar canal when investigating an infratemporal tumor. It is probable that if we had to deal with a tumor of this type, we would consider the use of Fisch's infratemporal route to approach it with all security.

Trigeminal Neurinomas

This again is a rare site, one encountered by us on 6 occasions, representing 2.2% of all the tumors of the cerebellopontine angle on which we have had to operate. We feel it right to distinguish them from the group of the other neurinomas because, though they may belong to the group of peripetrous tumors, they do not as a rule have any otoneurosurgical connotation.

The Problems

Morbid Anatomy

Since Jefferson [98], it has been classical to distinguish three pathologic types: A, supratentorial, B, subtentorial and C, astride the petrous ridge. Montaut [121], who reviewed all the reported cases at the time of his thesis, established that type A was the commonest (53% of cases), while type B accounted for 25% and type C 22%. Our own experience is different, since we record 1 type C and 5 type B. It is true that our otologic bias predisposes us to observe mainly tumors of the angle, but this very marked inversion of the trend leads us to think that a fair number of trigeminal neurinomas were previously taken for acoustic neurinomas.

Lazorthes and Bimes [110] have taught us that the cave of Meckel is a diverticulum of the posterior fossa which insinuates itself between the two layers of the dura which clothes the upper aspect of the petrous, and that the subarachnoid spaces extend into the cave as far as the exit of each of the three branches. Yasargil [188] has demonstrated that this is a prolongation, like a glove-finger, of the cerebellopontine cistern.

This arrangement explains why the neurinoma, which arises from the sensory root of the trigeminal ganglion, develops directly in this cistern without being covered in an arachnoid sheath as is the acoustic neurinoma. And doubtless, this explains why in our 5 type B cases we regularly found the acoustic and facial nerves in direct contact with the tumoral substance and very difficult to dissect and safeguard. On the contrary, in our first case, of type C, we were amazed at the ease with which we were able to strip these nerves and thus preserve hearing and expression. It is probable that the tumor, arising from a branch of the ganglion below the arachnoidal *cul-de-sac*, had pushed this ahead of it and become progressively sheathed in it while becoming embedded in the posterior fossa. This is only a supposition, which ought to be confirmed by careful study of the cases we may have to operate on subsequently.

Diagnosis

Our teacher, JE Paillas [134] had underlined after Montaut [121] how inconstant the trigeminal symptoms were, both subjectively and objectively, and even altogether absent in a quarter of the cases he had personally observed. Further, he found these symptoms quite often delayed in appearance in the course of the disease, since in half the cases they merely supplemented an initial audio-vestibular pattern. It should also be stressed that a number of the reported cases, of type A, progressed like tumors of the temporal fossa, diagnosis often not being made until the time of operation and sometimes even only after histologic examination of the operative specimen. This means that the clinician is often mistaken and that the diagnosis rests mainly on the neuroradiologic investigations.

A notch in the upper border of the petrous, as described by Fischgold [50], or the separation of the two cerebellar arteries, middle (AICA) and superior, with opposed concavities circumscribing the tumor are decisive signs; but analysis of the scanner images showing a tumor very internally, almost at the clivus and medial to the aperture of the IAM, is better placed to make the diagnosis (Fig. 118a).

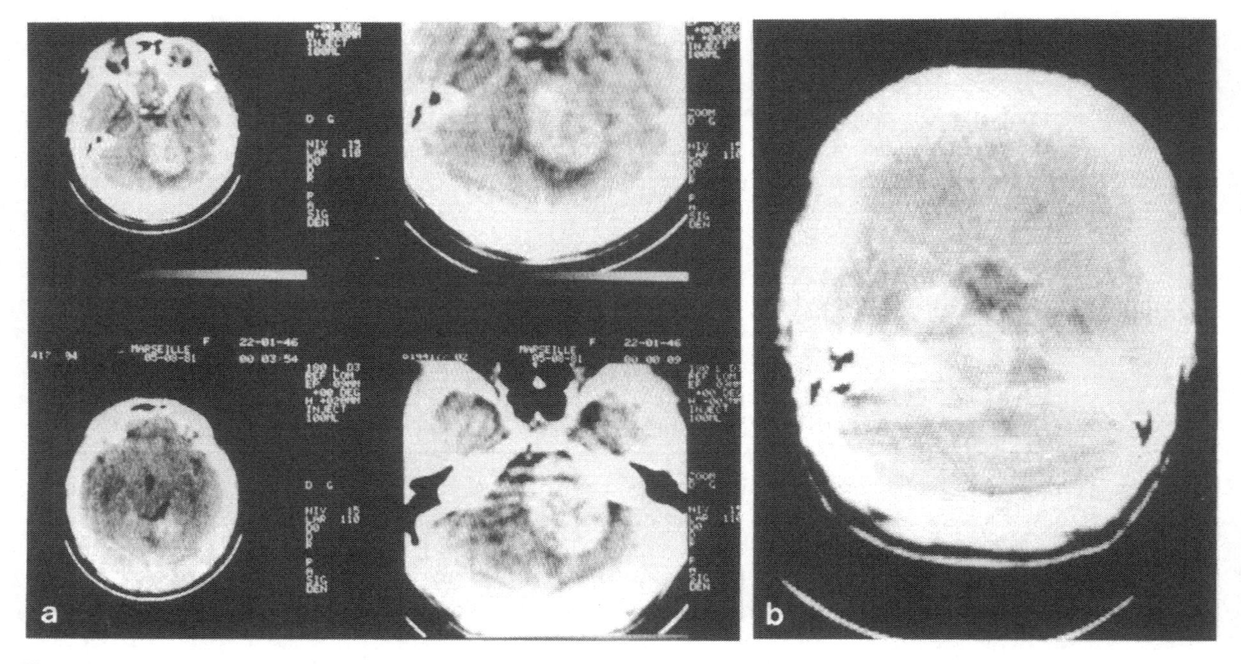

Fig. 118a, b. Trigeminal neurinomas. **a** Subtentorial neurinoma, type B (scans). **b** Tumor astride upper border of petrous, type C

When the tumor is evidently astride the upper border of the petrous (Fig. 118b) there is no longer any doubt. This diagnosis is important because the access route must in principle be very different and suited to the particular position of the tumor in relation to the facial nerve.

Treatment

The surgery of trigeminal neurinomas is naturally determined by the anatomic type, which decides whether the tumor is to be sought in the temporal or the posterior fossa. It is also determined by the paramedian situation of the tumor, near the clivus and the basilar trunk and its terminal branches and medial to the acoustico-facial bundle. There is often a retrocochlear type of auditory involvement, but this is usually relatively minor considering the size of the tumor. The result is that the translabyrinthine or transcochlear routes can as a rule be excluded. The problem is to access this tumor by attacking it internal to the IAM.

Indications

We feel that the only approach to be considered is a subtemporal route, possibly suprapetrous if the tumor remains localized in the cave of Meckel, but mainly subtemporal and then transtentorial [13] if there is an extension into the posterior fossa. This is a neurosurgical approach that we need not describe in detail here. We simply stress that, with microsurgical techniques and instrumentation, it is often adequate for removal of even large tumors descending quite low towards the mixed nerves, but that if control of the lower pole seems impossible from above it is always possible to go further and perform a small suboccipital craniectomy to look for the lower fragment by going behind and under the lateral sinus. We have had to do this on two occasions, with success. However, we cannot deny having on one occasion approached such a neurinoma by a translabyrinthine route because the clinical picture was essentially otologic and the conformation of the tumor in the scan with splaying of the aperture of the IAM had suggested an acoustic neurinoma. It was in this case that we had to abandon a fragment against the trigeminal, not because it was inaccessible but because during the dissection the patient exhibited serious and prolonged disorders of cardiac rhythm which led our anesthetist to ask for the operation to be halted. There were no serious consequences, but the access route required sacrifice of the facial nerve which constantly barred approach to the tumor and, after being contused, was finally divided.

Results

One patient died of a postoperative hematoma in the CPA. In the other 5, we have to report one section of the facial nerve in the case approached by the translabyrinthine route and 4 anatomic preservations of the nerve. Allowing for the difficulties of dissection referred to above, it is understandable that these 4 patients showed a postoperative facial paralysis. This regressed, in one case after 3 months with complete recovery and in one case after 12 months, with good resting tonus, good palpebral occlusion but moderate hemispasm mainly visible in expression (marked stage I of our classification). On the other hand, 2 operated cases were left with a facial paralysis and both had to undergo a hypoglossal-facial anastomosis, one after the failure of a transfacial anastomosis performed elsewhere.

We have always been surprised by the few sensory sequelae experienced by these patients, even when facial paralysis was persistent. All our patients claim to have preserved the sense of touch and only one complained of paresthesiae of formication type. Three even recovered corneal sensation, among them the patient who was left with facial paralysis and facial paresthesiae. It therefore seems that the tumor is related to the principal root and that the rootlets, which we were usually able to preserve, were intermediate rootlets. In view of the small number of cases studied, we are not certain that this is the correct explanation but it seems plausible.

Tumors of the Glomus Jugulare

As stressed by George [61], these are rare tumors. However, they are not exceptional and we have ourselves seen 41 cases since 1973. The limited approaches, adapted to the surgery of chronic suppuration, the only ones known before 1960, did not allow their proper exposure except when they were purely tympanic forms. This is why these tumors, dreaded by surgeons, were for long abandoned to the radiotherapists after Williams [193] and published his results. The progressive recurrences of irradiated tumors, reported in 1962 by Rossenwasser [154] rekindled surgical ardor, first for small tumors as recommended by McCabe and Fletcher [113] and then for larger ones under the stimulus of such pioneers as Gaillard and Rebattu [58], Shapiro and Neues [163], and Hilding and Greenberg [80]. We ourselves had made the attempt from 1972 [140]. It was mainly the work of House and Hitselberger [89] and of Fisch [47] that gave substance to the present-day surgery of these tumors of the glomus jugulare.

The Problems

Apart from the anatomic problems, already indicated in the two preceding chapters, we must discuss those inherent in the very nature of these tumors which are remarkable for their invasive, hemorrhagic and sometimes secretory character.

Invasive Features

It is the case that the forms localized to the tympanic cavity (type A) are the rarest. In the experience of Fisch and Kumar [107] they represent only 8.5% of cases. Usually, the tumor invades the pyramid to a greater or lesser extent. In one out of two cases, 47.8% exactly for Kumar and Fisch [107], it occupies the entire pyramid in more or less extensive fashion, eroding the labyrinthine massif and extending into the apex (type C). In 24.5% of cases, they state, it even sends extensions into the cranial cavity (type D) and in only one case in five (19.2%) does it remain rather limited, invading only the mastoid without affecting the labyrinthine massif (type B).

Reasons

There is no single explanation for this invasive property. A first factor may be the relative frequency of the various glomic structures scattered in the petrous. According to Guild [73], 55% are situated on the jugular dome while 25% are sited along the tympanic nerve and 20% along the nerve of Arnold-Cruweilhier. This may account for the higher incidence of type C tumors, but the rarity of the type A lesions does not fit well with the frequency of glomic structures at this site.

It may also be imagined that the pulsatile bruit produced by the tumor throbbing in the tympanic cavity is so unpleasant as to bring the patient for consultation sooner than the other forms, whose intraosseous development is much more silent and very slow and allows the occult invasion of the entire petrous before the tumor is discovered.

Finally, it may be supposed that only very rarely, as thought by Konheim and Ribbert [20], does the tumor originate from only one of these glomic structures, and that in fact the entire tympano-jugular glomic system can simultaneously give rise to a tumoral eruption, which would explain the diffuse nature of the lesion. In this context, the report by Buckingham et al. [20] of a tumor at the promontory, another at the hypotympanum and a third on the course of the nerve of Arnold-Cruweilhier argues in favor of a multicentric origin of this tumor. This last hypothesis of diffusion of the oncogenic process over the glomic system would explain the rare cases of bilateral tumors [7, 34, 59, 115], the somewhat commoner combination of a jugular glomus tumor with one of the carotid body [59-179] and also the very exceptional multiple tumors as in the case reported by Garcin et al. [59]. Again, these various combinations confirm the concept of a neural crest disorder comprising the pathologic conditions simultaneously involving several components of the APUD system, derived, as we know, from the neural crest. This concept makes it possible to understand the apparently confusing associations of glomus tumors with other neoformations throughout the organism, whether these be glomus tumors, now better known as paragangliomas than by the official name of chemodectomas, or adenomas at very varied sites (thyroid, parathyroid, pancreas, suprarenal, testis, ovary, pituitary) constituting a Werner type of polyadenomatosis, or yet again really malignant neoformations such as medullary thyroid carcinoma (the Sipple or Gorlin type).

Pathways

Spread is centrifugal, following the paths of least resistance – the orifices, the intrapetrous cavities and canals, and the loosely textured stretches of the temporal bone. Spector et al. [167] describe three possible directions:

a. Downwards via the posterior lacerate canal. The ramifications may take several paths: first intravenous, in the jugular vein, which justifies the routine stripping of the latter in types C and D; then perivenous and perineural, towards the infratemporal regions, which justifies a wide cervical exposure and the importance of the infratemporal or extended transcochlear routes.

b. Upwards, in the tympanic cavity, whether the tumor has arisen at this site on a tympanic glomus or has reached there after traversing the roof of the bulb. All paths are then open to it:

- The sinus tympani and the subfacial cellular tracts conduct it towards the intersinuso-facial space and the infralabyrinthine cells
- The cochlear window gives it access to the labyrinth and, across this structure, to the IAM.
- The aditus, once the epitympanic recess is filled, gives it access to the antrum and to the entire

mastoid, then the perilabyrinthine cells and also, after having crossed the tegmen tympani, freedom to develop in the temporal fossa
- The auditory tube and peritubal cells lead it to the skull base and the nasopharynx
- The carotid canal conducts it towards the anterior lacerate foramen, the paracavernous region and the cavernous sinus
- The EAM allows it to become embedded in the concha

c. Backwards, the tumor can spread into the posterior fossa, often extradurally, after having destroyed the cortex of the posterior face of the petrous, sometimes intradurally by filling the lumen of the sigmoid sinus or inferior petrosal sinus starting from the bulb, sometimes even intracranially, whether by destroying the dural barrier or crossing one of the natural orifices at this level - the aperture of the IAM from the labyrinth, the condylar canal from the bulb or one of the endo- or perilymphatic canals.

The result of all this is that it is not enough to confirm the existence of a tumor of the glomus jugulare. One must also endeavor to specify the various extensions and sometimes the combinations before making a valid plan of treatment.

The Problem of Malignancy

This is controversial and arises in three sets of circumstances:

Atypical Cells. It is indisputable that there sometimes exist histologic anomalies suspect of malignancy. As stressed by Lachard et al. [108], it is difficult to assess this precisely, especially as the incidence varies greatly, from 1 to 9%, for different authors. The criteria of malignancy are not at all formal, but are rather indications. They involve an increased number of mitoses, which are actually abnormal in rare cases, the presence of foci of centrolobular necrosis, and the invasion of vascular lumina and nerve sheaths. In practice, these histologic criteria are not sufficient since, as stressed by Romanet et al. [151], some tumors with atypical cells exhibit neither recurrence nor metastasis, whereas, conversely, histologically benign tumors may be complicated by metastases. In this field, therefore, the problem is far from solved.

Metastases. Cases have been reported where the glomus tumor was complicated by remote localizations, not only in the cervical or mediastinal nodes but also in numerous viscera: lung, bone, spleen, heart or kidney. The absence of atypical cells in a number of these cases may leave some doubt as to the true malignant nature of these lesions, especially as the concept of apudoma raises the question of the exact nature of these diverse localizations, i.e., whether they are all primary expressions of a neural crest pathology or else secondary foci from one primary tumor. The problem is the more difficult since, though cases are described which are truly malignant as proved by the rapid spread of the tumor, there are also cases of a tumor with metastases which remain stable over many years.

Local Malignancy. This is actually the commonest problem since this tumor usually gives evidence of a locally invasive capacity which would seem to argue in favor of malignancy. The bony destruction so characteristic of this type of tumor is indeed abnormal for a benign tumor and is regarded by some as evidence of a local malignancy, especially as there is no lack of cases where the tumor has also infiltrated the muscles, sometimes the cranial arteries or even the brain. However, the rarity of reported metastases (2.5% according to Zack and Lawson [195]) seems to confirm the strictly local nature of the invasive capacity of the tumor. However, this concept justifies the complete excision of the bone all around the tumor and also the routine inspection of the adventitia of the internal carotid artery and the epineurium of the facial and mixed nerves, as well as the dura, when the tumor has come in contact with them.

Hemorrhagic Features

Reasons

Its hemorrhagic features flow from the very nature of the glomus, which is essentially a vascular structure formed of a cluster of small vessels with a narrow lumen but a thick wall, with a very sinuous course, tangled but not branching nor anastomosed between themselves. These vessels are fed by a common artery and each drains into an adjacent capillary or vein. As stressed by Terracol et al. [178], the glomus may be considered in simplistic fashion as a collection of arteriovenous anastomoses. The tumor only reproduces to a greater extent the structure of a paraganglion and therefore has an extremely vascular structure, which is certainly mainly responsible for its particularly sinister reputation. Every otologist is aware of the hemorrhagic nature of a simple biopsy, often the cause of such abundant hemorrhage as to require prolonged packing and sometimes even an emergency mastoidectomy in order to obtain hemostasis. Excision likewise becomes extremely bloody as soon as the surgeon has penetrated the tumoral mass, which is not surprising since this arteriovenous structure is then open and discharging from either end of its network, the arterial afferents on one side and the dural sinuses on the other. Hence the importance of perfect exposure of these tumors, also the importance of methods suited to reducing the bleeding.

Reduction in Size

Various methods are to be considered. Hypothermia and hypotension were proposed at one time [186] but were soon abandoned because of their complexity. Preoperative ligature of the external carotid was used by many surgeons but its ineffectiveness, given the multiple anastomoses and the numerous tumoral pedicles, explains the search for techniques with a more distal effect in contact with the tumor. It is the case that careful coagulation of all the afferent pedicles and occlusion of the sigmoid and inferior petrosal sinuses before approaching the tumor are the best methods of operating without excessive blood-loss. They call for perfect exposure and a little practical ability which takes some time to acquire. Hence the attempt to develop preoperative methods capable of reducing the profusion of the tumoral blood-supply. In this context, radiotherapy may play a useful part. Spector et al. [166] have clearly shown that, while irradiation produces few cell changes and only some chromatin alterations, it leads to major intraparenchymatous sclerosis and a diffuse obliterative endovasculitis. These measures stabilize for a time the advance of the tumor, in any case known to be very slow, and seem to ensure a marked reduction of its hemorrhagic nature if they are operated soon after. This was the impression we gained in the 3 cases operated under these conditions. In fact, embolization appears to be the best method. It ensures the distal obliteration of the feeding pedicles. Unfortunately, it cannot be used for all the pedicles since some derive from arteries which are also functionally important for the brain, and therefore not to be embolized. Though considered from the early days of arteriography [19], this method only really became operational with Djindjian [40] and his work on selective catheterization. It does not much matter what material is embolized (muscle, spongel, absorbable or inert agents, cyanoacrylate, various beads); what is essential is to recognize the various pedicles that must be embolized to dry up the entire tumor.

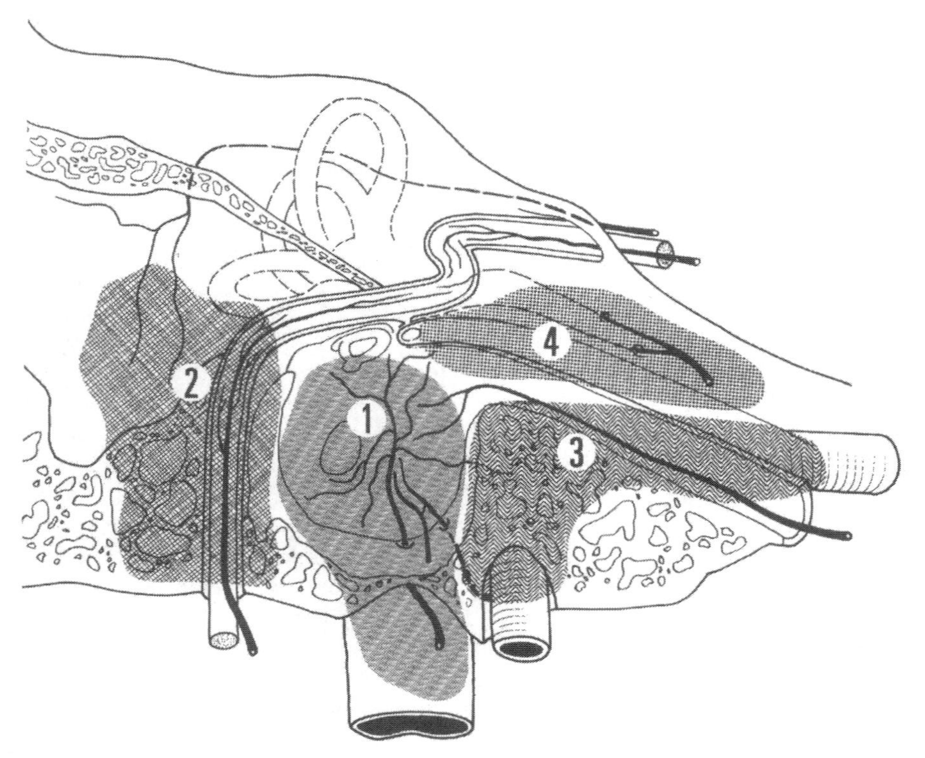

Fig. 119. Jugular glomus tumors. The vascular compartments (after Moret). Inferomedial *(1)*, posterolateral *(2)*, anterior *(3)* superior *(4)*

The Compartments

The concept of "compartments" is thus essential, and in this context the studies of Moret et al. [124] are of major importance. Studying the vascular architecture of the glomus tumors with a view to their systematic embolization, these authors established their compartmental structure, a very important concept since it presupposes that the selective embolization of each of these compartments, totally isolated from its neighbors, is the only means of complete success in this maneuver. Moret et al. [123] reported that 85% of glomus tumors were compartmented and that 4 compartments could be distinguished (Fig. 119):

- *The inferomedial compartment*. This is situated on the bulb and in the lower part of the tympanic cavi-

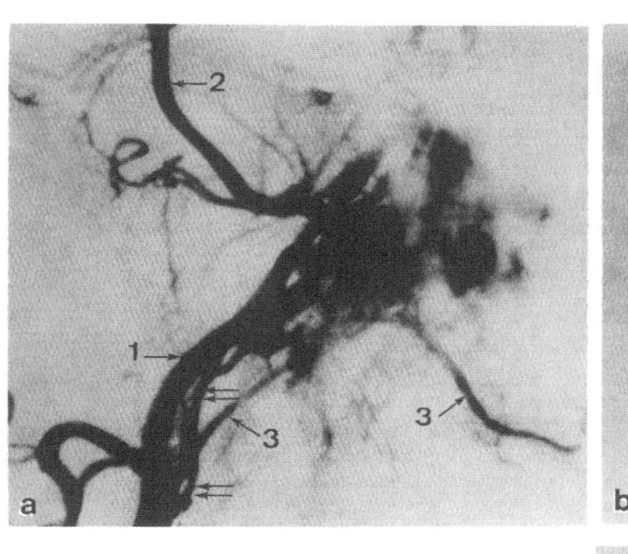

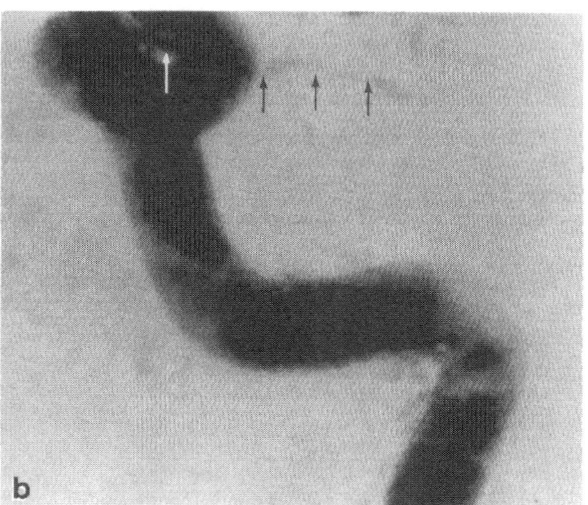

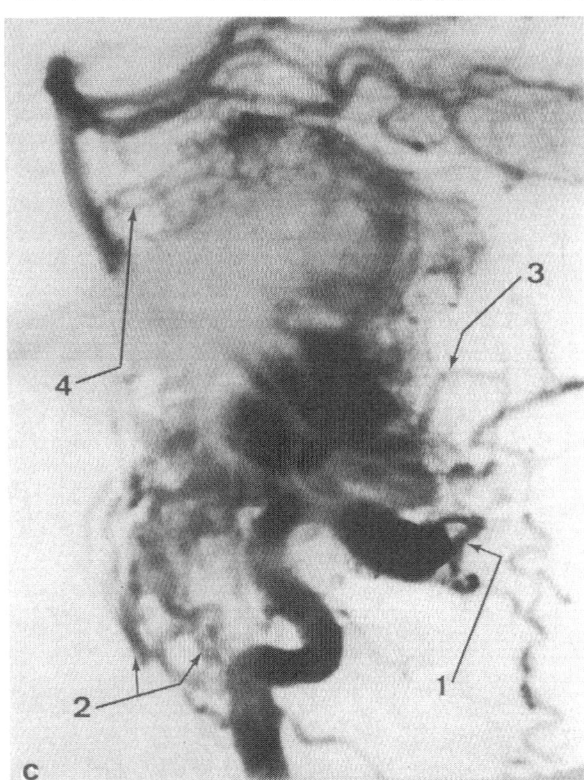

Fig. 120a–c. Inferomedial compartment. **a** External carotid (lateral view) *(1)*, ascending pharyngeal artery *(double arrow)*, internal maxillary artery *(2)*, occipital artery *(3)*. **b** Internal carotid (frontal view). Meningeal branch of siphon *(arrows)*. **c** Vertebral angiogram. Bulky tumor of jugular glomus. Posterior meningeal artery *(1)*, muscular branches *(2)*, PICA *(3)*, AICA *(4)*

ty. It is supplied by the ascending-pharyngeal artery, whose inferior tympanic branch, which accompanies the tympanic nerve, is normally distributed in the region of the promontory (Fig. 27). It extends mainly downwards, around or in the jugular vein and, across the condylar canal, towards the lower clivus and the region of the foramen magnum. As it progresses, it may collect the meningeal branches of the internal carotid, branches of the carotid siphon (Fig. 120), sometimes the meningeal branch of the

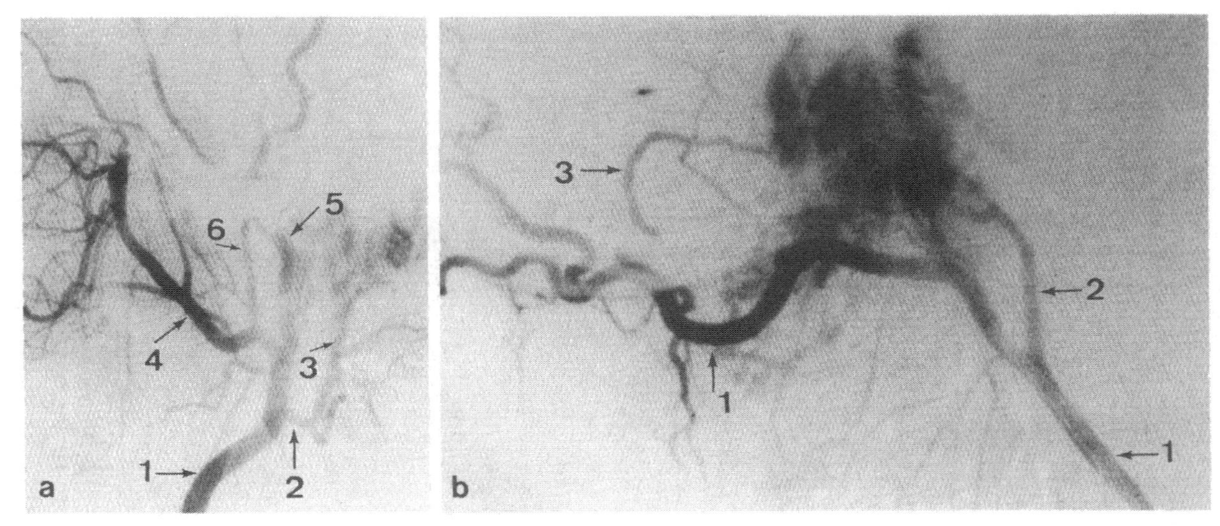

Fig. 121 a, b. Posterolateral compartment. **a** External carotid, left (lateral view) injected above origin of occipital artery. External carotid artery *(1)*, posterior auricular artery *(2)*, stylomastoid artery *(3)*, internal maxillary artery *(4)*, superficial artery *(5)*, middle meningeal artery *(6)*. **b** Selective injection right occipital artery *(1)*, stylomastoid artery *(2)*, mastoid meningeal branch of occipital artery *(3)*

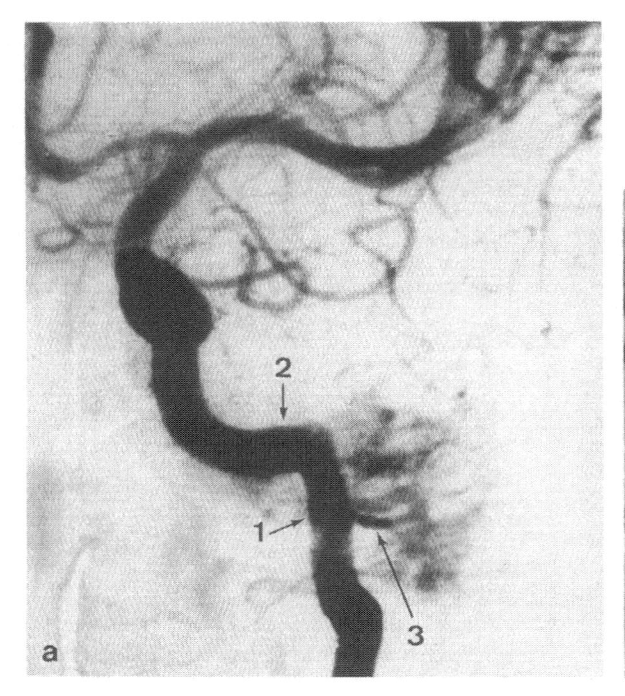

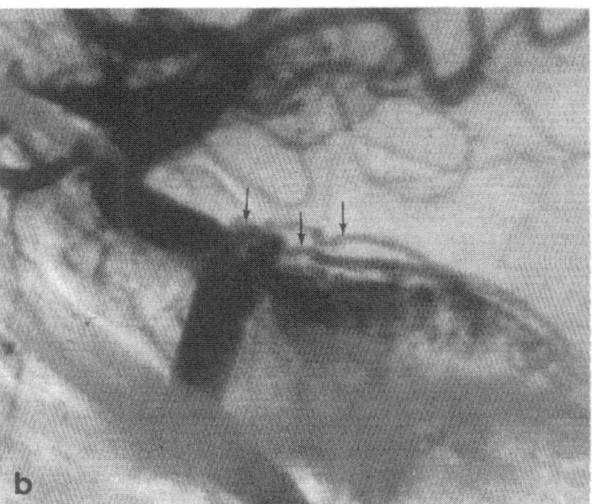

Fig. 122 a, b. Anterior compartment. **a** Internal carotid artery (frontal view). Intrapetrous/vertical portion *(1)*, horizontal portion *(2)*, carotico-tympanic branches *(3)*. **b** Internal carotid artery (frontal view). Meningeal branches of siphon *(arrows)*

vertebral artery, the posterior meningeal artery (Fig. 120) and sometimes even, when the tumor is large, the muscular branches of the vertebral artery (Fig. 120) and, if its extension is intradural, branches derived from the PICA and AICA (Fig. 120).

- *The posterolateral compartment.* This is situated in the posterior part of the tympanic cavity. It is supplied by the stylomastoid artery, a branch of the posterior auricular or occipital arteries, both branches of the external carotid (Fig. 27). This artery ascends in the facial canal to supply the facial nerve but is also distributed to the mastoid cells, to the posterior part of the tympanic cavity and to the labyrinth. As this compartment extends, it invades the mastoid cells and in particular the intersinusoidofacial space (Fig. 29). It may then capture the meningeal branch of the occipital artery, which enters the skull by the mastoid canal (Fig. 121), those of the vertebral artery via the posterior meningeal artery (Fig. 121) or even of the middle meningeal artery (Fig. 121).
- *The anterior compartment.* This is situated in the anterior part of the tympanic cavity. It is supplied by the carotico-tympanic branches of the internal carotid (Fig. 27) or the anterior tympanic artery, a branch of the maxillary artery (Fig. 27). It extends into the auditory tube, the peritubal cells and the carotid canal. It may then capture the meningeal branches of the carotid siphon (Fig. 122) or the tubal branch of the lesser meningeal artery (Fig. 123).

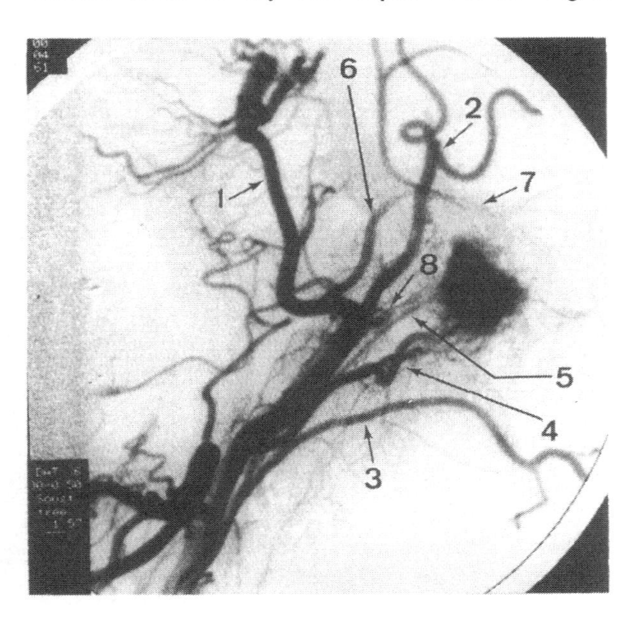

Fig. 123. Superior compartment (in a multicompartmental tumor). External carotid, left (lateral view) internal maxillary artery *(1)*, superficial temporal artery *(2)*, occipital artery *(3)*, posterior auricular artery *(4)*, ascending pharyngeal artery *(5)*, middle meningeal artery *(6)*, and its superior tympanic branch *(7)*, anterior tympanic artery *(8)*

It is essential to be familiar with each of these four compartments, and the differences between them for each particular case. This will allow the best performance of embolization, if this is done, but above all it allows the surgeon to routinely interrupt each of the afferents, which is still the best way to proper excision of these tumors.

Secretory Features

Histochemical studies and ultrastructural examination have shown that all the paraganglia, even those assigned as non-chromaffin and situated in the cervicothoracic region, as well as all the paragangliomas, form catecholamines. However, very few of the latter excrete enough of these agents to produce clinical effects. A review of the literature by Schwaber et al. [160] found only 21 cases of functional paraganglioma, to which they added three personal cases. To this series we can add only one further well-documented case, reported by Cantrell et al. [22]. Among these 25 cases, only 13 were tumors of the glomus jugulare: 12 of these secreted noradrenaline and only 1, case 1 of Schwaber et al. [160], secreted dopamine.

The diagnosis is often made only at the time of operation, when manipulation of the tumor provokes a hypertensive episode. But in fact a properly taken history will elicit suggestive evidence in the majority of cases: arterial hypertension, sometimes paroxysmal as in the case of Cantrell et al. [22], headaches, sweating attacks of pallor or flushing, nausea, tremor, excessive anxiety, recent loss of weight, palpitations or hyperglycemia. If any of these symptoms are present, or even routinely once the diagnosis of a glomus tumor has been suggested, the urine must be tested for levels of metabolites (metanephrine, vanyl-mandelic acid) and blood levels of noradrenaline should also be estimated. The secretory nature of the tumor can be confirmed by dynamic tests with rogitine - positive if a fall in systolic arterial pressure of at least 35 mm Hg or in diastolic pressure of 25 mm Hg lasts for over 4 minutes - or with histamine - positive when a rise in arterial pressure occurs during the two minutes following injection. It is then useful to make estimations of selective samples obtained by venous catherization of the jugular bulb, the subclavian vein, the superior and inferior vena cava and the renal and iliac veins so as to define the probable site of secretion.

In such cases the preparation of the patient for operation calls for certain precautions. Schwaber et al. [160] advise preventive treatment with alpha-blockers and, if tachycardia persists, the addition of a beta-blocker. In any case, the anesthetist should realize that he must avoid the use of atropine or halothane which potentiate histamine, of curare which liberates histamine, and of metaraminol (Aramine) which liberates catecholamines. In every case, even if there is no clinical evidence that the tumor is of a functional nature, the anesthetist must pay constant attention to the arterial pressure so as to correct any sudden rise.

Indications

Surgical excision is certainly the method of choice, radical as it may be, but it must be complete. Clearly, this is usually difficult and not devoid of risk; hence radiotherapy has an important role, In every case, the size of the tumor must first be precisely defined.

Assessment of Tumoral Extension

Clinical Picture

This sometimes gives an approach to the problem, since it may be imagined that each type of tumor corresponds to a suggestive clinical picture. A tumor strictly localized to the tympanic cavity (stage A, tympanic) might be expected to give rise only to otologic features: hypoacousia, throbbing buzzing, sometimes otorrhagia or otorrhea. When the tumor extends towards the mastoid and intersinuso-facial space (type B, tympanomastoid) facial paralysis may be added. When it occupies the posterior lacerate canal (type C, tympanojugular) it is natural for paralyses of the mixed nerves to be combined with the otologic signs. Finally, if the tumor has invaded the posterior fossa (type D, intracranial), signs of neurologic involvement and intracranial hypertension will appear. But in practice, while the presence of a sign has a formal value, its absence is of only relative significance. And it is well-known, especially for this type of very slowly developing tumor, that there is often a disparity between the pathologic and clinical features, so that a purely otologic picture, for example, may mask a tumor of type C or even D. Hence the importance of the paraclinical studies.

Otoradiologic Investigations

Radiography of the petrous bones, especially by tomography in multiple views and now by means of the scanner in bony window, afford a precise study of the pyramids. We must study in order:

- The posterior lacerate canal, enlargement of which is a symptom rather than a sign of extension. It must be remembered that the two orifices are usually asymmetric, the right usually being the largest (6 cases in 10). It is necessary to assess the exact importance of this enlargement, and in this context the index of Dichiro (sum of the length and the two widths, of the pars vasculosa and pars nervosa) is useful, since whenever the difference between the two sides exceeds 20 mm its value is unequivocal. Finally, the morphology of the outlines is also to be considered, aince a glomus tumor erodes the bone and makes the margins irregular and fuzzy (Fig. 124). The scanner, which is able to show simultaneously the bony changes and the tumor in the posterior lacerate canal, is essential to diagnosis (Fig. 124).
- The middle ear and mastoid must be carefully studied for evidence of filling of the tympanic cavity (Fig. 125), destruction of the outlines of the cavity or of the mastoid cells (Fig. 125), sometimes an enlargement of the third part of the facial canal (Fig. 125), anomalies which each have a certain localizing value.
- The inner ear, starting with the entire labyrinthine massif, must be assessed. Infralabyrinthine erosion, often labeled sublabyrinthine excavation (Fig. 126) is a pathognomic sign, especially when the borders are irregular and fuzzy. Destruction of the contours of the cochlea, vestibule, IAM or petrous apex are likewise signs of tumoral extension (Fig. 126). Indeed, while it is important to be aware of bony destruction as proof of intrapetrous extension when choosing and initiating treatment, even more important is the existence of peripetrous extension in the posterior fossa, in the supratentorial compartment and of course in the infratemporal region.

Neuroradiologic Investigations

These are essential. The scanner, using the parenchymatous window, perfectly shows the dimensions and boundaries of these possible extensions (Fig. 127).

MRI should certainly provide at least comparable results, but our own experience remains as yet inadequate (Fig. 127).

In our view, contrary to that of Merland et al. [120], angiography has retained its full place in the diagnosis of this type of tumor. More, since it is the basis of embolization techniques, we regard it as the fundamental investigation without which it would seem impossible, not to say reprehensible, to consider operation. Unlike the acoustic neurinomas, for which CT and in particular the scanner are es-

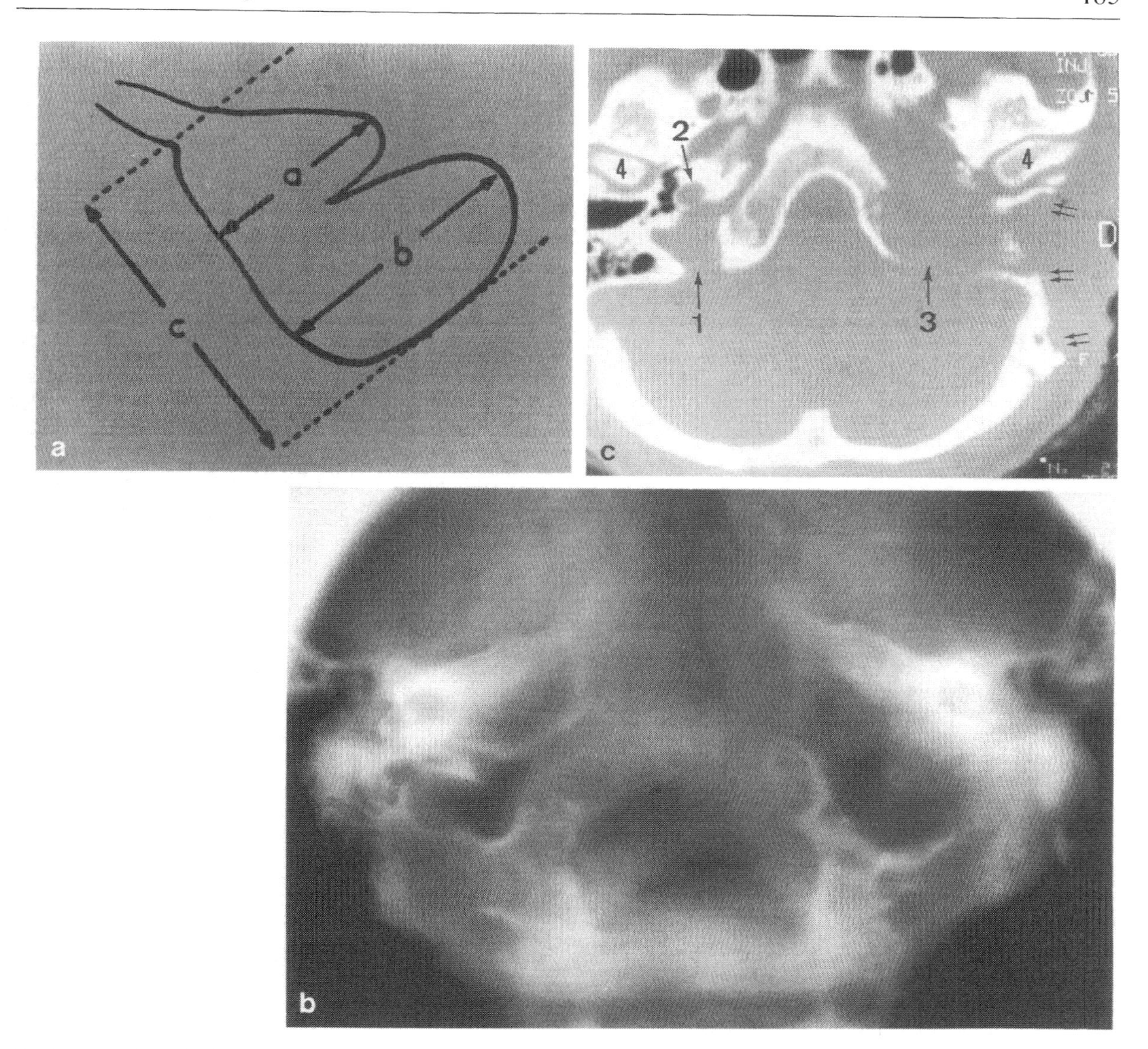

Fig. 124a–c. Posterior lacerate canal (tumor of glomus jugulare). **a** Di Chiro's index: the difference of the sums a (width of pars nervosa) + b (width of pars vasculosa) + c (length of internal orifice) is normally less than 20 mm in the tomograms. **b** Tomogram of posterior lacerate canal. Note on the left the blurred outlines of the internal orifice compared with the normal right side. **c** CT scan. Patient has undergone mastoid evacuation *(double arrows)* for "chronic otorrhea" but actually has a glomus tumor. Normal left posterior lacerate canal *(1)*, left carotid canal *(2)*, eroded right posterior lacerate canal *(3)*, condyles of mandible *(4)*

sential, the vascular nature of jugular glomus tumors and the importance of a complete knowledge of every afferent throughout operation, and of course in embolization, gives angiography a preponderant place. A neurinoma can be confidently operated on without angiography or scanner. A glomus tumor could not be operated without an angiogram, whereas this could be done without a scanner if the angiogram were of good quality. The identification of each of the afferent branches is essential when choosing operative tactics. If the surgeon is aware of the main pedicle of each of the four compartments and especially the plexuses that each of these is capable of capturing during its expansion, he knows exactly how to proceed when removing the entire tumor (Figs. 119 to 123).

Therapeutic Options

Surgery

The ideal treatment of tumors of the glomus jugulare is total *en bloc* excision. In view of the invasive,

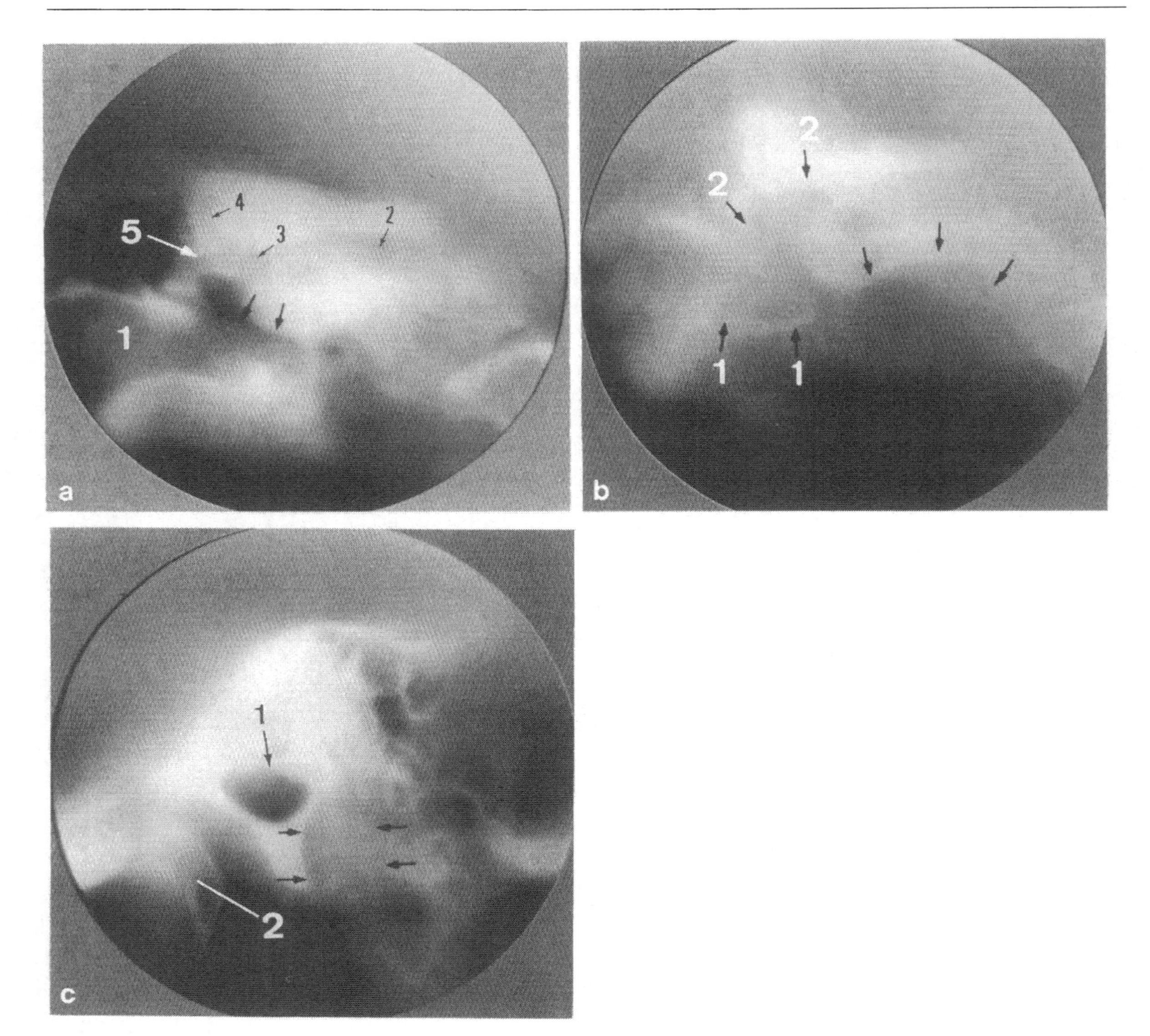

Fig. 125 a–c. Tomograms of petrous bones (tumor of glomus jugulare). **a** frontal view: EAM *(1)*, IAM *(2)*, vestibule *(3)*, anterior SCC *(4)*, lateral SCC *(5)*. Note the more opaque appearance of the lower half of the tympanic cavity *(arrows)* and EAM filled by the tumor. **b** frontal view: erosion of tympanal *(1)*, erosion of labyrinth *(2)*, infralabyrinthine erosion (arrows). **c** Lateral view: EAM *(1)*, condyle of mandible *(2)*, erosion of 3rd part of facial canal *(arrows)*

hemorrhagic and secretory features we have described, it is clear that if operation is to be successfully conducted certain principles must be resected:
- Perfect exposure of the tumor, so that one can get all around it without penetrating it
- Methodical devascularization by systematic interruption of each of the pedicles identified preoperatively, and then by occlusion of the sigmoid sinus, the inferior petrosal sinus and the accessory but sometimes dilated veins opening in or near the bulb especially the posterior condylar vein
- Excision *en bloc* of the tumor and the initial portion of the internal jugular vein, followed by careful inspection of its boundaries, especially at the dura mater, the internal carotid and the facial nerve.

The extent of the exposure obviously on the size of the tumor.

- For stage I (tympanic) cases, an endaural route is adequate provided the limits of the lesion are fully visible after elevation of the tympanic membrane. If this is not the case, a more extensive approach must be considered: a hypotympanotomy of Farrior's variant [44] of the infralabyrinthine route.

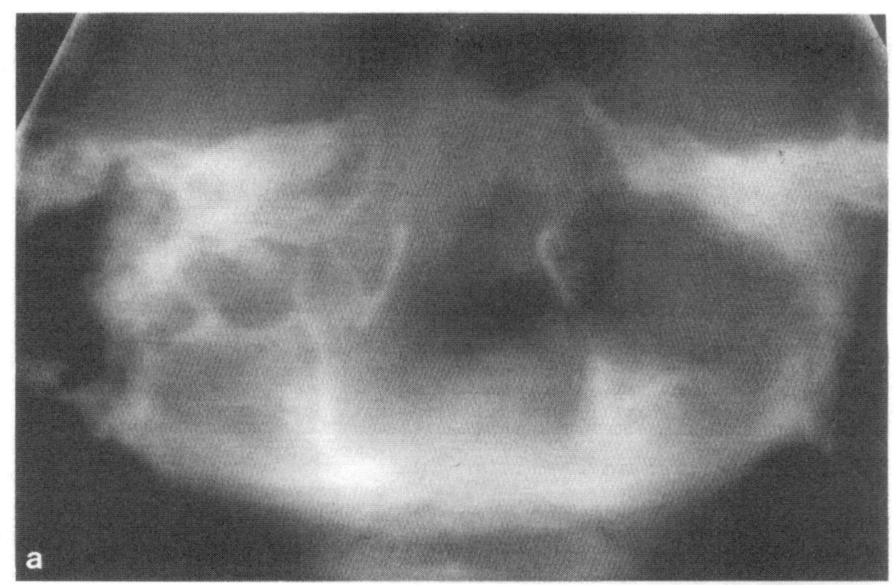

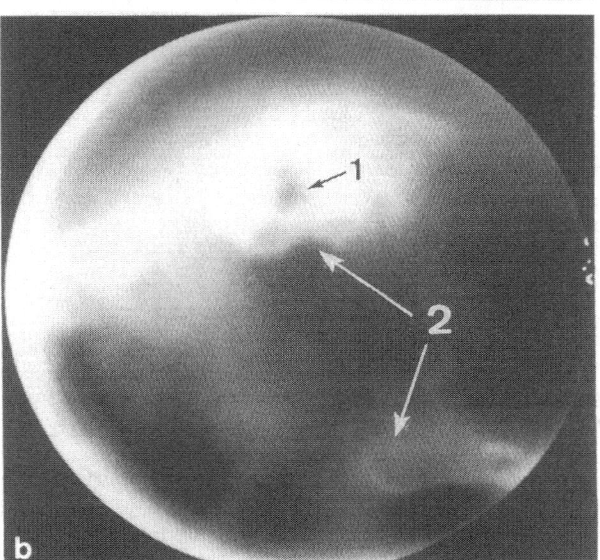

Fig. 126a, b. Tomograms of petrous bones (tumor of glomus jugulare). **a** Frontal view: very marked enlargement of left lacerate canal with irregular margins. **b** lateral view: sublabyrinthine erosion. Fundus of IAM *(1)*, expanded posterior lacerate canal *(2)*

- For stage II (tympanomastoid) cases, a transmastoid route must be used, comprising a wide posterior tympanotomy. This is the sublabyrinthine intersinusoido-facial route used by House [92] and advised by Charachon [24]. This approach, which is situated in the posterolateral retrolabyrinthine sector, is basically equivalent to performing the first stages of a retrolabyrinthine approach but without proceeding to expose the entire dura mater. In cases where the tumor is not too large, it gives excellent exposure. In contrary cases, one must decide on the infratemporal route of Fisch [47] or the variant proposed by Farrior [44].
- For stage II (tympanojugular) cases, the only route to be considered must be by an extended approach. When the bony destruction remains limited to the circumference of the posterior lacerate canal and the vertical portion of the carotid canal (Fisch's stage C1), the infratemporal route of Fisch [47] is perfectly satisfactory, possibly with Farrior's variant [44] if hearing is still useful. When the destruction extends to the labyrinthine massif (Fisch's stage C2), his infratemporal route is still indicated but is on the verge of becoming limited, and in our view an extended transcochlear route seems more suitable. When the hori-

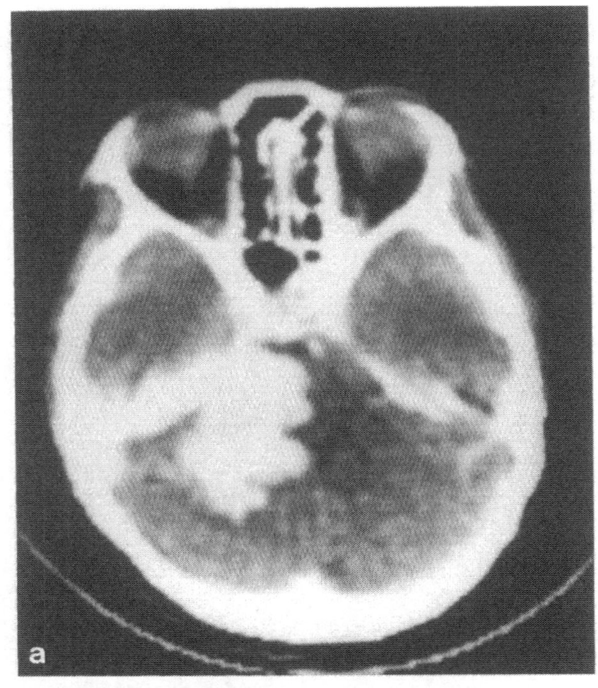

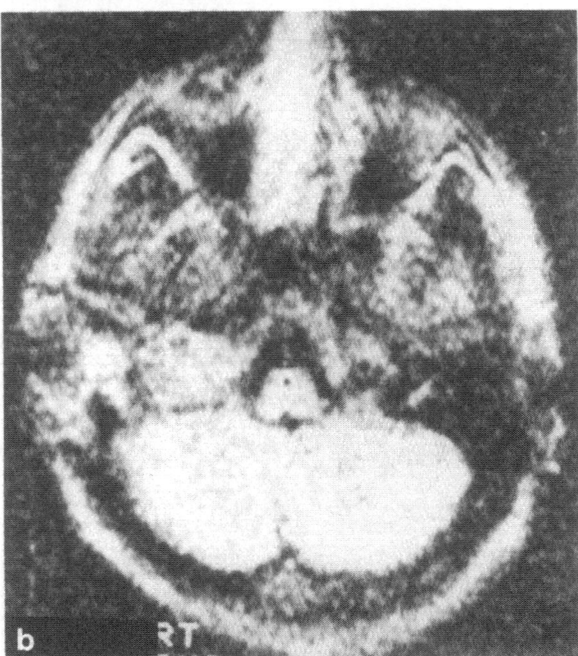

Fig. 127a, b. Glomus tumor. **a** CT scan: intracranial extension, type D3. **b** MRI: intracranial extension, type D2

zontal portion of the carotid canal is invaded (Fisch's stage C3), the extended transcochlear route seems the most reliable for complete excision, the more so since in these cases the labyrinthine destruction may well have opened the labyrinth and thereby settled the problem of hearing.

- For stage IV cases (with intracranial extension), one can certainly consider a two-stage procedure as advised by Fisch [47], i. e., an infralabyrinthine followed by a suboccipital approach, but it seems more logical to begin with an extended transcochlear approach, which has the advantage of even wide exposure of the tumor, so facilitating its excision and allowing one-stage treatment.

This leaves the problem of the possible necessity for ligaturing the internal carotid artery. We ourselves have never done this, but Fisch [48] states that the problem is easily resolved by the routine practice of a preoperative test of occlusion by means of an inflatable balloon. However, he does not specify what course he would take if occlusion were poorly tolerated and he was confronted with the necessity of sacrificing the carotid. Two attitudes seem possible: either to regard this situation as an absolute contraindication or to decide to abandon a fragment in contact with the artery.

Embolization

At first glance, this is an accessory treatment aimed at facilitating the operative stage by reducing the risk of hemorrhage, or at supplementing the effect of radiotherapy by adding to the thrombotic and necrotizing affect. However, as suggested by Merland et al. [120], it may be that improvements in embolizable material will one day will give it a curative role. This is certainly a possibility not to be ignored, since the technique of superselective embolization is so perfected that the only thing lacking is suitable materials. It is not utopian to imagine that it may be possible to develop agents capable of both interrupting the circulation and of impairing metabolism and cell mitosis once in place. For the moment, however, it is still a supplementary but not a secondary technique. Its chief employment is preoperative. Without denying its efficacy, we must point out nevertheless that full preoperative knowledge of the different pedicles allows the performance of a controlled operation in which the successive interruption of these afferents, facilitated by perfect exposure, leads to the same result. Moreover, it cannot be disputed that embolization of all the pedicles derived from the external carotid artery seriously impairs the vascularization of the superficial layers and thus imperils the healing processes, especially as the incisions of these superficial layers for an extended approach are often extensive and complex. Thus, the necrosis of some of our flaps has led us to reconsider the line of our incisions when embolization had been done previously (see the extended transcochlear route above). Some authors, like Merland et al. [120] have used embolization as a supplement to radiotherapy or even on its own, but it is still too soon to judge its efficacy. In cases of preoperative embolization, the release of fragments of muscle, spongel or aponeurosis seems the simplest technique, especially as the problem of absorption does not arise. On the other hand, when it precedes radiotherapy, and certainly when it is the only treatment considered, the nature of the embolized material becomes crucial since it is bound to persist. In current use are isobutyl cyanoacrylate, solid emboli of various shapes and sizes or liquid emboli, still in the experimental stage, but it is to he hoped that progress will reinforce the ischemic effects and tumoral necrosis. One major problem remains to be solved, that of the tumoral branches derived from the internal carotid and vertebral arteries in which embolization cannot be considered. This is the more unfortunate since it is just when the tumor is particularly large that these branches are precisely involved and the most could be expected from this technique. In our view, this is a serious handicap of the method at the present time.

Radiotherapy
This is recommended by Williams [193] and has a favorable effect on the functional symptoms but not, it seems, on the size of the tumor. Experience has shown that the effect is only temporary and that resumption of the progress of the lesion leads to a recurrence of symptoms, admittedly after a long interval. As we have already seen [166], it acts by producing intratumoral sclerosis and obliterative endovasculitis. It is not to be regarded as a curative treatment, but only as a palliative. To be sure of the result, the necessary dosage must not be less than 45 grays and, according to Bataini et al. [6], should even reach 50 to 60 grays without any problems of post-radiation necrosis though this risk cannot honestly be demied. Of course, such high dosage requires certain intervals and exact regulation as regards its action on the nervous system. Disappearance of the pulsating bruit, improvement in the disturbances of phonation and swallowing and the drying-up of aural discharge are so many indices of its effects; but the recurrence of symptoms, often after an interval of 10 to 12 years, is a clear proof that while this tumor is radiosensitive it is not curable by irradiation.

The Choice of Treatment

This is often difficult and cannot be stereotyped for every case. It lies between surgery and radiotherapy. At least three criteria are to be stressed.

Criteria of Choice
These are many and complex.

Age and general condition are to be considered primarily. It is obvious that the usual magnitude of the operative procedure and the slow development of the lesion will favor surgical abstention if the patient is rather elderly, especially since radiotherapy affords a hope of quite prolonged stabilization. This seems a sensible attitude, but it remains to fix the decisive age and the limit is not very precise. It must not be fixed too early, or else, at the time of recurrence of symptoma, often a dozen years later, the patient's age will now be such as to be a definitive contraindication and a cause of regret for not having intervened earlier. This is particularly the case since 80% of the patients are women, whose expectation of life is now around 75 years. It seems reasonable to accept that 65 years approaches a limit of age, but too struct an attitude should not be adopted. We oursleves have operated successfully on a woman aged 71 because her general state was excellent and the tumor not too large (stage B). Conversely, surgical abstention has seemed wiser for some younger patients with manifest systemic disorders (arterial hypertension, atheroma, coronary disease, diabetes, etc.). And it must be added that to fix the upper limit at around 65 years of age does not mean that every patient of a lesser age must be operated. Other factors have to be considered, such as the size of the tumor and its neurologic effects.

Tumoral size. For long, this was the determining criterion. Only the smaller tumors of stages A and B could reasonably be operated on, and larger tumors were to be managed by radiotherapy because of the great risk of hemorrhage and the much smaller chances of complete excision. Nowadays, the improvement in operative techniques and especially in access routes authorizes excision of the largest tumors, but of course the risks involved must be carefully weighed on each occasion. Every surgeon has become emboldened; we note, for instance, that Fisch, who in 1982 [47] categorized as D all tumors with an intracranial extension, seems to have revised the limits of this stage since in 1985 [48] he considered that intracranial but extradural extensions were still of type C and that only intradural extensions qualified as type D. We have seen that for each type of tumor there is now a very suitable route of access.

Neurologic involvement. This is the most debatable criterion. Several possibilities are to be distinguished. In certain cases there exist signs indicative of axial compression of the brain-stem or cerebellum. Since we know that radiotherapy hardly modifies tumoral size, it seems legitimate in these cases to adopt a more aggressive attitude, since the persistence of the compressive factor carries a serious risk of the rapid development of severe incapacity.

Other cases are dominated by symptoms indicative of hydrocephalus. The resolution of this problem leaves one in a much more comfortable situation without too great exertion, while avoiding any indication for a salvage procedure.

Finally, in other cases, and these are the commonest, the only problem is paralysis of the cranial nerves. The circumstances may be quite different, depending on whether these paralyses exist preoperatively (60% of our cases) or else are absent (40% in our series) but risk being provoked by the operation. In the former case, when they develop gradually and are already experienced by the patient, they will perhaps be more readily accepted. On the other hand, in the opposite situation these sequelae are much more distressing to tolerate. The actual handicap varies greatly with the nerve concerned. A dysphonia or a dysphagia may be quite easily improved by an injection of teflon into a vocal cord. There is the problem of preservation of hearing, which conditions the route of access: infratemporal if the cochlea is still functional, extended transcochlea if it has been destroyed. It may also condition the quality of the excision since the exposure is not the same in the two cases. Facial paralysis is not very common before operation (10% of our cases). On the other hand, the facial nerve bars operative access, is often related directly to the tumor and runs great risks during operation. It is quite common for it to need to be repaired postoperatively. This raises the problem of the hypoglossal nerve, which is often threatened by the tumor and also by the operation. Its paralysis is not of great importance in itself, but it raises the problem that hypoglossal-facial anastomosis is now impossible. Repair of the facial nerve can then only be effected by direct facio-facial anastomosis or by interposition of a graft from a peripheral nerve. It is fortunate that the facial nerve, well protected in the cerebellopontine cistern, is usually easily safeguarded in the angle during this operation, so facilitating repair techniques. The probability of there being problems of paralysis of the cranial nerves after operation is quite high, and the patient should be warned of this when the choice of treatment is under discussion.

Modalities of Choice

These vary depending on whether the tumor has just been discovered or, on the contrary, it is a matter of a recurrence or at least a return of clinical features which had been improved or stabilized by previous treatment.

- *The first indication.* This is certainly the most difficult and we have no "definitive" plan of treatment to propose. At best, we can discuss our own views.

• Operation seems to be indicated:

- Whenever the tumoral size is small (stages A and B), provided the patient is under 65 years of age
- whenever frank neurologic signs exist, particularly signs of compression of the brain-stem or cerebellum rather than paralyses of the cranial nerves, even and especially if the tumor is large. For a type C tumor, the infratemporal route is used if the cochlea is intact, the extended transcochlear route if it is damaged or even simply threatened. For a type D tumor, the extended transcochlear route. The following case is illustrative:

- Mr. SIM. aged 35. Early 1982, pulsating bruit in right ear. Investigation suggested a strictly tympanic glomus tumor but the retrotympanic approach used proved inadequate because of profuse hemorrhage and the manifest large size of the lesion. Postoperatively there developed gradually right deafness, originally of transmission and then mixed, and a posterior foramen lacerum syndrome. April 1984, development of right otorrhea, sometimes mixed with blood. When he consulted us, right otorrhea, virtual right cophosis, intermittent dysphonia and some difficulty in swallowing. On examination, a tumor filling the right EAM Radiologically: large stage D2 tumor. Extended transcochlear approach. Quasi-total excision; only the carotid adventitia was possibly not perfectly cleared at the end of operation. Postoperative facial paralysis which regressed in 6 months. Remains well after 3 years.

• Radiotherapy is more indicated:

- In patients over 65
- At whatever age, whenever the tumor, though large, produces no neurologic signs or when these are minor.

We illustrate this by two cases:

- Mrs. CEN, aged 70. For 10 years, progressive amyotrophy of the left trapezius, then dysphonia and dysphagia, symptoms established in 2 years. Condition stationary for 8 years without treatment and then, in April 1982, pulsating bruit in left ear and vertigo in September 1982. Moderate left kinetic cerebellar syndrome. Left mastoid systolic murmur. Radiologic findings: glomus tumor, D3. Radiotherapy (47 grays).

Disappearance of pulsatile bruit, improvement in swallowing and disappearance of vertigo and cerebellar syndrome. Condition stationary for 4 years.

- Miss NEG, aged 29. Pulsatile acouphenias and hypoacousis for 3 months. Otoscopy: tumor of external meatus, right. Radiography: glomus tumor, type D2. Surgery withheld. Radiotherapy: 60 grays. Disappearance of acouphenias. Condition stationary at 4 years.

- *Recurrence.* The decision depends greatly on the initial treatment.

• After radiotherapy, or surgery plus radiotherapy, it is necessary to operate or reoperate as far as possible. It is clear that a very elderly subject will run great risks during an operation whose only aim must be total excision. If the initial radiography had been administered in effective dosage, it would not be reasonable to embark on further radiation as the dose given would be either small and therefore ineffective, or else considerable again with an excessive risk of radionecrosis. In this context, there are only two alternatives: either to remove the tumor or to abandon the patient to his sad fate without proof to the contrary, embolization does not seem to be an efficient treatment once incapacitating neurologic signs exist. The following case well illustrates the situation in cases of recurrence:

- Mrs. BAR, aged 53. 1968, posterior lacerate foramen syndrome, diplopia, left deafness and pulsatile acouphenias on same side. Radiographic and arteriographic assessment suggests a tumor of the left glomus jugulare. Surgery withheld; radiotherapy 77 grays). Subsequent disappearance of diplopia and improvement in dysphonia. The patient retained the other symptoms unchanged until 1978. She then consulted us for return of diplopia. Neurologic examination showed major paralysis of all the cranial nerves from the trigeminal to the hypoglossal; only facial involvement remained minimal. The scanner and arteriogram revealed a large tumor with an intracranial extension of over 2 cm (D2). Excision by extended transcochlear route. Subsequent course without great problems apart from facial paralysis, for which hypoglossal-facial anastomosis was performed with a favorable result; the disorders of phonation and swallowing were improved after injection of teflon into the vocal cord. Condition still satisfactory 9 years later.

• After surgery without radiotherapy. It seems reasonable to begin by irradiating the tumor. If the irradiation is effective, there is nothing but to wait and monitor progress. If the symptoms persist or deteriorate, the only resort is surgery. Certainly, the involvement of many factors at the moment of choice makes the decision very difficult. What we would stress is that radiotherapy can only have a palliative effect, that only surgery can really cure the patient, but that the risks encountered, formerly uncontrollable but now more acceptable, still remain very great, at least at the functional level, when the patient has only very few postoperative symptoms.

Particular Choices

- *Surgery plus radiotherapy.* In the period when complete excision became impossible as soon as the tumor reached a certain size, this combination seemed very useful and was advocated routinely by some. Except in very special cases, this attitude can no longer be maintained. If surgery is undertaken, it must endeavor to be radical or at least to leave only very minimal fragments. In view of the slowness of growth of the tumor and the monitoring facilities provided by computed tomography, it seems legitimate to await recurrence before irradiating. To envisage the routine adoption of the combined treatment would amount to using up all one's ammunition at the outset. Only the abandonment of a major fragment seems to justify this combination, but we would regard this as an admissiion of failure. In our view, it is much more constructive to know how to assess the limitations of surgery and to dispense with it if the risks of incomplete excision are considerable. Irradiation in these cases would incur less danger to the patient and still leaves the possibility of a subsequent operation. It even seems to us that, for very large tumors, operation proved rather easier when the patient had been irradiated previously. This is why we think it preferable to irradiate, in a first stage, large tumors which have not yet produced major neurologic signs.

- *The place of embolization.* If this actually does render the tumor bloodless, there is every reason for carrying it out as it will certainly simplify the operation. This is possible when the tumor is supplied only by the external carotid. These are usually strictly tympanic tumors (type A) or else tympanomastoid tumors (type B) or tympanojugular tumors (type C) but of small size. If, however, the embolization is bound to leave persistent pedicles deriving from either the internal carotid or the vertebral arteries, its value seems illusory and essentially nil. This may seem paradoxical. However, it is not the ligature or coagulation of the branches of the middle meningeal artery which raise problems, it is that of the branches of arteries to the brain. Now, either these branches are fine and their coagulation raises no major problems, or else they are large, providing a major flow into the tumor, and the situation will not be changed by interruption of branches of the external carotid. As this technique is not devoid of disadvantages, particularly by endangering the blood-supply of the superficial layers, and is even capable of producing certain serious complications, we think it often unprofitable and even dangerous.

- *The place of CSF shunt.* Hydrocephalus is the cause of incapacitating clinical pictures, which may regress spectacularly after a shunt procedure. The ease of assessment of such hydrocephalus by means of CT allows the detection of patients with this lesion and their effective treatment. The clinical improvement then makes it possible to decide on the management of the case more calmly without being driven to a salvage procedure as the first examination might have suggested. This is illustrated by the following case:

- Mrs. VEI, aged 62 years. For 2 years, intellectual slowing, memory disturbance, headaches, then astasia-abasia. For 3 months, difficulty in swallowing and loss of visual acuity. On examination, bilateral Babinski. Neuroradiologic assessment showed a glomus tumor, stage D, on the left and triventricular dilatation. Ventriculo-peritoneal shunt spectacularly improved the astasia-abasia and the memory disorder. Activity virtually normal. Moderate left deafness, gait unsteady with deviation to left. Abolition of left corneal reflex. Examination otherwise normal. Occasional difficulty in swallowing.

Results

When we first discovered the possibilities offered by the extended transpetrous routes, we thought it feasible to approach every glomus tumor, of whatever type. For a time, our attitude was strictly surgical but the awkwardness of the operations and the functional sequelae they produce, particularly as regards the facial and mixed nerves, eventually moderated our enthusiasm. We then adopted our present policy, which is more conservative with larger tumors and older subjects. The results we report relate to our entire series, and their interpretation must therefore be tempered since they do not signify a fixed attitude but, on the contrary, a progressive evolution of our operative indications.

In 1984 [179] we reviewed a series which then amounted to 25 cases, of which 17 had been operated (68% of cases):

- 3 purely tympanic forms (stage A) had been excised with no problems or sequelae.
- 5 tympanomastoid forms (stage B) had been totally removed. One had been irradiated 10 years before and excision in this case was followed by a meningitis which caused the death of the patient. The other four had an uneventful course. One, who had a preoperative facial paralysis, needed a hypoglossal-facial anastomosis with a good functional result. The other three had no permanent facial sequelae, though one had a postoperative palsy which regressed in 2 months.
- 3 tympano-jugular forms (stage C) were extirpated, twice totally and once incompletely as residues seemed to have persisted in the pericarotid adventitia. This patient had supplementary irradiation (48 grays) and remains very well 8 years later. 2 patients had exhibited a postoperative facial paresis, rapidly and completely regressive in one case, persisting as a minor lower facial paresis in the other.
- 6 extensive forms (stage D) included two patients, aged 58 and 68 years, who died postoperatively, one from a Mendelson's syndrome and the other probably from brainstem softening. Two of the 4 survivors had a facial paralysis, one of which was treated by hypoglossal-facial paralysis. Excision was incomplete in two cases and was managed by postoperative irradiation in both; the follow-up was satisfactory in both, one at 9 years, the other at 13 years.

Since that period we have observed 16 further cases, only 6 of which (37%) were operated, bringing the number of our operated cases to 23 (56%); but it should be noted that, whereas before 1984 we operated on two patients out of three, since then we have only operated on one in three. These were a purely tympanic tumor whose excision was altogether simple and 5 extensive forms of which one, which was really aneurysmal (Fig. 120c) and could not be completely removed, died 5 weeks later after problems of CSF leakage and meningitis beyond control. The course of the other 4 patients had no great problems. Three showed a postoperative facial paralysis which regressed in each, and two had difficulty in swallowing and dysphonia, much improved after injection of teflon into a vocal cord. L8 patients were not operated but treated by radiotherapy alone. All benefited from a decrease of symptoms, the dysphonia and dysphagia having regressed in two out of three cases though without disappearing completely, so that the patients always needed to be very careful at meals and to drink between mouthfuls being subject to false passages. The pulsatile bruit lessened in intensity in every case and disappeared in two out of three cases, while the facial paralysis, when present, usually regressed. On the other hand, hearing continued to deteriorate and usually ended in unilateral cophosis. Only one patient died, 4 years after irradiation, but of intercurrent disorder (myocardial infarction). The results of the radiotherapy are certainly very interesting, but we are well aware that our treatment was only palliative and that there is every chance that these patients may return one day with a recurrence of symptoms and possible even a more serious neurologic picture. It is probable that the therapeutic possibilities will be limited, or even nonexistent, for the two-thirds of them who were over 60 years of age at the time of irradiation and who will probably be over 70 when they return.

Other Tumors

In our series, these represent only 10% of all the tumors that have come under our observation. In practically half the cases there was a meningioma, usually attached to the posterior aspect of the petrous. The next commonest were cholesteatomas, followed by a range of diverse tumors.

Meningiomas

These represent 6% of our series (19 cases out of 300 tumors). They were essentially meningiomas attached to the posterior aspect of the petrous. As first stated by Castellano and Ruggiero [23], this is the site of half the meningiomas of the posterior fossa.

Problems

Morbid Anatomy

Origin. The arachnoidal villi from which the meningiomas develop have a particular arrangement on the posterior aspect of the petrous, on the one hand in the vicinity of the various orifices of the cranial nerves, especially around the aperture of the IAM, the internal orifice of the posterior lacerate canal and the cave of Meckel, on the other hand along the superior petrosal sinus. According to Yasargil [189], Aoyagi and Kyumo [4] were the first to specify this fact. This entirely explains the site of attachment of the meningiomas developing on the posterior surface of the petrous, which, according to Yasargil [189], are situated in half the cases between the aperture of the IAM and the jugular foramen, in one case in ten between the aperture and Meckel's cave, and in one out of ten cases either above the aperture, between this and the superior petrosal sinus, i.e., on the supra-acoustic eminence, or at the inferior margin of the jugular foramen, between it and the condylar canal. Finally, in about one in twenty cases the meningioma is attached quite laterally, at the junction of the bend of the lateral sinus and the superior petrosal sinus.

It is not possible to ignore the problem of the so-called clival meningiomas, which, according to Yasargil [189], do not arise from the clivus itself but at its border, the petro-clival suture. The same author states that their extension makes it possible to distinguish purely clival, petro-clival and spheno-petro-clival forms. The distinction between the first of these and the other two forms is important. It is probable that the purely clival forms, median and, as it were, bilateral, must pose problems of access much more difficult than the petro-clival forms, whose evident lateralization assimilates them to the meningiomas attached to the apex of the petrous, medial to the IAM.

This zone of attachment to the petrous is usually limited, and even when the meningioma appears at first access, particularly by the scanner, to possess a wide foundation, the actual zone of adhesion to the meninges barely exceeds a square centimeter, sometimes much less. But the problem of attachment is not confined to the dura mater. The underlying bone is the site of changes which are often evidence of invasion. At the petrous it is rare for these to be shown in the radiographs or even the tomographs but, as Yasargil well remarks [189], despite the lack of visibility of these alterations before operation it is usual to observe them under the operating microscope. The scanner in bony window (Fig. 128) may be an excellent means for demonstrating them. Their presence calls for a very important stage of bony resection if a really complete resection of the tumor is to be done. At this zone of attachment the tumor receives its main blood-supply, what we call in our thesis [139] the attachment pedicle. This is important to bear in mind since it locates very well the origin of the tumor. One can picture how, starting from this point, the expanding tumor pushes back the cranial nerves, especially the acoustico-facial bundle, over its convexity. The position of the latter nerve may therefore be predicted before operation, as can the region to be first attacked, if possible, to devascularize the tumor before evacuating it. Thus, knowledge of the attachment pedicle is an important factor when selecting a route of access.

Situation. In general, in 9 cases out of 10 in our experience (17 of our 19 cases), meningioma develops in the posterior fossa. Despite the improvement in diagnostic techniques, it must be acknowledged that, as stated formerly by Russell and Bucy [157], these tumors remain large at the time of diagnosis, probably because of their very insidious mode of onset. In terms of the morbid anatomy, it seems important to distinguish (a) the meningiomas arising

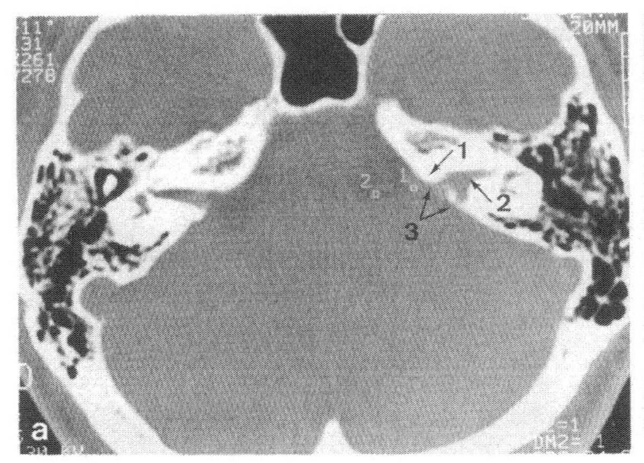

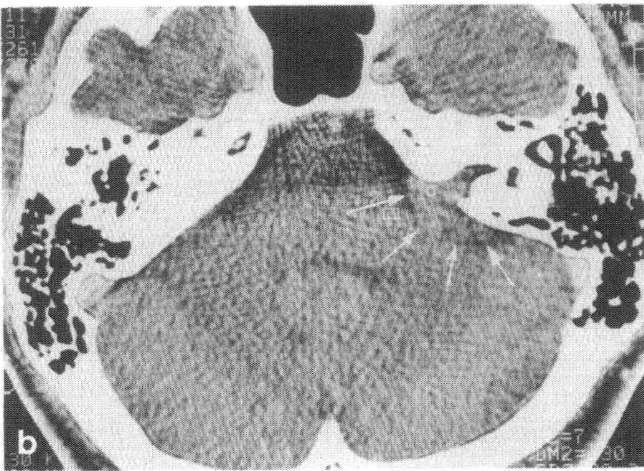

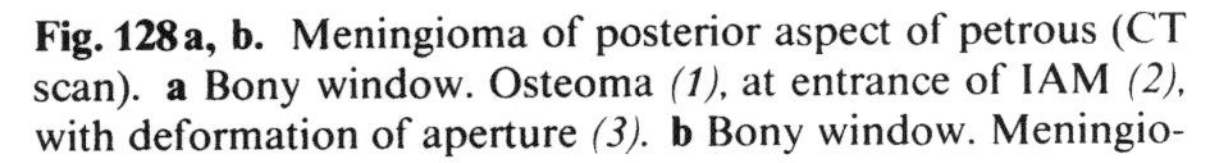

Fig. 128 a, b. Meningioma of posterior aspect of petrous (CT scan). **a** Bony window. Osteoma *(1)*, at entrance of IAM *(2)*, with deformation of aperture *(3)*. **b** Bony window. Meningioma spread out on posterior aspect of petrous around the aperture and penetrating the IAM

at the periphery of the aperture of the IAM and presenting at both the clinical and scanographic level (Fig. 129) like an acoustic neurinoma (11 of our cases) (b) the meningiomas developing medial to the aperture, appearing in the scan (Fig. 129) as internal manifestations spread out over the petrous apex (3 of our cases) and (c) the more external meningiomas, developing lateral to the aperture and occupying a lateral recess (Fig. 129) of the posterior fossa (3 cases). The fact that most of our team are otologists doubtless explains the very marked predominance of forms that may be qualified as "otologic", as against the "neurosurgical" forms, internal or external to the aperture. This distinction is important when the access route is being decided.

Classically, the meningiomas attached to the posterior aspect of the petrous do not invade the IAM, and the integrity of the latter in the presence of a tumor of the cerebellopontine angle is a diagnostic argument in favor of meningioma. However, some authors have reported the possibility of deformation of the meatus in cases of meningioma, suggesting that the tumor had invaded the meatus. Thus Maniglia [119], in a study of 220 meningiomas of the posterior fossa found in the literature, notes such changes in the IAM in 14% of cases. Dechaume [38], in his thesis, refers to 9% of cases. Nager [126] reported two cases of strictly intracanalicular meningioma found at autopsy, and Granick et al. [71] also report a meningioma that was both intra- and extracanalicular, identical with a neurinoma, also discovered at autopsy. Guiot and Bouche [74] seem to have been the first to operate on such an intracanalicular meningioma. We ourselves have operated on a case of an intracanalicular tumor by the suprapetrous route, in which all the investigations, including an opaque meato-cisternography, had led to the diagnosis of a strictly intracanalicular tumor. The diagnosis of acoustic neurinoma originally made was only corrected when the histology became available.

Since the now classic monograph of Nager [126], our knowledge of intrapetrous meningiomas has improved somewhat. In two-thirds of the cases, it only involves invasion of the pyramid by a meningioma attached to the temporal or cerebellar dura. It is easily imagined that the intrapetrous cavities, known often to be separated from the surface only by a slender cortical layer, may easily be invaded by an adjacent meningioma. To read Nager [126], it would seem that these are often meningiomas of the temporal fossa which have traversed the tegmen tympani, often known to be dehiscent, or else the roof of the geniculate ganglion compartment or a peritubal cell. Sometimes the tumor was situated on the posterior aspect and invaded a mastoid cell or a retrolabyrinthine cell. Finally, it is sometimes the roof of the jugular bulb which has been traversed by a meningioma arising at this site. But meningiomas exist whose origin really does lie within the petrous. Nager [126] believes that they arise from clumps of arachnoid cells to be found in the IAM, the posterior lacerate canal, on the geniculate ganglion and on the roof of the auditory tube. Thus Guzowski et al. [75], examining histologically 200 petrous bones removed and sectioned, found a small, strictly intrapetrous meningioma that seemed to derive from the dura mater across a dehiscence on the course of the tympanic nerve where it ascends towards the accessory hiatus, near the canal

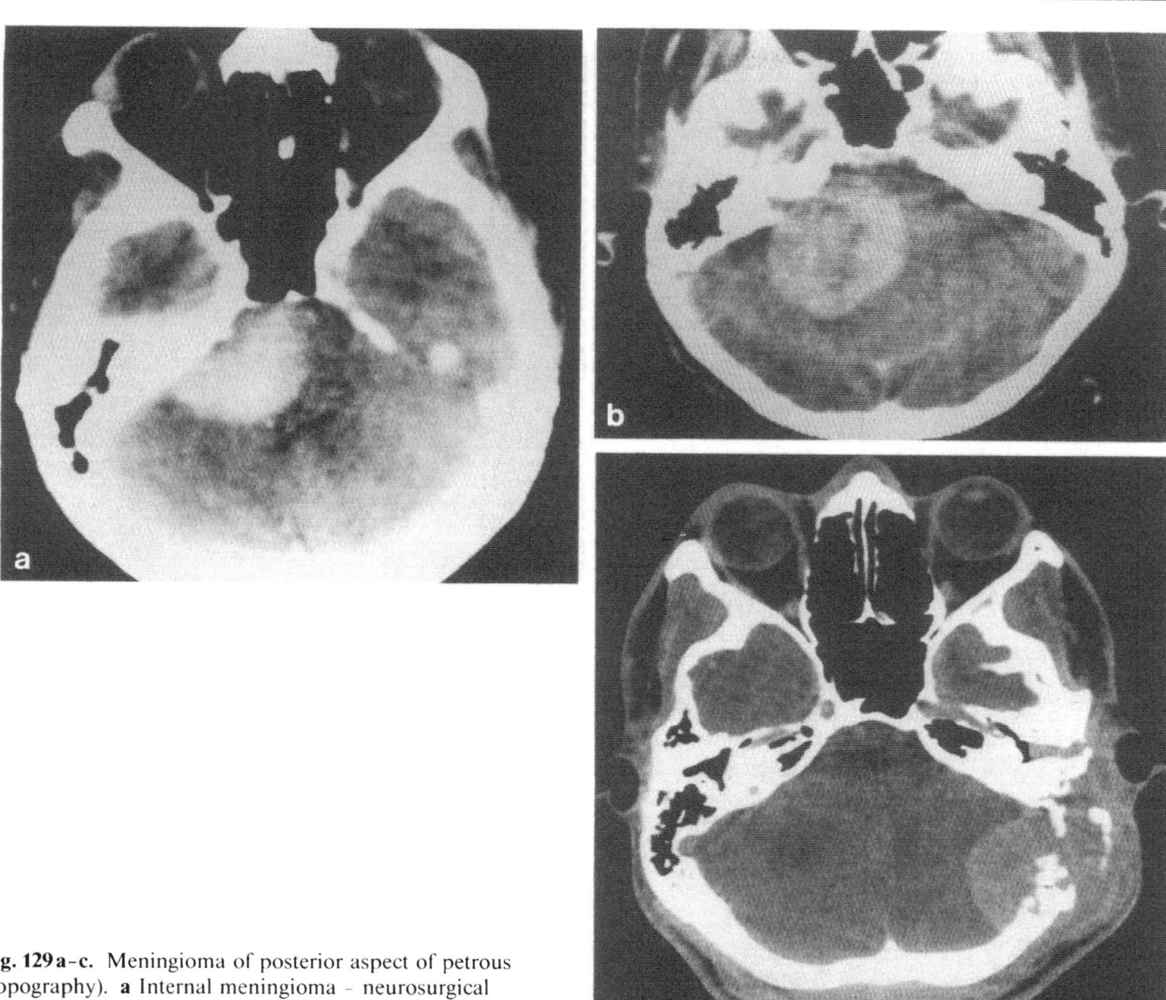

Fig. 129a–c. Meningioma of posterior aspect of petrous (topography). **a** Internal meningioma – neurosurgical form. **b** "Otologic" meningioma centered on IAM. **c** External meningioma

for the tensor tympani muscle. We have ourselves observed the case of a meningioma diffusely invading the petrous pyramid and having developed an infratemporal extension large enough to appear as a latero-cervical tumor, betraying the disease (Fig. 130).

Cisternal Relations. A meningioma attached to the posterior face of the petrous is by definition a subdural tumor. In view of the arrangement of the cisterns of the region, it can be seen that the nerves and vessels contained in these cisterns are always separated from the tumor by an arachnoid layer. Cautious extra-arachnoid dissection is thus the nest way of easily exposing the tumor while respecting these various structures. Yasargil [189], before us, had already stressed this important point. The best way to respect this arachnoid layer is again to approach the tumor by its base of attachment to the dura and not by its convexity. In this way, the arachnoid layer will not be torn at the outset. The translabyrinthine approach favors these tactics. The suboccipital approach, on the other hand, makes it necessary first to incise this layer before penetrating the tumor to evacuate it. This disadvantage is further aggravated by the fact that the acoustic and facial nerves are often spread out over the outer pole of the meningioma, which makes them particularly vulnerable when the tumor is attacked.

Diagnosis

– At the clinical level, it is very difficult to distinguish between an acoustic neurinoma and a meningioma. Classically, the latter would be more likely to affect the trigeminal and facial nerves, with relative respect for hearing. In fact, besides the total co-

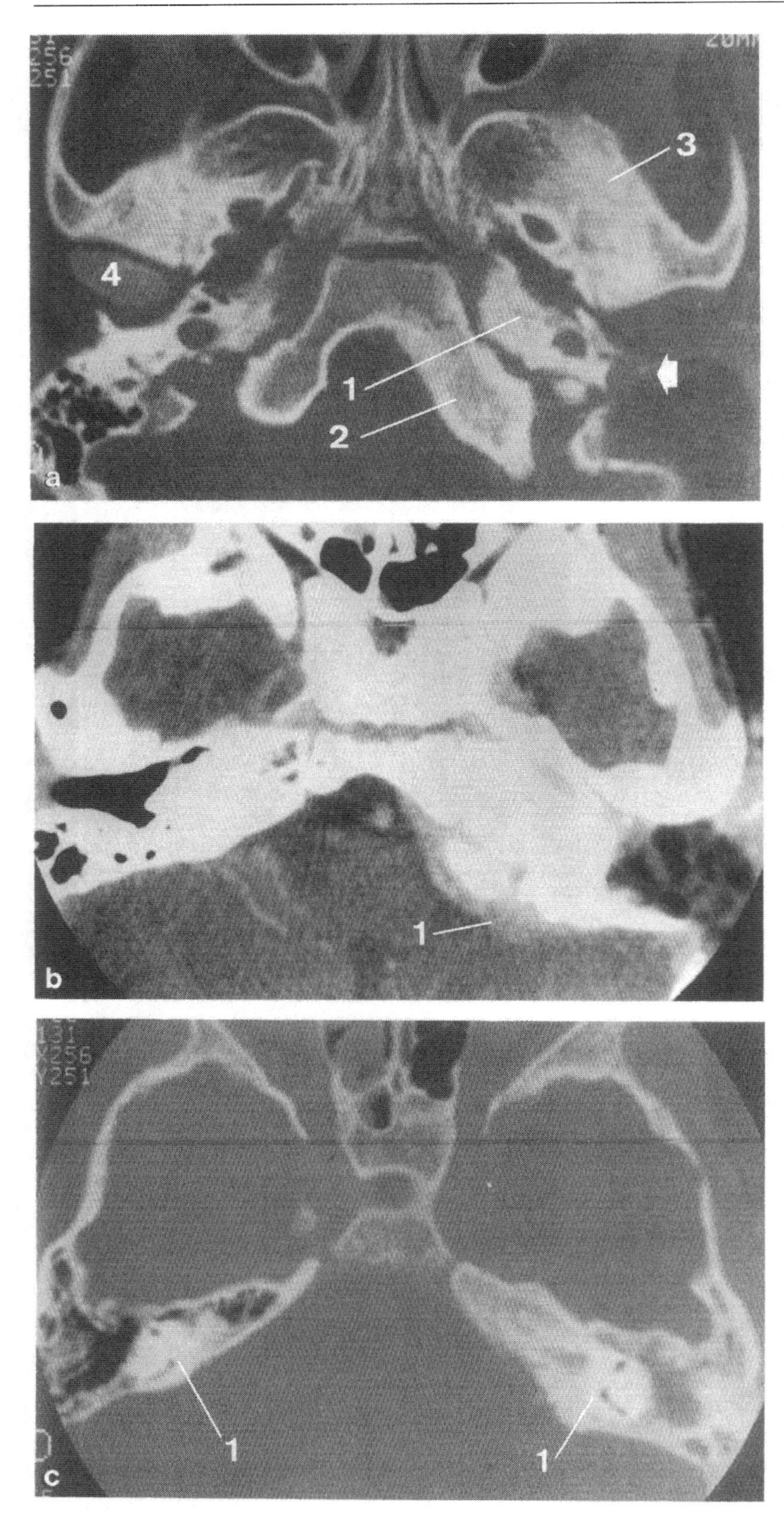

Fig. 130 a–c

phosis which was present in one case in trade in our series, a retrocochlear type of hypoacousia was noted in the other cases producing an average loss of 40 dbs, a considerably smaller figure than the average 74 dbs loss found in stage IV acoustic neurinomas. Now, 70% of our meningiomas were stage IV tumors. The extent of trigeminal involvement is usually modest and amounts to abolition of the corneal reflex (56% of Yasargil's cases, 62% of ours), which is quite trivial considering the size of the tumor. In our experience, facial nerve involvement was present in only 25% of cases, whereas it was present in 14% of our acoustic neurinomas.

- The clinical features mentioned were based on more longstanding and perhaps more advanced cases. The newer paraclinical investigations should lead to earlier diagnosis but are not determinant. Tomography of the IAM proved normal in 50% of cases, which is a useful feature. In the other 50%, however, this was a deceptive feature which was manifest in half the cases. Since 16% of our stage IV neurinomas had symmetric IAMs, one can understand the small value sometimes attached to this normality to the tomograms. Apart from the rare cases where there is a hyperostosis, it is the scanner that should attract attention because it classically shows a spontaneously hyperdense tumor (in 80% of our cases), taking up contrast strongly and attached more widely to the posterior face of the petrous. In fact, the presence of changes in the IAM (Fig. 130) has been responsible for errors in diagnosis. Of course, in doubtful cases it will be necessary to perform scintigraphy and angiography, which will settle the matter.

- The diagnosis may thus be erroneous. We have experienced this 6 times, always because we had not pursued enough investigations, but also because the tumor belonged to the forms we label "otologic" which are grossly evident in the scan and entirely resemble an acoustic neurinoma unless their nature becomes quite evident spontaneously. Moreover, this error may be perpetuated at operation, and in two cases the diagnosis was corrected only by the histologic findings. It has to be said that we have also experienced the converse situation, and that tumors regarded as meningiomas because of their texture, hemorrhagic nature and size of attachment to one of the margins of the vestibular aperture proved in fact to be neurinomas.

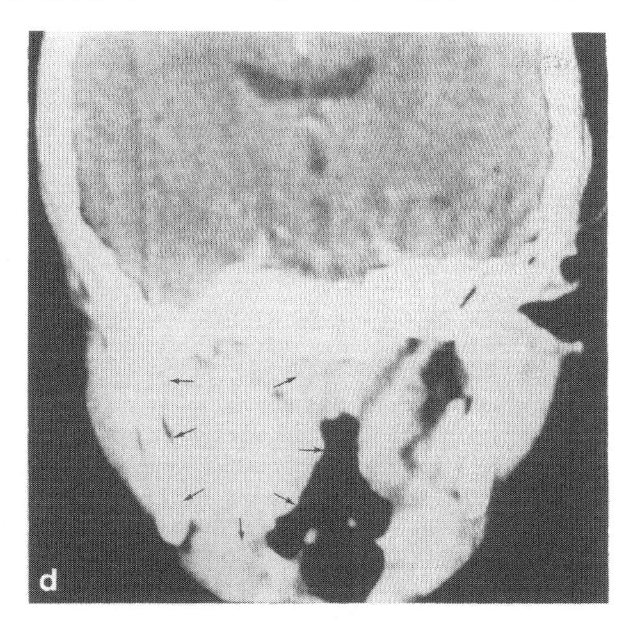

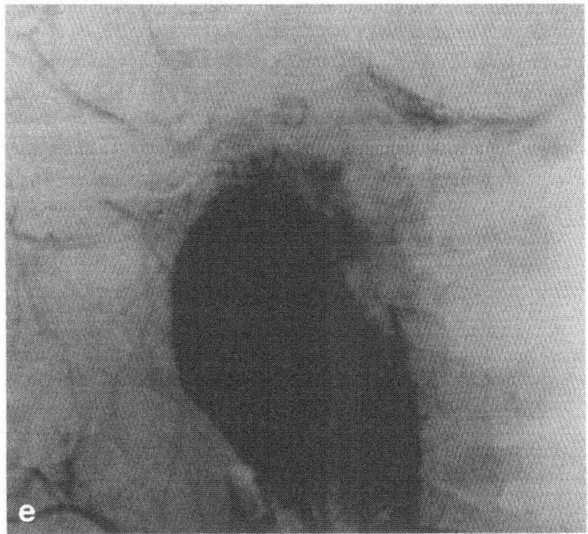

Fig. 130a–e. Intrapetrous meningioma. **a** CT scan (bony window), low section (after infratemporal approach: *white arrow*). Hyperostosis of petrous *(1)*, of lateral mass of occipital *(2)*, and of greater wing of sphenoid *(3)*, temporal condyle *(4)*. **b** CT scan (bony window), higher section showing hyperostosis of left pyramid. Labyrinthine massif and plane of lateral SCC *(1)*. **c** CT scan (parenchymatous window). Meningioma "en plaque" on posterior face of pyramid *(1)*. **d** Frontal CT scan. Infratemporal mass (border marked by arrows). **e** Arteriography. Injection en masse of cervical tumor

Treatment

- Meningiomas of the posterior aspect of the petrous closely resemble neurinomas where operation is concerned. The imperatives to be respected are the same. The extra-cisternal position of the tumor, provided one respects the arachnoid layer, greatly facilitates dissection of the different nerves, particularly the facial and cochlear, but there often exists around one or more orifices a zone of adhesion quite difficult to free without risking injury to the nerve(s) traversing this region. The same problem arises for the slender arteries accompanying these nerves, particularly those arteries to the internal ear whose fragility is ill-adapted to the manipulations inevitable with a tumor of the habitual size of a meningioma.

- In every case, total excision raises the problem of resecting the meninges and also the underlying bone, which, as Yasargil [189] has shown is often altered. It is a very conventional idea, where meningiomas are concerned, that one can never be sure of complete excision. Borovich and Doron [15] have shown that there is a regional pathology involved, and that the meningeal resection must extend well beyond the apparent limits of attachment of the tumor in order to resect numerous small meningoepitheliomatous clumps of cells infiltrating the dura mater. Unfortunately, this resection cannot usually be extensive at the posterior aspect of the petrous without wide resection of the pyramid and freeing of the intrapetrous course of the nerves. The problem is even more acute when, fortunately rarely, there is extensive invasion of the pyramid. To be "carcinologic", the resection then has to be extensive and this can only be done via an otoneurosurgical approach.
- Another difficult problem is that of the position of the acousticofacial bundle. It is very often displaced outwards, at the posterolateral pole of the tumor. During the resection of the latter, especially when this is undertaken by the suboccipital route, but also in the case of a translabyrinthine approach, the operative field is constantly barred by these nerves, which adds to the risks and in any case limits visibility at the innermost zones. The very fine operative photographs published by Yasargil [189] are very illustrative in this context.

Indications

Many routes have been proposed. Yasargil [189] lists six different approaches, some neurosurgical (infratemporal and transtentorial, supraoccipital and subtentorial, and classically suboccipital), the others otoneurosurgical (translabyrinthine, combined or not with a suboccipital approach, and translabyrinthine transtentorial as proposed by King and Morrison [104]). He himself uses only the classic suboccipital approach, but our attitude is much more eclectic. We use several approaches, the choice being determined mainly by the site of the tumor.

Otoneurosurgical Choices

The Extended Translabyrinthine Route. This is the route that we use most often (10 cases out of 19), an attitude shared by all those who hace acquired considerable experience of this approach. As stressed by Giannotta et al. [65], this approach makes it much easier to respect the facial nerve, which is one of the great problems in excision of this type of tumor, and by the very anterior access it provides to the posterior fossa also makes it much easier to respect the arachnoid layer. We would add that since, by definition, it involves a complete bony resection, it perfectly ensures eradication of the zone of attachment and thus offers the best conditions for a truly "carcinologic" operation. The extent of the bony resection performed by us ensures an exposure at least as large as that habitually offered by the suboccipital flap, especially since those who habitually use the latter approach endeavor to limit the size of their opening in order to reduce automatically the risk of traction on the cerebellum. We therefore entirely refute the comments of Rhoton added as an appendix to the article by Giannotta et al. [65]. This author is probably totally unaware of the extent of the exposure offered by a well-constructed extended translabyrinthine approach. Also, Rhoton advances the argument that there is a considerable risk of leakage of CSF after a translabyrinthine approach. If this was true at one time, it is no longer the case. The results we have reported in connection with acoustic neurinomas show that this risk of fistula is altogether comparable in the translabyrinthine and the suboccipital approaches, which destroys this argument. We use an extended translabyrinthine approach whenever the meningioma, because of its site, suggests an acoustic neurinoma and produces major hearing loss (above 40 dbs) and, of course, whenever it produces a cophosis, which is not exceptional since we have observed this sic times (35%).

Transcochlear Routes. An extended transcochlear route was used once for a large meningioma attached to the temporal margin of the posterior lacerate canal. This was in a woman aged 45 years with a very large meningioma (Fig. 131) fed by a large ascending pharyngeal artery (Fig. 131) and filling *en masse* like a tumor of the glomus jugulare (Fig. 131). Although the latter diagnosis was considered for a while, it was excluded by the findings of petrous tomography (Fig. 131) which showed absence of infralabyrinthine excavation, absence of erosion of the posterior lacerate canal and, on the contrary, a relative sclerosis of the margin of the in-

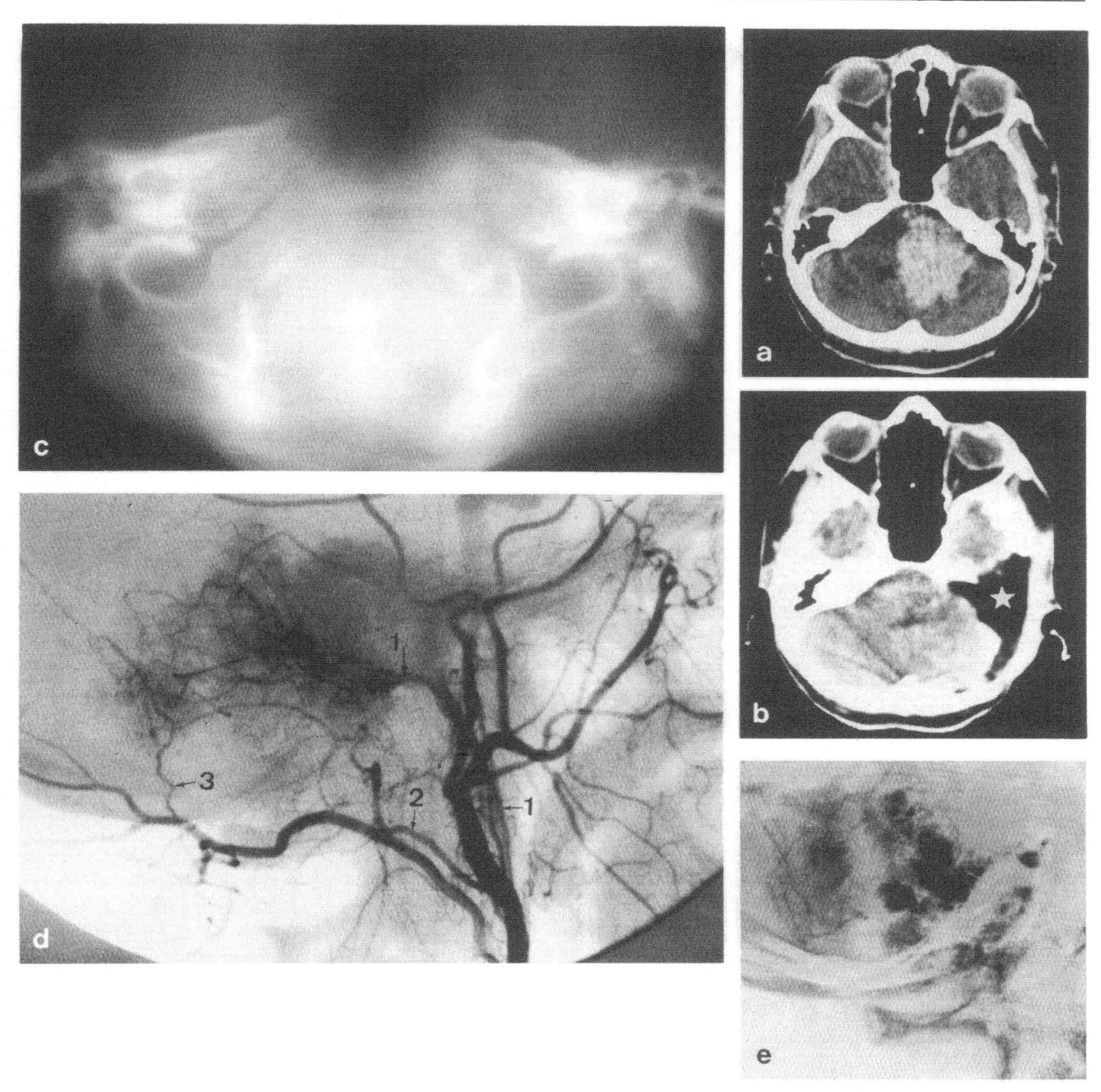

Fig. 131 a–e. Meningioma of foramen jugulare. **a** CT scan showing tumor. **b** Postoperative scan: *white star* marks bony resection of extended transcochlear approach. **c** Tomogram of petrous bones: no destruction of left lacerate canal, rather osteosclerosis of its margins. **d** External carotid arteriogram. Ascending pharyngeal artery *(1)*, posterior auricular artery *(2)*, and its stylomastoid branch, mastoid meningeal branch of occipital artery *(3)*. **e** Intense blush suggestive of jugular glomus tumor

ternal orifice, particularly of the jugular spine of the temporal bone. The absence of a pedicle derived from the carotid siphon demonstrated the very low attachment of the tumor.

This case is an excellent illustration of the indication for these routes of approach where a meningioma is concerned. They are reserved for meningiomas attached to the posterior lacerate canal. Based on the principle that excision of these tumors, which are usually of fair size, is a direct threat to hearing and the facial nerve, especially if it has been decided to perform a total resection including that of the base of attachment, we believe that only this route allows the best attainment of the main objectives. If the tumor is not too large, a simple transcochlear approach should suffice. In our case, the particularly large size and the vascularity led us to prefer an extended approach, which allowed complete excision of the tumor (Fig. 131).

The Suprapetrous Route. The limitation of the lesion to the confines of the IAM and the absence of major impairment of hearing are good enough reasons for considering this approach whatever the nature of the tumor. It is exceptional for this to be a meningioma. We observed one case of this type and treated it in this way. There was nothing unusual about the indication; it was only the nature of the tumor that proved surprising.

The Retrolabyrinthine Route. By resecting the posterolateral segment of the pyramid, this route gives a good exposure of meningiomas attached at the confluence of the superior petrosal sinus and the bend of the lateral sinus; moreover, it allows adequate resection of the zone of implantation of the tumor both on the dura and in the bone. The perfect expo-

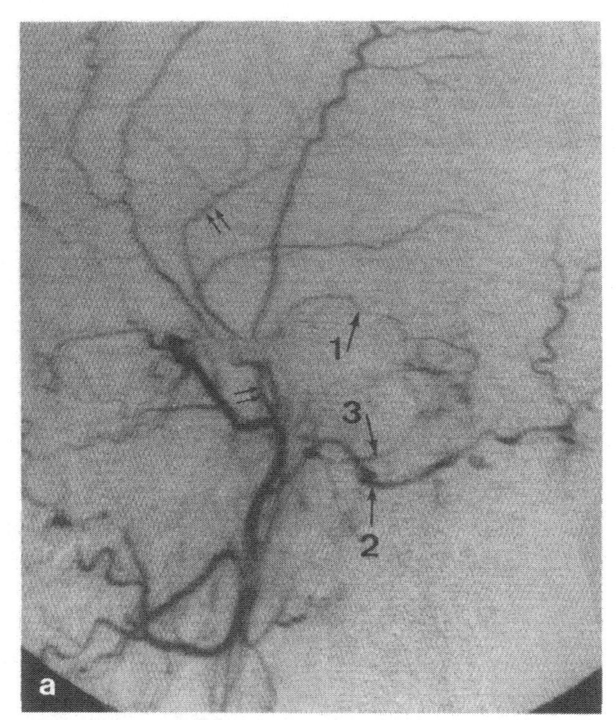

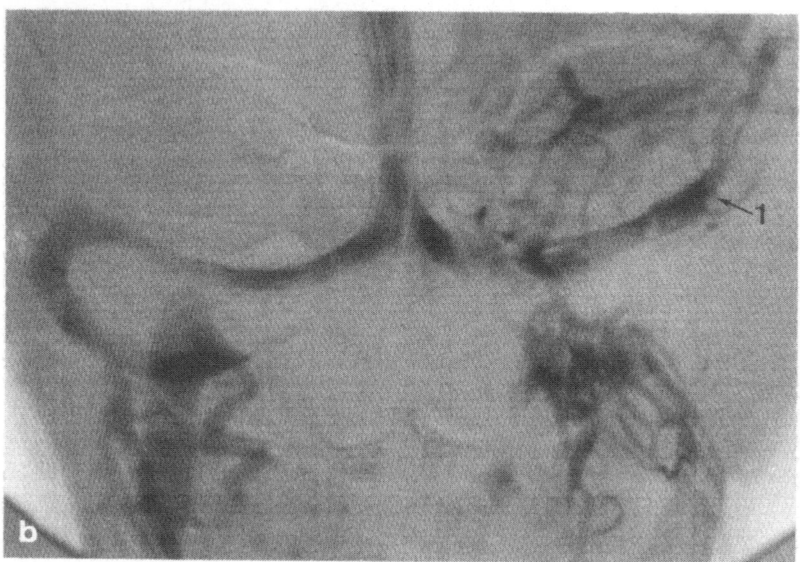

Fig. 132a, b. Meningioma of bend of lateral sinus. **a** External carotid arteriogram (lateral view). Middle meningeal artery and branches to vault *(double arrows)*. Superior tympanic artery *(1)*, occipital artery *(2)*, and its mastoid meningeal branch *(3)*. **b** Internal carotid arteriogram (frontal view): (venous phase) – obstruction of bend of left lateral sinus *(1)*

sure of the bend of the transverse sinus also allows resection of the invaded segment of the sinus if necessary and of the adjacent dura in the temporal and suboccipital zones. It may even be possible, as advised by Bonnal [14] to repair the lateral sinus since it is perfectly exposed on its three faces. We have successfully used this approach for a tumor of the bend of the lateral sinus that we considered to be a meningioma with complete obstruction of the sinus (Fig. 132). The planned operation went off very easily but the histology showed a cavernous angioma of the dura mater. We would have no hesitation in planning the same operation if faced with a meningioma situated in this region.

Neurosurgical Choices

It is not our intention to discuss these routes. Yasargil [189], with his enormous experience, has done so better than we can. In fact, we would stress that in certain situations it may be considered that the otoneurosurgical routes are contraindicated and that it is better to choose just a neurosurgical approach. But it has to be said that in these cases only the tumor and the dura mater are resected, and the bone is generally left intact. The osteoma, if it is obvious, will possible be resected, but with this technique there is a great risk of leaving in place some intrapetrous meningoepitheliomatous islets as a source of recurrence. This is the more likely if an attempt is made to preserve hearing, and this is why, when the tumor is attached to the immediate vicinity of the aperture of the IAM, we prefer to make a translabyrinthine approach. Certain tumors are attached manifestly more medially, at the petrous apex or at the petro-occipital suture. The presence of a stout pedicle derived from the carotid siphon (Fig. 133) is evidence of attachment at this site. If we know that the attachment is very medial, and that the laterally displaced facial and vestibulocochlear nerves form a permanent barrier to both a translabyrinthine and suboccipital approach, it seems preferable to use a subtemporal-transtentorial approach. This allows access to the upper pole of the tumor near its attachment. As a rule, interruption of the carotid pedicle can be managed successfully at the start of the operation, which greatly facilitates tumoral evacuation. A transcochlear approach might be considered for these meningiomas of the petrous apex, but this would seem to have two disadvantages in these cases. First it does not allow such direct access to the carotid pedicle, particularly when there is a large tentorial artery. Even if the resection is carried very high and far on the clivus, which is not without risk to the cavernous sinus and also the abducent nerve, it may well be impossible to reach the pedicle. On the other hand, as stressed by Yasargil [189], these meningiomas often have an extension on the temporal dura or the supper aspect of the tentorium in the vicinity of the posterior clinoid process which cannot be reached by this route, nor by the suboccipital route. Therefore, on three occasions we have performed sub-

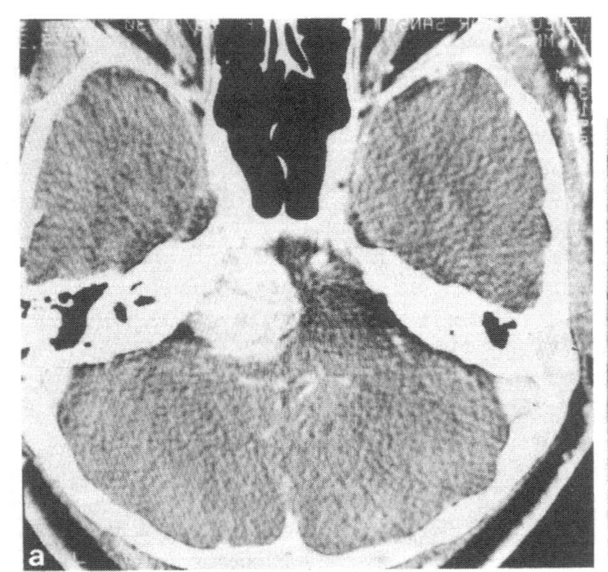

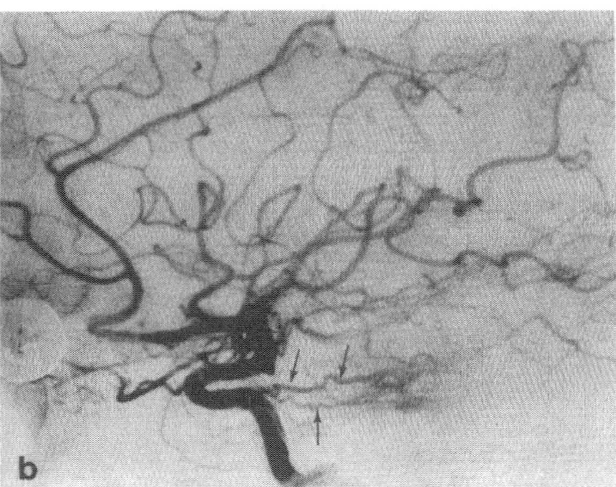

Fig. 133a, b. Meningioma of petrous apex. **a** CT scan. **b** Internal carotid arteriogram (lateral view). Very large meningeal artery of tentorium *(arrows)*

temporal-transtentorial access. Yasargil's criticism is that this does not allow for resection of the lowest fragments when the tumor descends very low in the posterior fossa. In our experience, admitedly limited, this criticism does not seem well-founded since tumoral evacuation allowed the lower pole to be eventually brought up and stripped without risk to the mixed nerves. In fact, this is possible only if the attachment is high enough, remaining at a distance from the posterior lacerate canal. In this context, the absence of a pedicle of attachment derived from the ascending pharyngeal artery is a useful diagnostic factor. On the contrary, in view of the position of the head in full rotation, we have felt that it was the external pole of the tumor that might be the more difficult to dissect, especially when its pressure on the acoustico-facial bundle from above has depressed these nerves downwards. This did occur to us on one occasion; but as we had foreseen this possibility it was easy, in the same operative stage, to expose the occipital squama and fashion a sub-occipital retrosigmoid access flap to reach this last fragment. Yasargil [189] makes a second criticism of this subtemporal-transtentorial approach. This concerns the need for considerable retraction of the temporal lobe, with all that this implies in direct cortical risks and risks to the veins, and this is the real nuisance with this approach. It may be that the principles laid down by Hongo et al. [85], i. e., routine emptying of the CSF and also retraction with narrow blades and the monitoring of retraction pressure, will lessen this problem.

Abstention

We do not intend to return to the general question of contraindications, but to discuss a particular situation in which, though operation is feasible, it seems excessive because of the risks of neurologic sequelae but mainly the risk of incomplete excision. This is the case of diffuse invasion of the petrous pyramid (Fig. 130). This is an eventuality that has come our way only once. The case was that of a young boy aged 13 years, with a cervical swelling under the angle of the mandible, clearly visible in the scan (Fig. 130). Biopsy had been reported as a chemodectoma. The absence of enlargement of the posterior lacerate foramen had then raised the diagnosis of a chemodectoma of the vagus, especially as arteriography (Fig. 130) injected *en masse* a tumor expanding the carotid fork. However, the operation performed necessitated resection of the lower part of the petrous in what might be called an infratemporal approach, since the tumor seemed to reach and even to adhere to the base of the skull. Histologic examination then revealed that it was a meningioma. Repetition of the tomographs of the petrous (Fig. 130) then showed that there was a diffuse hyperdensity of the pyramid. In particular, scanning by the bony window (Fig. 130) confirmed the diffuse sclerosis of the pyramid and its extension to the lateral mass of the occipital bone and also to the clivus and greater wing of the sphenoid. As this child clinically had only hypoacousia and minor disorders of swallowing and of taste on the posterior third of the tongue, and as it seemed that operation would require a complete petrectomy with adjacent resection of the lateral mass of the occipital bone, the clivus and greater wing of the sphenoid, and would not even then guarantee complete excision, especially at the cavernous sinus, the disproportion between the mutilating nature of such a procedure, its probable incomplete nature and the paucity of symptoms in this young patient led us to postpone intervention. It is clear that if this should be indicated in the future, only an oto-neurosurgical procedure could be considered.

Results

We have to report one death during the second postoperative week, probably due to difficulties in swallowing, in a patient who develoed secondary pulmonary infection with bilateral asphyxial pneumopathy.

The facial nerve was divided twice, during extended translabyrinthine procedures. This was at the start of our series. The patients were improved by hypoglossal-facial anastomoses. The continuity of the facial nerve was preserved 16 times. Facial motility was normal after operation in 6 patients, while the other 10 had a facial paralysis. In 5 cases this rapidly regressed with eventual normal motility; in 4 recovery was prolonged, ending with normal palpebral occlusion but with hemispasm (type 2 A or 1 B). One patient did not recover and was treated by hypoglossal-facial anastomosis. In this respect our results are comparable with those of Yasargil [189], but as regards audition they have of course been very different. We were able to preserve hearing only after a suprapetrous approach, performed for an intracanalicular meningioma, and after 2 subtemporal transtentorial approaches. Unfortunately, towards the 8th postoperative month one of these patients developed progressive cophosis, probably due to progressive cochlear sclerosis. This course clearly shows that one must wait before claiming success as regards hearing, and that the results reported by Yasargil [189], which are the dream every neurosurgeon must hope for, are in fact very difficult to attain. The sacrifice of hearing is counterbalanced by the complete excision of the bony base of the tumor, except in the 3 cases of subtemporal-transtentorial approach. This is one of the advantages of an otoneurosurgical approach where meningioma is concerned. Apart from the one death, the complications consisted of 2 laryngeal paralyses, much improved by injection of teflon into the vocal cord, and 2 CSF fistulae which dried up rapidly and spontaneously.

Cholesterol-Containing Tumors

We are perfectly well aware that the use of this term means adopting an altogether heterodox attitude, one which groups under the same heading disorders differing as much in their pathogenesis as in their clinical features and treatment. Their common feature is that they contain crystals of cholesterol, which suggests an apparent kinship that is open to dispute, but the point of grouping them in the same chapter is precisely so that they can be better differentiated. From the neurosurgical aspect, there are two expanding lesions containing cholesterol: the craniopharyngioma and the cholesteatoma or epidermoid cyst. The site of craniopharyngioma is in the infundibulo-hypophyseal region whereas cholesteatoma is most often situated in the cerebellopontine angle but also sometimes in the chiasmatic or parasellar regions, in the fourth ventricle and sometimes in the diploë of the skull vault. These two lesions are of dysembryoplastic origin, the craniopharyngioma deriving from ectoblastic remnants in the pouch of Rathke while the cholesteatoma develops from ectoblastic fragments carried along by the rudimentary vessels on their way to the primitive neural tube.

From the otologist's viewpoint, there are three entities. First, the cholesteatoma, which is an acquired disorder, rarely very expansive, and with multiple mechanisms (epithelial invagination through a gap, proliferation of epithelial cells in the subepithelial layers) but in which the chief etiologic role is that of infection. Then there is the epidermoid cyst, or primary cholesteatoma, which is of the same nature as the neurosurgical cholesteatoma. This arises from ectoblastic fragments which have migrated into the mesoderm that will give ride to the petrous. It is therefore intrapetrous and extradural, unlike the cholesteatoma of the neurosurgeon which is intradural. Lastly, there is the cholesterol-containing granuloma, which is a quite special otologic lesion. According to Nager and Vanderveen [127], this is "simply a tissue reaction in response to the presence of a particular foreign body made of cholesterol crystals". These crystals are derived from the breakdown of blood shed into a blocked petrous cavity. Chronic obstruction, by preventing the ventilation and drainage of the intrapetrous cavities, produces a mucosal edema and then a petechial extravasation; the cholesterol crystals than give rise to a nonspecific reaction with giant-cells and vascular proliferation leading to further effusions and further deposition of crystals, so maintaining the cycle. The obstruction may be anywhere: in the auditory tube, the tympanic cavity or the intrapetrous cell-system. The nature of the obstructive lesion is various (polyp, inflammatory granulation, sclerosis, cholesteatoma). When the granulomatous reaction develops within the tympanic cavity, it produces the classic cholesterin otitis resulting in a picture of idiopathic hemotympanum. When this develops in the mastoid it gives rise to a quite special type of lesion, the "chocolate" cyst. And when it develops in the cells of the petrous apex, it produces the very gradual development of such cysts which may reach a considerable size before becoming manifest: the only recently recognized giant cholesterin cyst. Thus, in the otoneurosurgical field, it is necessary to be aware of the three entities capable of producing an expansive petrous or peripetrous lesion.

Cholesteatoma of the Cerebellopontine Angle

This is a lesion wellknown to neurosurgeons since the report of Lepoire and Pertuiset [112]. It is uncommon and we have only seen 3 cases. At the outset, at any rate, it is more likely to give rise to facial neuralgia and so lead to consultation with those of our colleagues dealing with pain problems. In view of the specialized nature of our team, this doubtless explains the small number of our own cases. At a more advanced stage, it produces a clinical picture more typical of the CPA, involving several nerves but still dominated by the trigeminal, as stressed by Fischer et al. [52].

The scan, with its image of extra-axial hypodensity not taking up contrast (Fig. 134), is characteristic, more perhaps than MRI, which shows a hypersignal in spin-echo.

The approach must be neurosurgical, and mainly by the suboccipital route, since the relative integrity of the acoustico-facial bundle and the ease with which the nerves are separated contraindicate a translabyrinthine approach. When the tumor has a

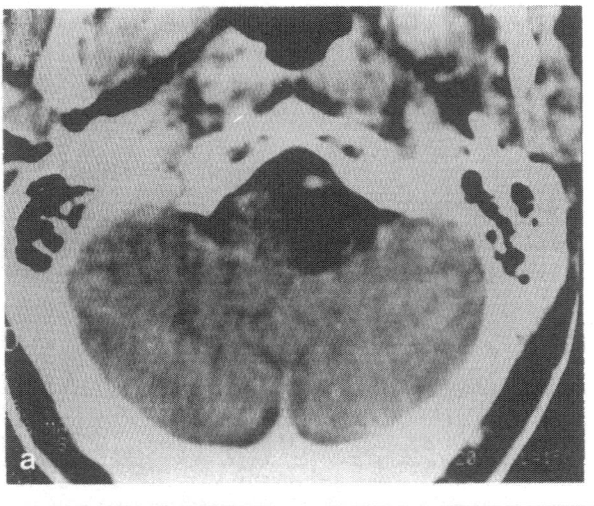

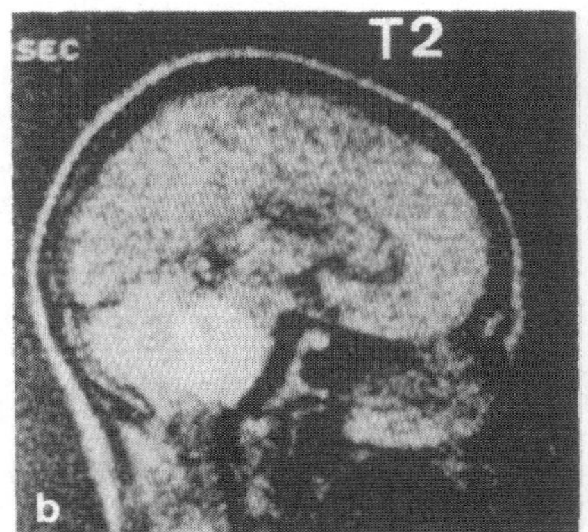

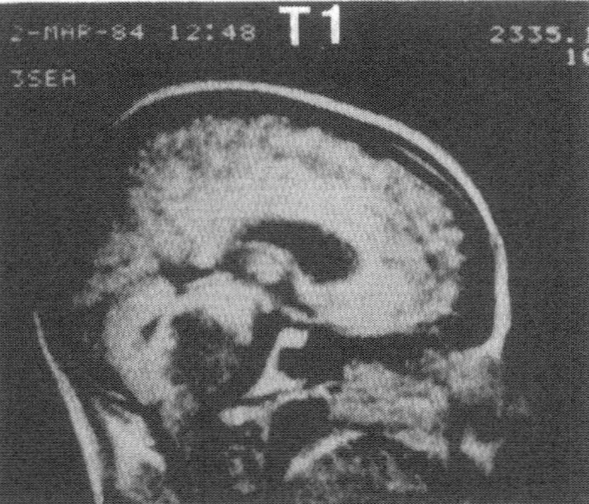

Fig. 134 a, b. Cholesteatoma of cerebellopontine angle. **a** CT scan: characteristic hypodensity. **b** MRI: hyporesonance in Tl and hyperresonance in T2

major extension towards the petrous apex and the foramen of Pacchioni, a subtemporal transtentorial approach may be considered. Its pearly-white appearance, characteristic from the outset, earns it is very picturesque name of pearly tumor.

Primary Cholesteatoma of the Petrous

This is an otologic tumor. To clearly distinguish it from a secondary cholesteatoma, otologists join with House and Sheehy [92] in defining it as a benign tumor of embryonic origin, "developing behind a normal tympanum and without previous ear infection". It is a relatively rare tumor, well studied by Derlacki [39], McDonald et al. [181] and recently in France by Tran Ba Huy et al. [181]. It is a pearly tumor, but extradural. It is a tumor commoner in men (70% of cases) around the age of 40, usually manifested by a gradual facial paralysis, sometimes by a "cold-induced" paralysis but whose failure to regress should attract attention. At the initial examination there is a perceptive deafness and a deep mixed deafness. These findings are quite consistent with the concept of Fisch [45], who believes that primary cholesteatoma arises at the geniculate ganglion just above the cochlea, soon threatening these two structures, "strangulating the facial nerve near the geniculate ganglion and perforating the cochlea". It arises within the labyrinthine massif and remains for long confined within the petrous apex, but ends by invading the perilabyrinthine cells, sometimes becoming embedded in the middle ear at the level of inner part of the epitympanic recess. This arrangement is quite different to that of the secondary cholesteatoma, which usually arises in the middle ear and remains confined to the posterolateral part of the pyramid, only exceptionally crossing the labyrinthine massif. Our own observations are altogether consistent with this concept of Fisch [45]. In half of our cases, the primary choles-

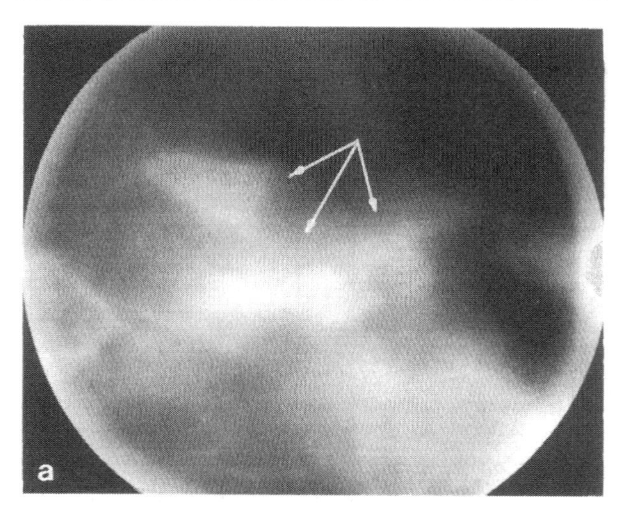

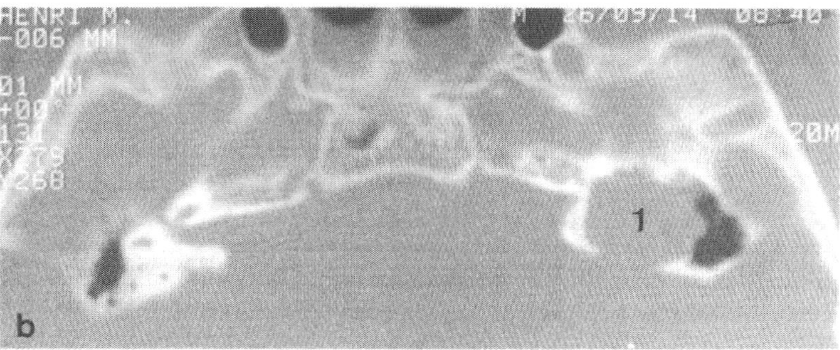

Fig. 135 a, b. Primary cholesteatome of petrous. **a** Tomogram (Pöchl's view). Loss of substance with sharp borders in region of geniculate ganglion *(arrows)*. **b** CT scan. Punched-out loss of substance in region of geniculate ganglion with fine cortical reaction on posterior border *(1)*

teatoma was situated at the geniculate ganglion and invaded the perilabyrinthine cells, but did not appear in the cavities of the middle ear. The other half of the cases were in patients who had been operated some years previously for a cholesteatoma of the middle ear and now reported with a large tumor having invaded and more or less destroyed a large part of the petrous apex to spread out against the dura mater just beneath the trigeminal ganglion. It is probable that all that the first operation had done was to remove that part of the tumor protruding into the tympanic cavity. The deeper nucleus, left in place, then had plenty of time to silently erode the inner portion of the pyramid, progressively destroying the facial nerve and the cochlea. The diagnosis is based on the radiographs (Fig. 135), which show a punched-out appearance of the labyrinthine massif in the region of the geniculate ganglion, and especially on the scan (Fig. 135) which shows a hypodense but not negative lesion (0 to 30 HU), without take-up of contrast, while the bony window view shows destruction in the region of the ganglion and sometimes an extension over the upper face of the petrous. In MRI, this is a hyper-resonant lesion in spin-echo; this is suggestive enough, but the non-visibility of the bony structures makes this examination, in our view, less useful than it might be for the assessment of bony spread.

Thus is an otosurgical tumor, as its extradural site means that the help of the neurosurgeon is not usually required. Our team has operated on 10 cases and, depending on the extent and size of the tumor, a suprapetrous approach was required on two occasions and, more often, a transpetrous approach indicated by destruction of the labyrinthine massif. A translabyrinthine approach was used 3 times and a transcochlear approach 5 times. We report only one section of the facial nerve, which was repaired by a facio-facial anastomosis. 3 patients had a postoperative paralysis which fully regressed only once, the other two patients being dealt with by hypoglosso-facial anastomosis.

The Cholesterol Giant Cyst

This is a new entity, promoted, as Sterkers et al. [173] have suggested, by antibiotics. Just as these have transformed the course of acute otitis and increased the incidence of seromucous otitis, so they have modified the character of infections propagated to the tip of the petrous which produce what the otologists call petrositis or apicitis. These cases of acute osteitis were formerly responsible for the clinical picture described by Gradenigo, which combines an acute otitis with otorrhea, abducent paralysis and trigeminal neuralgia. The inhibition of ear infections doubtless favors obstruction of the ventilation of the apical cells, and this now sometimes results in the development of a cholesterol granuloma whose slow course and situation in a relatively silent zone allow considerable enlargement before producing symptoms suggestive of a tumor of the cerbellopontine angle. Wyler et al. [194] described such a lesion in 1974, as did Gaeck [57] in 1975 and Montgomery [122] in 1977, but interest in this pathology was not really aroused until the paper by House and Brackmann [93] in 1982. Since then, Lo et al. [112 a] have reported the first 10 cases of the otologic group of the St. Vincent Hospital at Los Angeles and stressed the identity of this disorder. In 1985, Gherini et al. [64] discussed the therapeutic attitudes of the same group. Occasional cases have been reported by Graham et al. [70], Sterkers et al. [173], Tran Ba Huy et al. [181] and by ourselves [138]. A review of the literature makes one wonder whether certain lesions reported under other headings, such as the dermoid cyst of Behnke and Shindler [10] or the mucocele of Osborn and Parkin [133] were not in fact of the same nature. Macroscopically, this is a cystic lesion with a well-organized fibrous wall, containing a yellowish-brown viscous substance with sparking flakes, like a craniopharyngioma, and semi-solid curds. The chocolate color derives from deposits of hemosiderin, the yellow from lipids, and the flakes from crystals of cholesterol. The microscope shows that the fibrous wall is devoid of the keratinized squamous epithelium which lines the walls of primary cholesteatomas. In the interior there are cholesterol crystals surrounded by multinucleate giant-cells indicative of a foreign-body reaction. The whole is embedded in fibrous tissue studded with macrophage cells loaded with hemosiderin and traversed by blood-vessels. Frequently, streaks of fresh blood are found adjacent to zones of manifestly old hemorrhage in process of absorption. All these findings are completely consistent with the ideas of pathogenesis discussed above and derived from the experimental studies of Friedman [54], Beaumont [8], Main et al. [116] and others.

Clinically, these patients are usually referred for unilateral hypoacousia, often retrocochlear, frequently associated with vertigo and tinnitus and sometimes with signs of involvement of other cranial nerves: facial hemispasm or paresis dysphonia and disturbance of swallowing in one of our cases. The tympanum is normal and there is no evidence of antecedent otitis. The audiometric involvement is variable. On the contrary, it is the scanner that makes the diagnosis. It shows a cystic lesion at the petrous apex that does not take up contrast, about isodense with the brain, that has variably excavated the petrous apex and projects to a varying degree into the cerebellopontine angle, from which it remains separated by a well-outlined shell (Fig. 136). The bony destruction is often impressive, involving the petrous apex but also extending on to the clivus. It has sharp curved outlines (Fig. 136).

Treatment is surgical, and consists of evacuating the cyst and ventilating the cavity by means of a silastic drain connecting it to the mastoid or the tympanic cavity. The approach depends on the state of the hearing. If this is markedly compromised, or destroyed, the translabyrinthine route is the best as the most direct and the most "permeable". If the hearing is still worth saving, one must skirt round the labyrinth, either by using the infratemporal route if the jugular bulb is not too high, or by the subpetrous route if the bulb is in the way. After emptying, drainage is then installed between the cyst and the mastoid in the first case and between the cyst and the epitympanic recess in the other. The important thing is to identify the lesion before operation, as simple emptying of the cyst with drainage calls for a less mutilating procedure than complete excision of the cyst-wall. For example, in our two cases we undertook a wide exposure by an extended transcochlear route. True, eradication was complete and the risk of recurrence by obstruction of drainage impossible, but in both cases the patient had a facial paralysis and one had to undergo a hypoglossal-facial anastomosis. It is certain that a translabyrinthine approach, justified in both these cophotic patients, would have allowed perfect drainage while ensuring the safety of the facial nerve.

Other Tumors

These are rare, and they all constitute curiosities discovered during exploration for audiovestibular disorders, sometimes elaborate but all evocative of

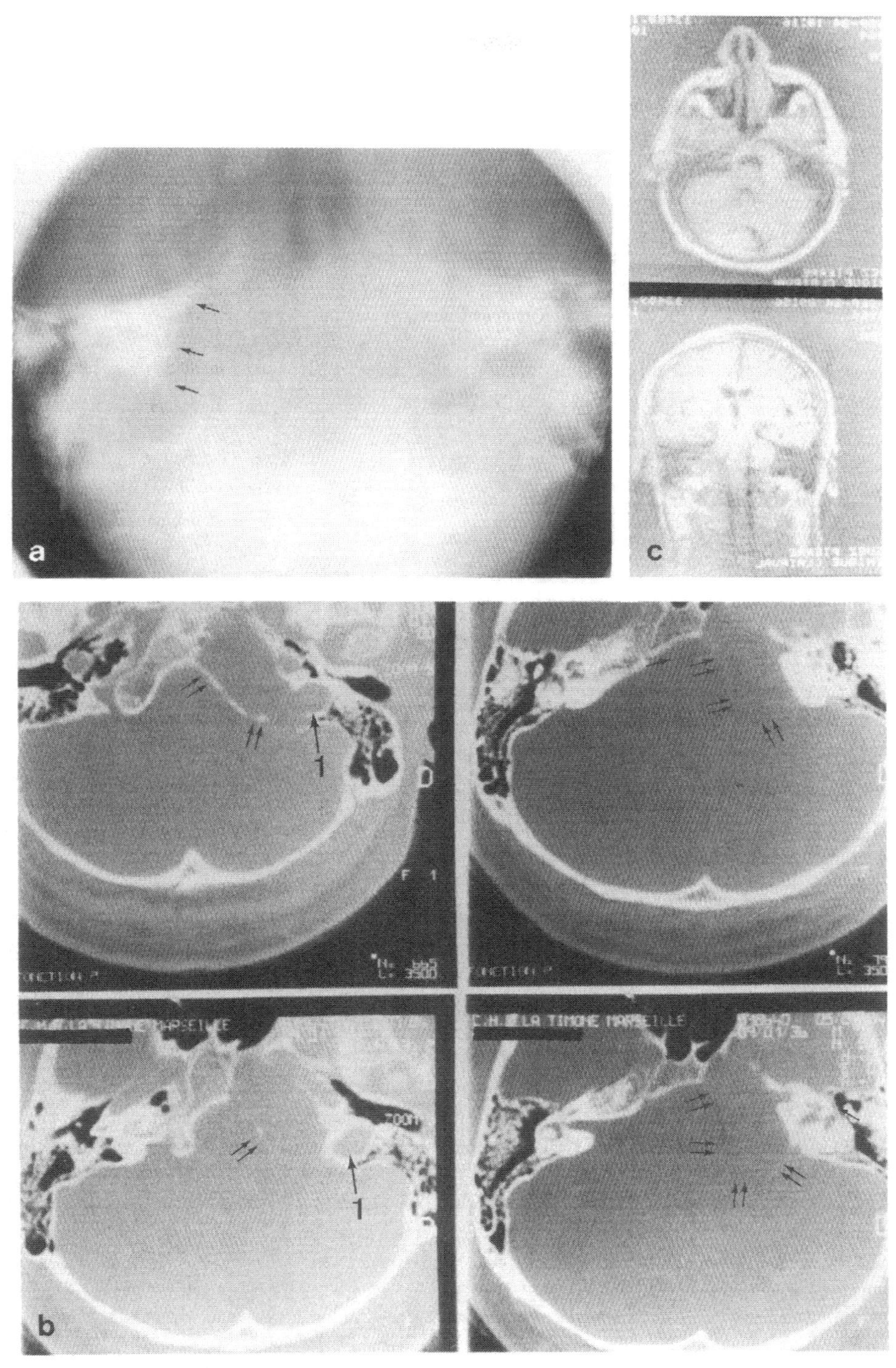

Fig. 136a–c. Giant cholesterol cyst. **a** Tomogram. Sharp-edged erosion of petrous apex *(arrows)*. **b** CT scan. The cyst has destroyed the entire apex. It is bounded by a fine cortical layer *(double arrow)*. Jugular bulb *(1)*. **c** MRI. Hyper-resonant extra-axial, lateral, pontobulbar lesion

a lesion of the cerebellopontine angle. Clinical diagnosis is difficult because the clinical picture is very much the same. One feature that might attract attention is the rapidity of development (in malignant lesions), but a rapid or even very abrupt clinical course is sometimes known to occur with authentic neurinomas, to be explained by a vascular factor or, when growth is manifestly accelerated, by the operation of hormonal or humoral factors (growth factor?). It is difficult to provide a coherent classificatin of all the tumors or expansive lesions that have been reported. It seems best to make a catalogue based on the frequency of the various etiologies, as done by Brackmann and Bartels [18].

Arachnoid Cysts

Arachnoid cysts of the cerebellopontine angle are rare. Brackmann and Bartels however observed 7 cases, while Rousseaux et al. [156] collected 10 cases over 20 years, which we find very strange as we have never seen one ourselves. One wonders whether the infective factor they adduce in determining this lesions may not take a special course in the Lille region for reasons unclear to us. They state that the clinical history begins with various otologic symptoms (especially acouphenia and vertigo). Examination shows mainly a hypoacousia, which, so they say, is likely to be bilateral. Brackmann and Bartels [18] stress the retrocochlear features which direct investigation towards the cerebellopontine angle. A dilatation of the IAM indicative that the lesion is of long standing is not exceptional. The scan shows a well-circumscribed fluid image that suggests the diagnosis, but there may be confusion with cholesteatoma, lipoma, certain cystic neurinomas or meningiomas. If a scan made for some other condition incidentally reveals the lesion, simple surveillance may be all that is required but the presence of symptoms of the CPA calls for operation, which should not involve the labyrinth. Of course, a classical suboccipital approach is indicated but it would not be unreasonable, in an otoneurosurgical setting, to use a retrolabyrinthine route which would allow excellent management of the lesion or, if another lesion were discovered, could be extended as required into a translabyrinthine approach. Obviously, such an extension can only be considered if major hearing loss already exists.

Vascular Tumors

Brackmann and Bartels [18] report 5 personal cases: 4 hemangiomas and one hemangioblastoma. We have ourselves seen a cavernous angioma of the tentorium cerebelli, growing in bilocular form with a supratentorial subtemporal mass and a subtentorial mass at the upper part of the CPA, and also a cavernous angioma arising at the bend of the lateral sinus which we originally regarded as an angioblastic meningioma (Fig. 132). According to Brackmann and Bartels [18], the relative rapidity of the clinical course should be indicative. The scanner, by sometimes showing a marked uptake of contrast, suggests the vascular nature of the tumor. The existence of a cyst with a mural tumor is suggestive of a hemangioblastoma.

Glial Tumors

The "bell-clapper" shape of gliomas of the brainstem is classical. It might be thought that this anatomic shape determines a cerebellopontine angle syndrome, but there is no direct causal relationship and there are known to be intra-axial forms of tumor of the brain-stem, with few symptoms, capable of suggesting a tumor of the CPA. However, with the scanner it seems that there should no longer be any diagnostic error. However, we were deceived on one occasion by a scanner image (though of the first generation) (Fig. 137) and Brackmann and Bartels [18] report two cases similar to ours where a translabyrinthine approach revealed an obvious glial tumor. These bell-clapper forms must be compared with the exceptional but possible glial tumors arising from heterotopic foci of glial cells situated in the meninges. These are embryologic anomalies which may eventually give rise to a glial tumor. Fewer than 30 cases have been reported, and recently Sceats et al. [158] reported a case arising in the CPA and perfectly simulating an acoustic neurinoma in terms of its morphology and scan images.

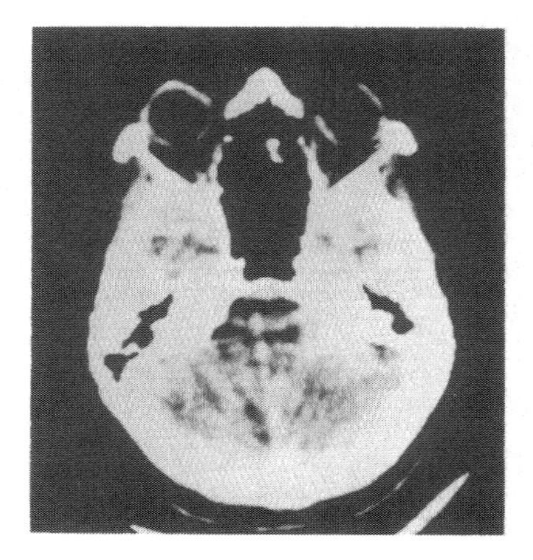

Fig. 137. Glioma of brainstem. CT scan

In fact, the absence of auditory signs and of expansion of the IAM could have suggested this diagnosis with such a fine CT image. It should also be recalled that several cases of medulloblastoma developing in the cerebellopontine angle have been reported. Brackmann and Bartels [18] published 2 personal cases and Coux et al. [28] collected 11 other cases from the literature and added two of their own. Usually, there was an extension of the tumor towards the angle from the cerebellum or sometimes the embedding of a tumoral fungation through a foramen of Luschka (lateral opening of the IVth ventricle). These anatomic forms may be compared with the rare but classic extraventricular papillomas of the choroid plexus: Kalangu et al. [100] have recently reported a case and reviewed the literature. These observations, even the most recent, not quite complete clinical pictures combining in different degrees intracranial hypertension, a cerebellar syndrome and signs of involvement of the cranial nerves, i.e., "neurosurgical" pictures. The absence of expansion of the IAM might possibly suggest the diagnosis, but the clinical pictures as well as the radiologic and in particular the scanographic images thrown open every hypothesis. It is certain, in these cases, that the absence of expansion of the IAM and the relatively benign auditory involvement should indicate a classical suboccipital approach rather than a translabyrinthine route.

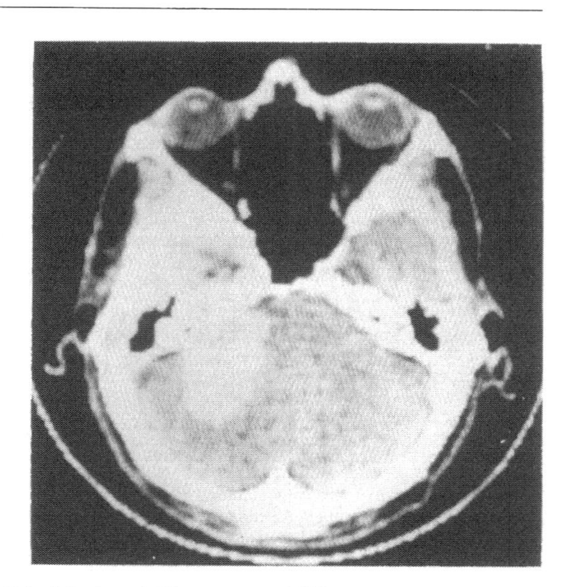

Fig. 138. Meningeal fibrosarcoma. CT scan

Malignant Tumors

We have observed 3, each different in nature.

– A fibrosarcoma in a man of 61, who had been complaining for 1–2 months of severe hemicrania justifying at that time a preliminary assessment by tomography and scanography which proved normal. 6 months later, the development of a facial paralysis and the persistence of the headaches led him to consult us. The facial paralysis was partial but marked. There was a hypoacousia with a loss of 70 dbs and the evoked auditory potentials reached only to the cochlear I wave. The scan (Fig. 138) showed a tumor of somewhat irregular but dense texture before injection with moderate uptake of contrast. Arteriography showed no abnormal vascularization. An extended translabyrinthine approach allowed the removal of a yellowish hemorrhagic tumor which filled the IAM to the fundus. The patient died 3 weeks later from the rupture of a pulmonary bulla.

– A leptomeningeal nevo-carcinoma in a woman aged 52, with gradually progressive paralysis of the IXth, Xth and XIth nerves over a year. Clinically, there was a cophosis of the same side. The IAMs were normal, the lacerate canal not eroded. The scan showed a stage II tumor, spontaneously dense, not taking up contrast. A translabyrinthine approach allowed removal of a friable, blackish, hemorrhagic tumor extending over the posterior aspect of the petrous from the IAM to the jugular foramen. Histologic examination confirmed nevo-carcinoma. The absence of any other cutaneous or choroidal lesion indicated the primary nature of the lesion.

– A metastasis in a man aged 58, without previous history, complaining of hypoacousia, unsteadiness, vertigo, otalgia and then facial paralysis, all having evolved over 3 months. Audiometry showed a virtual cophosis with absence of evoked auditory potentials. The scan showed a hyperdense tumor of the angle, taking up contrast (Fig. 139). Arteriography showed no abnormal vascularization. By way of an extended translabyrinthine approach, a tumor filling the IAM was completely excised without injury to the facial nerve. Histology showed the metastatic nature of the tumor, probably from a thyroid carcinoma, which was confirmed and operated on. A year later the patient seemed in perfect health, but had meanwhile been irradiated for a metastasis in the dorsal spine. 6 months later he returned with severe pain due to metastases in the sacroiliac and hip regions requiring the placement of a morphine pump. He died 5 weeks later.

To our knowledge, malignant primary tumors other than gliomas have been reported only by Brackmann and Bartels [18] (1 fibrosarcoma) and

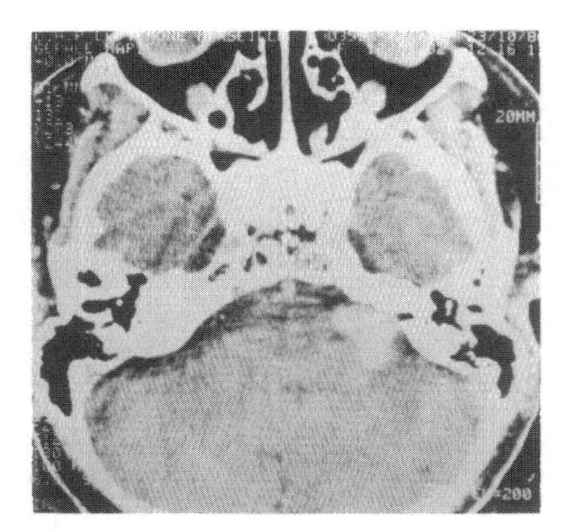

Fig. 139. Metastasis of cerebellopontine angle. CT scan

House [88] (2 hemangiosarcomas). The rapid onset of the symptoms, the development of a facial paralysis and the rapid growth of the lesion in successive scans made the best contribution to the diagnosis. The course, always fulminating, attested to the malignancy of these tumors.

The fact that the melanoblasts can derive from the neural crest explains why, during their migration from the neural crest, the Schwann cells or the meningoblasts may sometimes bring melanoblasts along with them. This would account for the presence of melanic lesions on the nerves or meninges. These lesions vary greatly in their extent and course, from simple melanic pigmentation of the meninges to the existence of true leptomeningeal nevo-carcinomas. There have also been described schwannomas or meningiomas containing melanin cells, the color of these tumors then being due, according to Dastur et al. [37], to phagocytosis of melanin pigments by the tumoral cells, in full proliferation and capable of macrophagic activities. Brackmann and Bartels [18] have observed 3 metastases developing in the CPA, all occurring after treatment of the primary tumor, situated in the lung, breast and larynx respectively. The rapid onset of symptoms in a patient with such a history strongly suggested the diagnosis. In our case the metastasis predominated. The rapid onset of the symptoms might have suggested the diagnosis of "atypical pathology"; but in no way do we regret the operation, which was both radical and a solution to the etiologic diagnosis.

Lipomas

When an intracranial lipoma is present, it is usually situated in the pericallosal cisterns. Probably because it is associated with other anomalies of the nervous system (microgyria, microcephaly, etc.) responsible for neurologic disorders, particularly epileptic attacks, it is discovered almost incidentally in the course of one of these manifestations. Moreover, it is now discovered quite by chance by the scanner in subjects without any abnormal manifestations. The site is then very variable: on the basal cisterns, particularly in the midline, in the lateral cerebral fissures or in the ventricles. All these sites are usually asymptomatic, and perhaps only lesions in the cerebellopontine angle are the source of a frank clinical picture usually based on a perceptive deafness. In a recent study, Rosenblum et al. [153] collected only 11 cases, including their own, to which we can add the case of Steimle et al. [168], the 2 cases of Pensak et al. [144] and the 2 cases of Brackmann and Bartels [18], which bring the total of known cases to 16. The radiographs may show a regular erosion with a definite sclerosed border (case of Fukui [56]), but it is mainly the scanner which provides the diagnosis by revealing the existence of a frankly hypodense lesion (25 to 45 HU), provided this is large enough to be visible, which is not always the case. The fact that it is a hamartoma and that its excision is usually only partial is no argument for routine intervention, nor for a translabyrinthine approach. Thanks to the use of evoked auditory potentials and CT scans which allow easy surveillance of these patients, one can wait until deterioration of auditory function is confirmed before deciding on a suboccipital approach.

Chondromas

Recently, Dany et al. [36] have reported a case of chondroma of the angle and collected 12 other cases from the literature. This is an extradural tumor, at least at the start of its development. It arises from cartilaginous embryonic vestiges persisting at the synostoses and, in this particular case, at the temporo-occipital suture. It is a slowly developing tumor which often produces a progressive paralysis of several cranial nerves. Though a cartilaginous tumor, it is calcified in only a quarter of the cases, and in half the cases produces only an image of erosion of the pyramid. The scan shows a hypodense tumor not taking up contrast, while the intrathecal injection of contrast demonstrates its extra-axial position. At operation, it is a well-circumscribed, multilobular, extra-arachnoidal tumor. We have never met such a case but, in view of its intraosseous origin and the risk of recurrence, we should be inclined to consider operation by an otoneurosurgical route so as to resect the entire zone of attachment if possible.

Other Tumors

Those reported include:

2 teratomas, one by Brackmann and Bartels [18], the other by Waters et al. [192]. This is a tumor of childhood, often large, in which the scan shows a non-uniform structure, sometimes with calcifications

1 dermoid cyst, by Brackmann and Bartels [18]: the presence of a germinative membrane made up of a stratified pavement epithelium seems formally to distinguish this tumor from a cholesterol giant cyst, which has no such layer

1 craniopharyngioma, by Altionörs et al. [3]. In the absence of an intra- or suprasellar lesion, the site of this tumor is rather surprising and leads us to think that it may be possible to compare this lesion with the preceding.

References

1. Albin MS, Benegin L, Dujovny M, Bennet MH, Jannetta PJ (1975) Brain retraction pressure during intra-cranial procedures. Surg Forum 26: 499-500
2. Albin MS, Babinski M, Maroon JC, Jannetta PJ (1976) Anesthetic management of posterior fossa surgery in the sitting position. Acta Anes Scand 20: 117-128
3. Altinörs N, Senveli E, Erdogan A, Arda N, Pak I (1984) Craniopharyngioma of the cerebellopontine angle. J Neurosurg 60: 842-844
4. Aoyagi T, Kyuno K (1912) Über die endothelialen Zellzapfen in der dura mater cerebri und ihre Lokalisation in derselben, nebst ihrer Beziehung zur Geschwulstbildung in der dura mater. Neurologia 11: 1-12
5. Atkinson WJ (1949) Anterior inferior cerebellar artery: its variation, pontine distribution and significance in surgery of the cerebello-pontine angle tumors. J Neurol Neurosurg Psychiatry 12: 137-151
6. Bataïni J, Pontvert D, Jaulerry C, Brunin F, Gaboriaud G, Scherrer A (1985) Tumeurs du glomus jugulaire. Posibilités de la radiothérapie. Neuro-chirurgie 31: 377-380
7. Bartels JP (1949) De tumoren van hut glomus jugulare. Doctorate Thesis. Groningen. Van gorsum et com
8. Beaumont GD (1967) Cholesterol granuloma. J otolaryngol Soc Aust 2: 28-35
9. Bébéar JP (1973) Apport des techniques récentes dans le diagnostic et le traitement du neurinome de l'acoustique. Thèse médecine Bordeaux N° 227
10. Behnke EE, Schindler RA (1984) Dermoid of the petrous apex. Laryngoscope 94: 779-783
11. Belal A, Lynthicum FH, House WF (1982) Acoustic tumor surgery with preservation of hearing. A histopathologic report. Amer J Otology 4/1: 9-16
12. Bennett MH, Bunegin L, Albin MS, Dujovny M, Hellstrom HR, Jannetta PJ (1977) Evoked potential correlates of graded brain retraction pressure. Stroke 8: 487-492
13. Bonnal J, Louis R, Combalbert A (1964) L'abord temporal trans-tentoriel de l'angle ponto-cérébelleux. Neuro-Chir (Paris) 10/1: 3-12
14. Bonnal J (1982) La chirurgie conservatrice et réparatrice du sinus longitudinal supérieur. Neuro-chirurgie 28: 147-172
15. Borovich B, Doron Y (1986) Recurrence of intra-cranial meningiomas: The role played by regional multicentricity. J Neurosurg 64: 58-63
16. Brackmann DE, Hitselberger WE (1979) In Neurological Surgery of the Ear, Vol II, Silverstein H and Norrel H. Edit Aesculapius Publ CO (Birmingham, Alabama) 173-177
17. Brackmann DE (1979) Acoustic neuroma surgery: Otologic medical group results. In: Neurological Surgery of the Ear, Vol II, Silverstein H and Morrel H Edit. Aesculapius Publ Co (Birmingham, Alabama) 248-259
18. Brackmann DE, Bartels LJ (1980) Rare tumors of the cerebello-pontine angle. Otolaryng Head Neck surg 88: 555-559
19. Brooks B (1930) The treatment of traumatic arterio-venous fistula. Sth Med J (Birmingham) 23: 100-106
20. Buckingham RA, Kenji Amii Perrelli SL (1959) Multicentric origin of glomus jugulare tumors. Arch Otolaryng 70: 104-107
21. Cannoni M, Pech A, Zanaret M, Thomassin JM, Goubert JL (1982) Notre expérience de l'anastomose hypoglosso-faciale. Ann Oto-Laryng (Paris) 99: 299-302
22. Cantrell RW, Kaplan MJ, Atukn O, Winn HR, Jahrsdoerferra (1984) Cathecolamine secreting infratempora fossa paraganglioma. Ann Otol Rhinol Laryngol 93: 583-588
23. Castellano F, Ruggiero G (1953) Meningiomas of the posterior fossa. Acta Radiologica, Suppl 104: 1-177
24. Charachon R, Junien, Lavillauroy C (1975) Intérêt de la voie intersinuso-faciale sous-labyrinthique. Ann Otolaryng (Paris) 92: 445-449
25. Charachon R, Roux O, Dumas G (1978) Tumeur du nerf facial. A propos de trois observations. Ann Oto-Laryng (Paris) 95: 777-784
26. Charachon R, De Rougemont J, Chirossel J, Gratacap B (1986) Notre expérience de la chirurgie des neurinomes du VIII. Ann Oto Laryng (Paris) 103: 495-499
27. Chobaut JC (1985) Intra-temporal facial nerve tumors. In Facial nerve. Portmann M Edit, Masson, New-York 417-423
28. Choux M, Léna G (1982) Le médulloblastome Neuro-chirurgie 28 suppl I: 1-229
29. Clark WC, Moretz WH, Acker JD, Gardner LG, Eggers F, Robertson JH (1985) Nonsurgical management of small and intra-canalicular acoustic tumors. Neurosurgery 16: 801-803
30. Clemis JD, Mastricola PG, Schuler, Vogler M (1982) Sudden hearing loss in the contralateral ear in post-operative acoustic tumor: three cases reports. Laryngoscope 92/1: 76-79
31. Clemis JD (1984) Hearing conservation in acoustic tumor surgery: pros and cons. Otolaryng Head and Neck Surg 92/2: 156-161
32. Cohen NL (1979) Acoustic neuroma surgery with emphasis on preservation of hearing. Laryngoscope 89: 886-896
33. Cohen NL, Ransonoff J (1984) Hearing preservation, posterior fossa approach. Otolaryng Head and Neck Surg 92/2: 176-183

34. Conley JJ (1956) Multiple paragangliomas in the head and neck. Ann Otol 65: 356-360
35. Dandy W (1934) Effects on hearing after subtotal section of cochlear branch of auditory nerve. Bull John Hopkins Hosp 55: 240-243
36. Dany A, Vidal J, Dumas M, Ravon R, Bokor J (1980) Les chondromes de l'angle ponto-cérébelleux. A propos d'une observation personnelle. Neurochirurgie 26: 355-357
37. Dastur DK, Sinh G, Pandyas K (1967) Melanotic tumor of the acoustic nerve. Case Report. J Neurosurg 27: 166-170
38. Dechaume JP (1966) Les méningiomes de la fosse postérieure. A propos de 28 observations. Thèse médecine Lyon 1: 253
39. Derlacki EL (1977) Congenital cholesteatoma of the middle ear and mastoid: a fourth report. In Shambaugh GE, Shea JJ Eds, Proceeding of the Shambaug Fifth International Workshop on middle ear microsurgery and fluctuant hearing loss. Huntsville (Ala), Strude Publishers 156-161
40. Djindjian R (1975) Arteriographie supersélective de la carotide interne et embolisation. Ann Med Int 126: 569-572
41. Dolan EJ, Tuckers WS, Rotenberg D, Chui M (1982) Intracranial hypoglossal schwannoma as un unusual cause of facial nerve palsy. J Neurosurg 56: 420-423
42. Domb GH, Chole RA (1980) Anatomical studies of the posterior petrous apex with regard to hearing preservation in acoustic neuroma removal. Laryngoscope 90: 1769-1776
43. Donaghy RMP, Numoto M, Wallman LJ, Flanagan ME (1972) Pressure measurement beneath retractors for protection of delicate tissues. Am J Surg 123: 429-431
44. Farrior JB (1984) Anterior hypotympanic approach for glomus tumor of the infratemporal fossa. Laryngoscope 94: 1016-1021
45. Fisch U, Rüttner J (1977) Pathology of intratemporal tumors involving the facial nerve. In facial nerve surgery. Fisch U Edit Aesculapius pub Co (Birmingham, Alabama) 448-456
46. Fisch U (1978) Otoneurosurgical approach to acoustic neurinomas. Prog neurosurg 9: 318-336
47. Fisch U (1982) Infra-temporal fossa approach for glomus tumors of the temporal bone. Ann Otol Rhinol Laryngol 91: 474-479
48. Fisch U (1985) La voie d'abord infra-temporale pour les tumeurs du glomus jugulaire. Neurochirurgie 31: 367-376
49. Fisch U (7 et 8. 11. 1986) Reconstruction intra-crânienne du nerf facial après exérèse du neurinoma de l'acoustique. Communication colloque "Neurinomes de l'acoustique. Acquisition et controverses". Toulouse (in press)
50. Fischgold J, Metzger J (1961) Precisions sur la radiographie du rocher dans les neurinomes du trijumeau. Rev Neurol 104: 308-311
51. Fischer G, Constantini JL, Mercier P (1980) Improvement of hearing after microsurgical removal of acoustic neurinoma. Neurosurgery 7/2: 154-159
52. Fischer G, Bret Ph, Hor F, Pialat J, Massini B (1984) Les kystes épidermoïdes de l'angle ponto-cérébelleux, 6 observations. Neurochirurgie 30: 365-372
53. Fischer G, Morgon A, Fischer C, Bret Ph, Massini B, Kzaiz M, Charlot M (1987) Exérèse complète des neurinomes de l'acoustique. Preservation du facial et de l'audition. Neuro-chirurgie 33: 169-183
54. Friedmann J (1959) Epidermoïd cholesteatoma and cholesterol granuloma. Ann Otol Rhinol Laryngol 68: 57-79
55. Fuentes JM, Uziel A (1983) Neurinomes intra-pétreux du nerf facial et de ses branches. A propos de deux observations. Neurochirurgie 29: 197-201
56. Fukui M, Tanaka A, Kitamura K, Okudera T (1977) Lipoma of the cerebello-pontine angle. Case report. J Neurosurg 46: 544-547
57. Gacek R (1980) Evaluation and management of petrous apex cholesteatoma. Otolaryng Head Neck surg 88: 519-523
58. Gaillard J, Rebattu JP (1963) Deux nouvelles observations de tumeurs glomiques tympano-jugulaires opérées par voie cervico-mastoïdienne avec déroutation du facial. J Fr d'O. R. L. 12/1: 83-92
59. Garcin M, Toy Riont J, Cagnol C, Huguet JF, Olivier JC (1968) Tumeurs glomiques multiples. Glomus jugulaire bilatéral associé à un glomus carotidien, à propos d'un cas. J Fr d'O. R. L. 17/5: 399-405
60. Gardner G, Robertson JM, Clark WC (1983) 105 patients operated for cerebello-pontine angle tumors. Experience using combined approach and CO2 laser. Laryngoscope 93: 1049-1055
61. George B (1985) Les tumeurs du glomus jugulaire. Presentation générale. Neurochirurgie 31: 333-341
62. German WJ (1961) Acoustic neurinomas. A follow up. Chir Clin Neurosurg 7: 1-20
63. Gueurkink NA (1977) Surgical anatomy of the temporal bone posterior to the international auditory canal: an operative approach. Laryngoscope 87: 975-986
64. Gherini SG, Brackmann DE, Lo WM, Solti-Bohmann LG (1985) Cholesterol granuloma of the petrous apex. Laryngoscope 95: 659-664
65. Giannotta SL, Pulec JL, Goodkin R (1985) Translabyrinthine removal of cerebello-pontine angle meningiomas. Neurosurgery 17/4: 620-625
66. Glasscock ME, Hays JW, Murphy JP (1975) Complications in acoustic neuromas surgery. Ann Otol Rhinol Laryngol 84: 530-540
67. Glasscock ME, Hays JW, Jackson CG, Steenerson RL (1978) A one stage combined approach for the management of large cerebello-pontine angle tumors. Laryngoscope 88/10: 1563-1576
68. Glasscock ME, Hays JW, Miller GW, Drake FD, Kanok MM (1978) Preservation of hearing in tumors of the internal auditory canal and cerebello-pontine angle. Laryngoscope 88/1: 43-55
69. Glasscock ME, Kveton JF, Jackson CG, Levine SC, McKennan KX (1986) A systematic approach to the surgical management of acoustic neuroma. Laryngoscope 96: 1088-1094
70. Graham MD, Kemink JL, Latack JT, Kartush JM (1985) The giant cholesterol cyst of the petrous apex. A distinct clinical entity. Laryngoscope 95: 1401-1406
71. Granick MS, Martuza RL, Parker SW, Ojemann RG, Montgomery WW (1985) Cerebello-pontine angle meningiomas: clinical manifestations and diagnosis. Ann Otol Rhinol Laryngol 94: 34-38
72. Greiner CF, Phillipied M, Collard C, Conraux C, Buccheit F, Maitrot D (1971) Hypoacousie contro-latérale dans le neurinome du VIII. Evolution post-opératoire. Rev Oto-Neuro-Ophtalmo 43/1: 54-58
73. Gild SR (1941) A hitherto unrecognized structure, the glomus jugularis in mass. Anat 79 [suppl 2]: 28
74. Guiot G, Bouche J (1965) Meningiomes et neurinomes intra-pétreux. Neuro-chirurgie 11: 361-362

75. Guzowski J, Paparella MM, Rao KN, Hosmino T (1976) Meningiomas of the temporal bone. Laryngoscope 86: 1141-1146
76. Hagen PT, Scholz DG, Edwards W (1984) Incidence and size of patent foramen during the first decades of life: an autopsy study of 965 normal hearts. Mayo Clin Proc 59: 17-50
77. Hakuba A, Hashi K, Fujotani K, Ikuno H, Nakamura T, Inoue Y (1979) Jugular foramen neurinomas. Surg Neurol 11: 83-94
78. Harker LA, McNabe BF (1978) Iowa results of acoustic neuroma operations. Laryngoscope 88: 1904-1911
79. Harner SG, Laws ER, Onofrio BM (1984) Hearing preservation after removal of acoustic neurinoma. Laryngoscope 94: 1431-1434
80. Hilding DA, Greenberg A (1971) Surgery for large glomus jugular tumor. Arch Otolaryng 93: 227-231
81. Histace B (1979) La microchirurgie du neurinome de l'acoustique. 53 interventions par abord sous-occipital. Thèse Médecine Lyon, N° 369
82. Hitselberger WE, Hugues RL (1968) Bilateral acoustic tumor and neurofibromatosis. In monograph II. Acoustic neuroma, House WF. Edit Arch Otolaryng 88: 700-711
83. Hitselberger WE, Gardner G (1968) Other tumors of the cerebello-pontine angle. In Monograph II. Acoustic neuroma, House WF. Edit Arch Otolaryng 88: 712-714
84. Hitselberger WE, House WF (1979) Partial versus total removal of acoustic tumors. In Acoustic Tumors, Vol II: Management House WF, Luetje CM. Edit University Park Press (Baltimore) 265-268
85. Hongo K, Kubayashi S, Yokon A, Sugita K (1987) Monitoring retraction pressure on the brain. J Neurosurg 66: 270-275
86. Horrax G, Poppen JL (1939) Experiences with the total and intracapsular extirpation of acoustic neuromata. Ann Surg 110: 513-524
87. House WF (1964) Evolution of trans-temporal bone removal of acoustic tumors. Arch Oto-Laryngol 80: 731-741
88. House WF (1968) Acoustic neuroma. Monograph II. Arch oto-laryng 88: 575-715
89. House WF, Hitselberger WE (1976) The transcochlear approach to the skull base. Arch Oto-Laryng 102: 334-342
90. House WF, Luetge CM (1979) Acoustic Tumors: I Diagnosis. II Management. University Park Press (Baltimore)
91. House WF (1979) Cochlear effects. In: Acoustic Tumors, Vol II. Management. House WF, Luetge MC. Edit University Park Press (Baltimore) 209-211
92. House JW, Sheehy JL (1980) Cholesteatoma with intact tympanic membrane - a report of 41 cases. Laryngoscope 90: 70-75
93. House JL, Brackmann DE (1982) Cholesterol granuloma of the cerebello-pontine angle. Arch Otolaryngol 108: 504-506
94. House JW (1985) Facial nerve grading systems, in M Portmann Edit Facial nerve, Masson Publ, New-York Paris 34-41
95. Hughes GB, Sismanis A, Glasscock ME, Hays JW, Jakson CG (1982) Management of bilateral acoustic tumors. Laryngoscope 92: 1351-1359
96. Ishii S, Hayner R, Kelley WA, Evans JP (1959) Studies of cerebellar swelling. II. Experimental cerebral swelling produced by supra-tentorial extra-dural compression. J Neurosurg 16: 152-166
97. Iwagana M, Yamamoto E, Yamauchi M, Fukumoto M, Uchno R, Sawada S (1984) Facial nerve neurinoma. Two cases located in the horizontal portion. Laryngoscope 94: 938-941
98. Jefferson G (1955) Trigeminal neurinomas with some remarks on the malignant invasion of the gasserian ganglion. Clin Neurosurg 1: 11-54
99. Jung TT, Jun BH, Shea D, Paparella MM (1986) Primary and secondary tumors of the facial nerve. A temporal bone study. Arch Otolaryng Head neck Surg 112: 1269-1273
100. Kalangu K, Reznik M, Bonnal J (1986) Papillome du plexus choroïde de l'angle ponto-cérébelleux. Présentation d'un cas et revue de la littérature. Neurochirurgie 32: 242-247
101. Kasantikul V, Netsky MG, Glasscock ME, Hays JW (1980) Acoustic neurilemmoma. Clinicoanatomical study of 103 patients. J Neurosurg 52: 28-35
102. Kaye AH, Hahn JF, Kinney SE, Hardy RW, Bay JW (1984) Jugular foramen schwannomas. J Neurosurg 60: 1045-1053
103. Kimura R (1976) Experimental blockage of the endolymphatic duct and sac and its effect on the inner ear of the guinea pig. Ann Otol Rhinol Laryngol 76: 664-669
104. King TT, Morrisson AW (1980) Translabyrinthine and transtentorial removal of acoustic nerve tumors. J Neurosurg 52: 210-216
105. Koos WTh, Spetzler RF, Böck FW, Salah S (1976) Microsurgery of cerebellopontine angle tumors. In clinical microsurgery. Koos WTh, Bock FW, Spetzler RF. Edit G Thieme Publ Stuttgart 91-112
106. Koos WTh, Perneczky A (1983) Pathomorphologie et pathophysiologie des neurinomes de l'acoustique. In diagnostic et traitement des neurinomes de l'acoustique. Lazorthes Y, Clanet M, Fraysse B Edit (Toulouse) 37-59
107. Kumar A, Fisch V (1983) The infratemporal fossa approach for lesion of the skull base. Advances. Technical standard in Neuro-surg 10: 187-220
108. Lachard A, Hassoun J, Charpin C, Toga M (1984) Les paragangliomes de la tête et du cou (chémodectomes). Aspects anatomo-pathologiques, in Les chemodectomes cervico-céphaliques, Leroux, Robert J, Pech A. Edit Masson Publ Paris 3-16
109. Laha RK, Dujovny M, Rao S, Barrionuevo PJ, Bunegin L, Hellstrom HR, Albin MS, Taylor FN (1979) Cerebellar retraction: significance and sequelae. Surg Neurol 12: 209-215
110. Lazorthes G, Bimes C (1947) Remarques sur la constitution du cavum de Meckel. CR Ass Anat 312-319
111. Lechat Ph, Guggiari M, Lascaut G, Fuschiardi M, Evans J, Drobinski G, Klimczak K, Grosgugeat Y (1986) Detection par échographie de contraste d'un foramen ovale perméable avant neurochirurgie. Presse Med 30: 1409-1410
112. Lepoire J, Pertuiset B (1957) Les kystes épidermoïdes cranio-encéphaliques. Masson Edit, Paris, 1-106
112 a. Lo WW, Solti-Bohmann LG, Brackmann DE, Grunskin P (1984) Cholesterol granuloma of the petrous apex: CT Diagnosis. Radiology 153: 705-711
113. Mc Cabe BF, Fletcher J (1969) Selection of therapy of glomus jugulare tumors. Arch Otolaryngol 89: 156-159
114. Mc Donald TJ, Cody DTR, Ryan RE (1984) Congenital cholesteatoma of the ear. Ann Otol Rhinol Laryngol 93: 637-640
115. Mc Neill KA, Milner GAW (1955) Bilateral tumor of the glomus jugulare. J Laryngol Otol 69: 430-431

116. Main TS, Shimada T, Lim DJ (1970) Experimental cholesteral granuloma. Arch Otolaryngol 91: 356-359
117. Malca S (1985) Le neurinome des nerfs mixtes. Formes intra-crannienes et du foramen jugulaire. Thèse Marseille 1-468
118. Malis LI (1975) Microsurgical treatment of acoustic neurinomas, in Microneurosurgery, Handa H. Edit Igaku Shoin Publ 105-120
119. Maniglia AJ (1978) Intra and extra-cranial meningiomas invalving the temporal bone. Laryngoscope Suppl 12: 88-89
120. Merland JJ, Reizine D, Guimaraens L, Ruffenacht D, Melki JP, Riche MC, George B (1985) L'angiographie diagnostique et thérapeutique dans le bilan et le traitement des tumeurs du glomus jugulaire. A propos de 32 cas. Neurochirurgie 31: 358-366
121. Montaut J (1962) Les neurinomes du trijumeau. Thèse Medecine Nancy
122. Montgomery WW (1977) Cystic lesion of the petrous apex: trans-sphenoidal approach. Ann Otolaryngol 86: 429-435
123. Moret J, Delvert JC, Lasjaunias P (1982) La vascularisation de l'appareil auditif. Normal, variantes, tumeurs glomiques. J Neuroradiol 9: 213-260
124. Morrisson AW, King TT (1982) Translabyrinthine tumoral of acoustic neuromas. In: Neurological Surgery of the Ear and Skull Base. Brackmann DE Edit. Raven Press Publ 227-233
125. Murata T, Akuba A, Okumura T, Mori K (1985) Intra petrous neurinomas of the facial nerve. Report of three cases. Surg Neurol 23: 507-512
126. Nager GT (1964) Meningiomas involving the temporal bone: clinical and pathological aspects. ChC Thomas Edit. Springfield, ill
127. Nager GT, Vanderveen TS (1976) Cholesterol granuloma involving the temporal bone. Ann Otol Rhinol Laryngol 85: 204-209
128. Neely JG, Hough J (1986) Histologic findings in two very small intracanalicular solitary schwannomas of the eighth nerve. Ann Otol Rhinol Laryngol 95: 460-465
129. Neff W (1947) The effects of partial section of the auditory nerve. J Comp Physiol Psychol 40: 205-215
130. Numoto M, Donaghy RMP (1970) Effects of local pressure on cortical vessels in the dog. J Neurosurg 33: 381-387
131. Ojeman RG, Levine RA, Montgomery WM, Mc Gaffigan P (1984) Use of intra operative auditory evoked potentials to preserve hearing in unilateral acoustico neuroma removal. J Neurosurg 61: 938-948
132. Olivecrona H (1967) Acoustic tumors. J Neurosurg 26: 6-13
133. Osborn AG, Parkin JL (1979) Mucocele of the petrous temporal bone. Ann J Radiol 132: 680-681
134. Paillas JE, Grisoli F, Farnarier Ph (1973) Neurinomes du trijumeau. A propos de 8 cas. Neurochirurgie (Paris) 20: 41-54
135. Palva T, Troupp H, Jauhiainen T (1985) Hearing preservation in acoustic neurinoma surgery. Acta otolaryngol (Stockholm) 99: 1-7
136. Palva T, Johnsson LG (1986) Preservation of hearing after removal of the membranous canal with a cholesteatoma. Arch Otolaryngol, Head Neck Surg 112: 982-985
137. Pech A, Cannoni M, Thomassin JM, Abdul S, Zanaret M (1981) Le diagnostic d'extension des tumeurs du glomus jugulaire. Ann Otolaryng (Paris) 98: 215-222
138. Pech A, Cannoni M, Pellet W, Triglia JM, Zanaret M, Thomassin JM (1986) Les tumeurs de l'angle ponto-cérébelleux à l'exception des neurinomes de l'acoustique. Ann Oto-Laryng (Paris) 103: 293-301
139. Pellet W (1968) La vascularisation des méningiomes. Thèse Médecine Marseille 1-238
140. Pellet W, Cannoni M, Lavieille J, Lehmann G, Vacherat S, Mouren P (1973) Volumineuse tumeur du glomus jugulaire: voie d'abord combinée oto-neuro-chirurgicale. Neurochirurgie (Paris) 19: 567-579
141. Pellet W, Cannoni M, Pech A (1979) La collaboration oto-neuro-chirurgicale dans l'abord trans-labyrinthique des neurinomes de l'acoustique. Neurochirurgie 25: 84-90
142. Pellet W, Cannoni M, Pech A, Triglia JM (1987) Le neurinome de l'acoustique. Approche oto-neuro-chirurgicale. Rev Neurol (Paris) 143: 614-619
143. Pennybacker J, Cairns H (1950) Results in 130 cases of acoustic neurinomas. J Neurol Neurosurg Psychiat 13: 272-277
144. Pensak ML, Classcock ME, Gulya AJ, Hays JW, Smith HP, Dickens RE (1986) Cerebellopontine angle lipomas. Arch Otolaryngol head Neck surg 112: 99-101
145. Perlman H, Kimura R, Fernandez C (1959) Experimentation on temporary obstruction of the internal auditory artery. Laryngoscope 69: 591-599
146. Pertuiset B (1970) Les neurinomes de l'acoustique développés dans l'angle ponto-cérébelleux. Neurochirurgie 16: suppl 1, 1-147
147. Pool JL (1966) Suboccipital surgery for acoustic neurinomas: advantages and disadvantages. J Neurosurg 24: 483-492
148. Portmann M, Bébéar JP, Francois JH (1984) Notre expérience des tumeurs glomiques tympano-jugulaires. In les chemodectomes cervico-céphaliques. Leroux, Robert J, Pech A Edit. Masson Paris Publ 131-134
149. Pulec J (1969) Facial nerve tumors. Ann Otol-Rhinol-laryngol 78: 962-982
150. Rasmussen AT (1940) Studies of the VIIIth cranial nerve of man. Laryngoscope 50: 67-83
151. Romanet PH, Haguenauer JP, Gaillard J (1978) Chemodectomes de la tête et du cou. Enceph Med Chir (Paris) ORL., Fasc 20955, A-10-12
152. Rosenbloom SB, Carson BS, Wang H, Rosenbaum AE, Uduarhelyi GB (1985) Cerebello-pontine angle lipoma. Surg Neurol 23: 134-138
153. Rosenblum WI, El-Sabban F (1978) Platelet aggregation and vasoconstriction in undamaged microvessels on cerebral surface adjacent to brain traumatized by a penetrating needle. Microvascular research 15: 299-307
154. Rossenwasser H (1969) Glomus jugulare tumors. Arch Otolaryng 89: 186-192
155. Rougerie J, Guyot JF (1964) Essai de conservation du nerf facial dans l'ablation des neurinomes de l'angle ponto-cérébelleux. Neuro-chirurgie 10: 13-21
156. Rousseaux M, Lesoin F, Petit H, Jomin M (1984) Les kystes arachnoïdiens de l'angle ponto-cérébelleux. Neurochirurgie 30: 119-124
157. Russel JR, Bucy PC (1953) Meningiomas of the posterior fossa. Surg gynéc Obstet 96: 183-192
158. Sceats DJ, Quislingr, Rhoton AL, Ballinger WE, Ryan P (1986) Primary leptomeningeal glioma mimicking an acoustic neuroma: case report with review of the literature. Neurosurgery 19: 649-654
159. Schuknecht HF (1977) Pathology of vestibular schwannoma (acoustic neurinoma). In Neurological Surgery of

the Ear, Sylverstein H, Norrel HA, Vol I. Aesculapius Publ 193–197
160. Schwaber MK, Glasscock ME, Nissen AJ, Jakson CG, Smith PG (1984) Diagnosis and management of catecholamine secreting glomus tumors. Laryngoscope 94: 1008–1015
161. Schwartz HE, Morgan DE, Calcaterra TC (1978) Recovery of eighth nerve function after cerebello-pontine angle surgery. Arch Otolaryngol 104: 231–233
162. Shambaugh GE, Arenberg IK, Barney PL, Valvassori GE (1969) Facial neurilemmonas. Arch Otolaryng 90: 742–755
163. Shapiro MJ, Neues DK (1964) Technique for removal of glomus jugulare tumors. Arch Otoryng 79: 219–224
164. Shea MC, Robertson JT (1979) Acoustic neuroma. A comparative study of translabyrinthine and suboccipital approach. Amer J Otol 1: 94–99
165. Sheptak PE, Jannetta PJ (1979) The two stage excision of huge acoustic neurinomas. J Neurosurg 51: 37–41
166. Spector GJ, Maisel R, Ogura JH (1974) Glomus jugulare tumors. A clinicopathologic analysis of the effect of radiotherapy. Ann Oto-laryngol 83: 26–32
167. Spector GJ, Sobol S, Thawley SE, Maisel RH, Ogura JH (1979) Patterns of invasion in the temporal bone. Laryngoscope 89: 1628–1639
168. Steimle R, Pageaut G, Jacquet G, Bourghli A, Godard J, Bertaud M (1985) Lipoma in the cerebello-pontine angle. Surg Neurol 24: 73–76
169. Sterkers JM Removal of bilateral and unilateral acoustic tumors with preservation of hearing, in Neurological Surgery of the Ear. Silverstein H, Norrel NA Edits. Aesculapius Publi, Vol II. 269–277
170. Sterkers JM, Hamann KF (1979) Neurinomes de l'acoustique bilatéraux à propos de 8 observations. Ann Otolaryng 96: 623–635
171. Sterkers JM, Batisse R, Gandon J, Cannoni M, Vaneecloo JM (1984) Les voies d'abord du rocher. Librairie Arnette Publ (Paris) 1–195
172. Sterkers JM, Desgeorges M, Sterkers O, Corlieu P, Viala P (1986) Chirurgie du neurinome de l'acoustique et autres tumeurs du conduit auditif interne et de l'angle ponto-cérébelleux à propos de 602 cas. Ann Oto-Laryng (Paris) 103: 487–492
173. Sterkers JM, Corlieu P, Viala P, Riviere F, Chobaut JC (1986) Les pétrosites actuelles. Ann Oto-Laryng (Paris) 103: 515–521
174. Sterkers O, Viala P, Riviere F, Sterkers JM (1986) Neurinomes du nerf facial intra-temporal. Classification anatomo-clinique de 12 cas. Ann oto-laryng (Paris) 103: 501–508
175. Silverstein H, Mc Daniel A, Norrel H, Wasen J (1985) Conservative management of acoustic neuroma in the elderly patient. Laryngoscope 95: 766–770
176. Tarlov E (1980) Total one stage suboccipital microsurgical removal of acoustic neuromas of all sizes. Surg Clinics North Amer 60: 565–591
177. Tator CH, Medzelski JM (1985) Preservation of hearing in patients undergoing excision of acoustic neuromas and other cerebellopontine angle tumors. J Neurosurg 63: 168–174
178. Terracol J, Guerrier Y, Guibert HL (1956) Le glomus jugulaire. Masson Edit (Paris) 1–119
179. Thomassin JM, Zanaret M, Sarrat P, Cannoni M, Pech A (1984) Les problèmes diagnostiques et thérapeutiques des chémodectomes tympano-jugulaires. In les chémodectomes cervico-céphaliques, Leroux Robert J, Pech A, Edit. Masson Publ (Paris) 140–148
180. Tos M, Thomsen J (1982) The price of preservation of hearing in acoustic neuroma surgery. Ann Otol Rhinol Laryngol 91: 240–245
181. Tran Ba Huy P, Levy C, Bensimon JL, Cophignon J (1968) Cinq cas de kyste épidermoïde (ou cholestéatome primitif) de la pyramide pétreuse. Aspect clinique radiologique, pathogénique et thérapeutique. Intérêt de la R. M. N. dans le diagnostic et la surveillance post-opératoire. Ann Oto-laryng (Paris) 103: 363–371
182. Triglia JM (1983) Le neurinome de l'acoustique. Son diagnostic et son traitement (à propos de 130 observations). Thèse Médecine Marseille 160
183. Ulso C, Sehested P, Overgaard J (1981) Intracranial lypoglossal neurinoma: Diagnosis and post-operative care. Surg Neurol 16: 65–68
184. Uziel A, Benezech J, Frerebeau Ph (1986) Les possibilités de conservation de l'audition dans la chirurgie des neurinomes de l'acoustique. Ann Oto-laryng (Paris) 103: 127–132
185. Vaneecloo FM, Jomin M, Janssen B, Burny A, Lejeune E, Allouche F (1985) Le traitement oto-neurochirurgical des tumeurs de l'angle ponto-cérébelleux à propos d'une série des 93 cas. Ann Oto-Laryng (Paris) 102: 157–161
186. Vourch G, Guilmet D, Guiot G, Bouche J, Frèche C (1969) Intervention d'exérèse d'une volumineuse tumeur glomique sous circulation extra-corporelle et hypothermie profonde. Ann Oto-laryng (Paris) 86: 466–473
187. Wigand ME, Rettinger G, Haid T, Berg M (1985) Die Ausräumung von Octavus Neurinomen des Kleinhirnbrückenwinkels mit trans-temporalem Zugang über die mittlere Schädelgrube. HNO 33: 11–16
188. Yasargil MG, Smith RD, Gasser JC (1977) Microsurgical approach to acoustic neurinomas. Adv Technical Standards in Neurosurg, Krayenbuhl H, Edit, Vol 4. Springer Verlag (Wien New-York) 93–129
189. Yasargil MG, Mortara RW, Curcic M (1980) Meningiomas of basal posterior cranial fossa. Adv Technical Standard in Neurosurg, Krayenbulh H, Edit, Vol 7. Springer Verlag (Wien New-York) 1–110
190. Yliloski J, Palva T, Collan Y (1978) Eighth nerve in acoustic neuromas. Arch Oto-Laryngol 104: 532–537
191. Wade PJ, House WF (1984) Hearing preservation in patients with acoustic neuromas via the middle fossa approach. Otolaryngol Head Neck Surg 92: 184–193
192. Waters DC, Venes JL, Zis K (1986) Case report: childhood cerebello-pontine angle teratoma associated with congenital hydrocephalus. Neurosurgery 18: 784–786
193. Williams IG (1957) Radiotherapy of tumors of the glomus jugulare. J Fac Radiol 8: 325–338
194. Wylder AR, Leech RW, Reynolds AF, Ojemann GA, Mead C (1974) Cholesteral granuloma of the petrous apex. Case report. J Neurosurg 41: 765–768
195. Zack FG, Lawson W (1982) The paraganglioma chemoreceptor system. Physiology-pathology and clinical medicine. Springer Verlag Edit, New-York Heidelberg Berlin 15–23
196. Zlotnik EI, Smeyanovitch AF, Tyappo EP (1982) Method of eyelid closure in facial nerve paralysis. Technical note. J Neurosurg 57: 722–723

Subject Index

Italicized page numbers indicate major references.